Technician Electronics 2

This book covers the standard unit in Electronics at level 2 (TEC U75/76/010) of the Technician Certificate in Electronics and Communications Engineering (TEC Programme A2). The learning structure of the syllabus and content are closely followed and the book is written with the clarity and simplicity of Dr. Meadows' other works. There are numerous worked examples and questions and answers and the book contains many line illustrations.

Dr. Meadows is Head of the Department of Electronic and Communications Engineering at the Polytechnic of North London and author of *Technician Physical Science* 1, *Technician Engineering Science* 2 (Cassells Technician Series), *Problems in Electrical Circuit Theory* 1 *and* 2, *Electrical Network Analysis, Electrical Communications* and *Advanced Pure Mathematics.*

Technician Electronics 2

R. G. MEADOWS

M.Sc., Ph.D., C.Eng., M.I.E.E.,
M.Inst.P., A.R.C.S.

CASSELL
LONDON

CASSELL LTD.
35 Red Lion Square, London WC1R 4SG
and at Sydney, Auckland, Toronto, Johannesburg,
an affiliate of Macmillan Publishing Co. Inc., New York

First published 1978

ISBN 0 304 30006 3

Printed and bound in Great Britain at
The Camelot Press Limited, Southampton

Preface

This text is written specifically for students taking the level 2 unit Electronics 2, although I hope it will be also useful as an introductory course in Electronics for Technician, Engineering and Science students.

Electronics 2 is a standard unit for Technician Certificate Courses in the A2 subject area of Electronics and Communications Engineering.

The text follows exactly the same order as the published TEC syllabus and an identical numbering sequence of the specific learning headings is used. I hope that this provides a rapid aid for reference purposes. I have also included many worked examples within the text and problems with answers at the end of each chapter.

I should like to express my thanks to my colleagues for many helpful discussions, with particular thanks to Paul Clark and Mike Brinson; very special thanks to June Lomas and my wife Lynn for their help in preparing the manuscript; and to Dr. H. W. French for his extremely helpful and valuable comments.

Finally, I hope students will find the book interesting and helpful in their studies.

RICHARD MEADOWS

March 1978

Dedicated to 'Medium son' Alexis Dominic.

Contents

B Thermionic Valves

A Elementary Theory of Semi-conductors

Semi-conductor solid-state devices have transformed the field of electronics over the past two decades. Solid-state diodes, transistors, and integrated circuits have now replaced thermionic valves as the basic electronic building block in the majority of electronic equipment. All solid-state electronic devices depend for their operation on controlling the flow of charge carriers in semi-conductor materials.

We thus start our study of electronics by looking into the properties and structure of semi-conductors to see how these materials conduct electricity. Subsequently, we will consider the behaviour and applications of two basic semi-conductor devices: *p-n* junction diodes and bipolar transistors.

Section 1: *The expected learning outcome of this section is that the student should understand the simple concept of semi-conductors.*

1.1
Properties of semi-conductors in relation to conductors and insulators

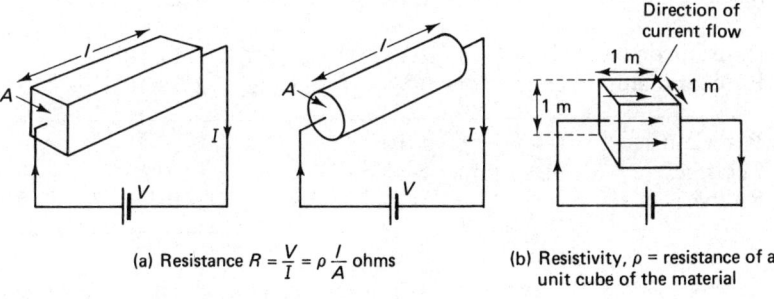

(a) Resistance $R = \dfrac{V}{I} = \rho \dfrac{l}{A}$ ohms

(b) Resistivity, ρ = resistance of a unit cube of the material

Fig. 1.1.

A semi-conductor material is one whose ability to conduct electricity lies roughly midway between that of a good conductor and a good insulator. We quantify the ability of material to conduct electricity by means of two terms: resistivity and conductivity.

1

The electrical resistance R of a material, see fig 1.1(a), is given by

$$R = \frac{V}{I} = \rho \frac{l}{A} \text{ ohms}$$

where V = voltage across the material in volts (V),
$\quad I$ = current flowing in the material in amperes (A),
$\quad l$ = length of material in metres (m),
$\quad A$ = cross-sectional area of material in square metres (m^2),
$\quad \rho$ = resistivity of material, units ohm-metres (Ωm).

The definition of the resistivity of the material follows from the above formula for resistance and is defined as the resistance between the opposite faces of a one-metre cube of the material when the current flows perpendicularly between its faces, as shown in fig. 1.1(b). The conductivity, denoted by σ (Greek symbol sigma), of the material is the reciprocal of resistivity, so $\sigma = 1/\rho$. Conductivity has the units of siemens per metre (S/m). The siemen (S) is the SI unit of conductance, conductance normally denoted by the symbol G being the reciprocal of resistance R, i.e. $G = 1/R$ siemens.

Table 1.1 *Resistivities of some good conductors, semi-conductors, and good insulators*

Material	Class	Resistivity ρ (Ωm)	Resistivity relative to copper, $\rho/\rho_{\text{copper}}$
Copper	Good	$1\cdot7 \times 10^{-8}$ at 20°C	1
Aluminium	conductors	$2\cdot8 \times 10^{-8}$ at 20°C	1·64
Silver		$1\cdot6 \times 10^{-8}$ at 20°C	0·95
Pure silicon	Semi-	$2\cdot3 \times 10^3$ at 27°C	$1\cdot4 \times 10^{11}$
Pure germanium	conductors	0·45 at 27°C	$2\cdot6 \times 10^7$
Porcelain	Good	$>10^{11}$	$>10^{18}$
Ceramics	insulators	$>10^{12}$	$>10^{19}$
Rubber		$>10^{11}$	$>10^{18}$

Table 1.1 lists the resistivities of some metals which are excellent conductors of electricity, together with those for the semi-conductor materials of pure silicon and germanium and some good insulating materials. In practice small amounts of certain elements are added to pure silicon and germanium to produce semi-conductor materials suitable for fabricating practical devices. The addition of such impurities considerably reduces the resistivity. The resistivities of semi-conductors

are also very temperature-dependent, in some cases increasing with increasing temperature, whilst in others falling with increasing temperature. The latter characteristic contrasts with a metal whose resistivity normally increases with a temperature rise.

Finally, to summarize, we can state that at room temperatures semi-conductors have resistivities of the order of 10^{-3} to 10^3 Ωm, very much higher than metals which have resistivities of the order of 10^{-7} to 10^{-8} Ωm, but very much lower than insulating materials where resistivities range from 10^4 to 10^{15} Ωm. The really important property of a semi-conductor is not so much its intermediate value of resistivity but the fact that it can conduct current in two independent ways, by negative charges (free electrons) and positive charge (holes), and that the conduction of electricity by means of these carriers can be effectively controlled. In a metal, conduction occurs by the movement of free electrons. In an insulator, there are virtually no charge carriers free to conduct electricity.

1.2
The two most common types of semi-conductor material are silicon and germanium

The definition of a semi-conductor as a material whose conductivities lie between metallic conductors and insulators covers a wide range of materials. This range includes silicon, germanium, gallium arsenide, selenium, cadmium sulphide, indium antimonide, and the oxides of copper, iron, zinc, cobalt, manganese, and nickel. However, the best known and most widely used material is silicon, followed to a lesser extent by germanium. The majority of solid-state devices—diodes, transistors, and integrated circuits—are now made from silicon. Germanium is used to make diodes and transistors, and other materials—particularly gallium arsenide—are now used for more specialized applications.

1.3
The structure of *p*-type and *n*-type semi-conductors

This sub-section contains:

(a) A simple review of atoms and crystal structure.
(b) A simple model of a metal and the conduction processes in a metal.
(c) The structure of pure silicon and germanium semi-conductors.
(d) The structure of *n*-type and *p*-type semi-conductors.

It therefore contains a fair bit more than is required at first sight to achieve the specific learning objective. (a), (b), (c) supplement Sub-sections 1.1, 1.4, and 1.5 as well as helping us in our understanding of the structure of *p*-type and *n*-type semi-conductors.

(a) A simple review of atoms and crystal structure

In atomic theory, we regard matter as composed of atoms, atoms of a given element being identical (except in some cases where isotopes occur). The atom itself can be pictured as consisting of a positively charged central core, known as the nucleus, surrounded by tiny negatively charged particles rotating around the nucleus in a series of orbits (in a way analogous to those in which the planets rotate around the sun). These negatively charged particles are known as electrons. Each carries the smallest amount of electric charge which can exist; this is denoted by e and is equal to 10^{-19}C. The electrons orbiting closest to the nucleus are bound tightly by attractive forces exerted by the positively charged nucleus. In certain materials, all the orbital electrons are held securely, and hence such materials act as insulators. There are no free electrons available to conduct electricity. In a metal, however, the outermost electrons are bound very weakly and can break free. These free electrons provide the mechanism for the conduction of electricity in a metal.

In the solid state, the atoms of a substance are very densely packed together, and in the case of many materials, including metals, semi-conductors, and some insulators, the atoms are arranged in regular arrays. Such materials are said to be crystal in structure and the regular array of atoms is known as a crystal lattice.

(b) A simple model of a metal and the conduction processes in a metal

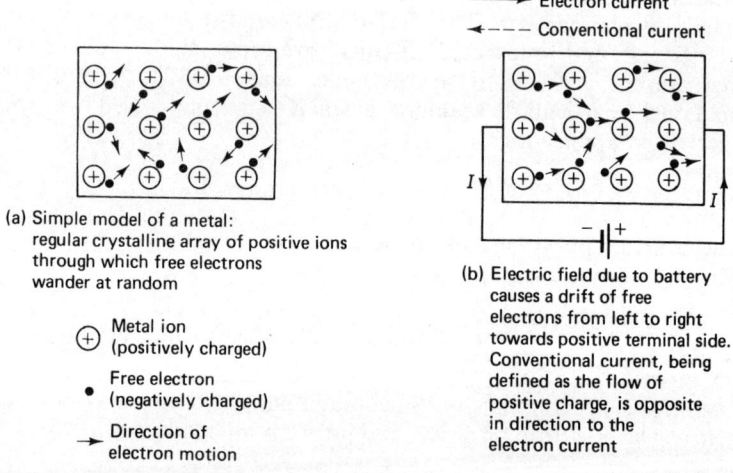

(a) Simple model of a metal: regular crystalline array of positive ions through which free electrons wander at random

⊕ Metal ion (positively charged)

• Free electron (negatively charged)

→ Direction of electron motion

(b) Electric field due to battery causes a drift of free electrons from left to right towards positive terminal side. Conventional current, being defined as the flow of positive charge, is opposite in direction to the electron current

Fig. 1.2.

In a metal, the outermost electrons of each atom in the crystal lattice are very weakly bound and may easily break away from the atomic core, leaving the latter positively charged. These electrons are known as free electrons, 'free' as they are able to wander through the crystal lattice. In an electrically neutral atom, the total positive charge of the nucleus is equal to the total negative charge of the surrounding electrons. If one or more electrons is removed from an atom, the remaining atomic core becomes positively charged, whilst if an atom gains one or more electrons, it becomes negatively charged. Atomic cores with a net positive or a net negative charge are known as ions. Thus, it would be more accurate to describe the crystalline structure of a metal as consisting of a regular array of positive ions through which free electrons may drift randomly.

Fig. 1.2 shows a simple model of a metal: a crystal lattice composed of positively charged metal ions with a number of free electrons exactly equal in number to those lost by the ions. When the metal is acted upon by an electrical force, for example by applying a battery across the material, a uniform drift is superimposed on the previously random free electron paths. This drift of electrons constitutes the electrical current in a conductor and the process is illustrated diagrammatically in fig. 1.2(b). When an electron reaches the right-hand side of the metal in the Figure, it flows via a conducting wire to the positive terminal of the battery. At the same time, an electron is released from the negative battery terminal and enters, via the conducting wire, the left-hand side of the metal. Thus, at all times the metal remains uncharged. Note, also, that the conventional current flow, taken as the flow of positive charge, is in the opposite direction to the electron flow.

The model described above provides a simple explanation for the good conduction properties of a metal. It will also help us to see the basic difference between the properties and structure of metals and semi-conductors.

(c) *The structure of pure silicon and germanium semi-conductors*

The crystal lattice structure of pure silicon is shown diagrammatically in fig. 1.3. The structure of germanium is exactly similar, so the following discussion applies equally to germanium. Each silicon atom has four outer electrons and these electrons are known as valence electrons. The four valence electrons each form bonds with an electron from an adjacent atom so that each atomic core is effectively surrounded by eight electrons, four of its own and four from adjacent atoms. Atoms with eight outer electrons are stable and the bonds formed by the pairing of electrons serve to bind one atom to the next. The bonds are known as covalent or electron-pair bonds and are represented in the diagram of fig. 1.3 by dashed lines between the silicon ions. The latter each carry an effective positive charge of +4 electron units since each atomic core has 'lent' 4 electrons to form covalent bonds with its 4 neighbours.

At very low temperature, close to absolute zero, 0 K or −273°C, the

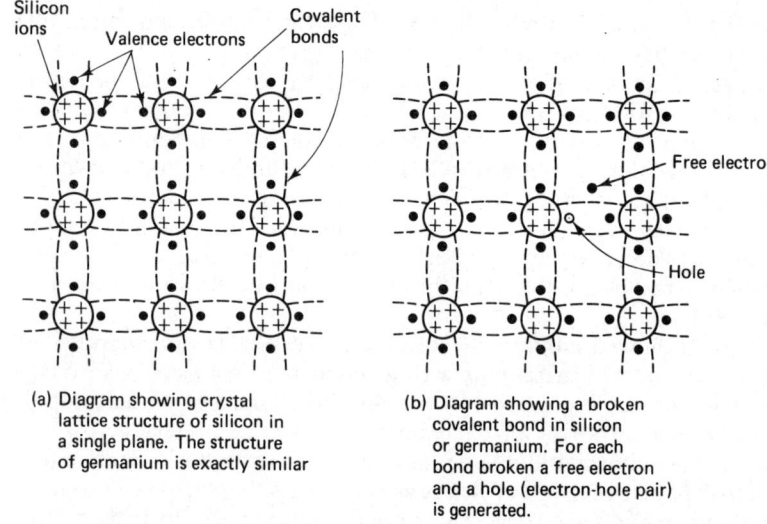

(a) Diagram showing crystal lattice structure of silicon in a single plane. The structure of germanium is exactly similar

(b) Diagram showing a broken covalent bond in silicon or germanium. For each bond broken a free electron and a hole (electron-hole pair) is generated.

Fig. 1.3.

bonds hold virtually all valence electrons rigidly in place. There are therefore no free electrons available to conduct electricity and hence at very low temperatures silicon and germanium act as perfect insulators. However, as the temperature is raised the semi-conductor gains thermal energy and some electrons acquire sufficient energy to break their covalent bond and become free electrons. In physical electronic terms we say that such an electron has gained sufficient energy to move into the conduction energy band of the semi-conductor. These free electrons provide one of the two conduction processes in a pure semi-conductor. At room temperature the number of bonds broken and hence the number of free electrons is $1 \cdot 6 \times 10^{16}$ electrons per cubic metre for silicon and $2 \cdot 5 \times 10^{19}$ electrons/m^3 for germanium. Although these numbers appear at first sight immensely large they show in fact that only one bond in 3×10^{10} silicon atoms is broken, and one in $1 \cdot 8 \times 10^9$ germanium atoms, whereas in a metal there are one or more free electrons per atom giving the order of 10^{29} free electrons per cubic metre.

For every electron breaking its covalent bond and entering the conduction band a 'hole' is left behind. The term hole is the name given to the absence of an electron in a covalent bond and is represented diagrammatically in fig. 1.3(b). A hole created in one part of the crystal lattice can be filled by a valence electron jumping into it from a nearby atom. This in turn leaves another hole which can be filled (subsequently) by another valence electron breaking its bond, and so on.

We can thus identify two carriers of current in a semi-conductor: free electrons (or conduction band electrons as they are frequently called), formed by valence electrons gaining sufficient heat energy (or other energy such as light), and holes which act as effective positive charges and move through the semi-conductor by valence electrons jumping from one hole to another. When holes and free electrons are generated in a semi-conductor by heat or light energy they are always generated in pairs, so that in a pure semi-conductor the numbers of holes and free electrons are always equal.

(d) *The structure of n-type and p-type semi-conductors*

In the production of semi-conductor crystals for use as diodes, transistors, and integrated circuits, certain elements are deliberately added to pure silicon and germanium. This process is known as doping and the foreign atoms added are known as impurities. The process requires very sophisticated control since the amount of impurity atoms added is typically 1 impurity atom to 10^8 silicon or germanium atoms. Two main types of impurities are used: pentavalent atoms having 5 valence or outer electrons and trivalent atoms having 3 valence electrons. A list of the most commonly used 'dopants' for silicon and germanium semi-conductor device fabrication is given in Table 1.2.

Table 1.2 *List of* n-*type and* p-*type impurities used to dope silicon and germanium*

Pentavalent impurity atoms used for production of *n*-type semi-conductors		Trivalent impurity atoms used for production of *p*-type semi-conductors	
Phosphorus	(P)	Boron	(B)
Arsenic	(As)	Aluminium	(AL)
Antimony	(Sb)	Gallium	(Ga)
		Indium	(In)

Pentavalent impurity atoms added during the doping process are known as donor atoms, since they donate current carriers. The semi-conductor so formed is known as *n*-type, since the donated current carriers are free electrons, i.e. negatively charged carriers. Fig. 1.4 shows a diagram of the structure of an *n*-type semi-conductor formed by adding pentavalent impurity atoms to silicon. Four of the five impurity atom's valence electrons form covalent bonds with four valence electrons from four adjacent silicon atoms. The fifth electron is effectively unpaired and, at all but the very lowest of temperatures, is free to move, and thus to contribute to a current of electricity. For diode and transistor applications the order of 1 impurity atom to 10^8 silicon atoms is introduced and since silicon contains about 5×10^{28} atoms/m^3 the

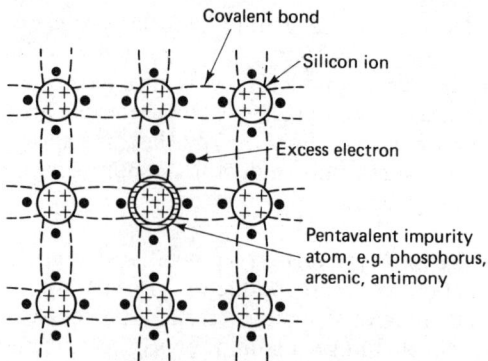

Diagram showing structure of *n*-type silicon
produced by adding atoms with 5 valence electrons.

Fig. 1.4.

number of free electrons donated in this case would be 5×10^{20} electrons/m³, still very much smaller than a metal which has the order of 10^{29} electrons/m³. It is of course possible to introduce very much larger numbers of impurity atoms and increase the conductivity of a semi-conductor much further. Indeed this is done in some applications (e.g. temperature sensitive resistors) but a doping concentration of the order 1 in 10^8 is typical for diode and transistor fabrication.

Trivalent impurity atoms added during the doping process are known as acceptor atoms because they create holes that can accept electrons. The semi-conductor so formed is known as *p*-type, the *p* denoting that the current carriers are positively charged, i.e. holes which effectively act as positively charged carriers. Fig. 1.5(a) shows a diagram of the structure of a *p*-type semi-conductor formed by adding trivalent impurity atoms to silicon. The three valence electrons of an impurity atom form covalent bonds with three valence electrons from three adjacent silicon atoms. However, the fourth bond is left deficient by one electron, so a trivalent atom creates an electron vacancy and we refer to this vacancy as a hole. The simple diagram of fig. 1.5(b) is intended to help explain how a hole moves through a semi-conductor. The energy to break a covalent bond and thus release a valence electron to fill a hole is small in a *p*-type semi-conductor and may be easily supplied by the battery. Thus when the battery is connected across the *p*-type semi-conductor some covalent bonds are broken and the electrons released move to fill the holes present owing to the addition of *p*-type impurity atoms. The net result is that some valence electrons breaking their bonds move towards the positive battery terminal side and the holes formed by the vacation of the valence electrons appear to drift towards the negative battery terminal side, just as if they were positive charges. It has now

8

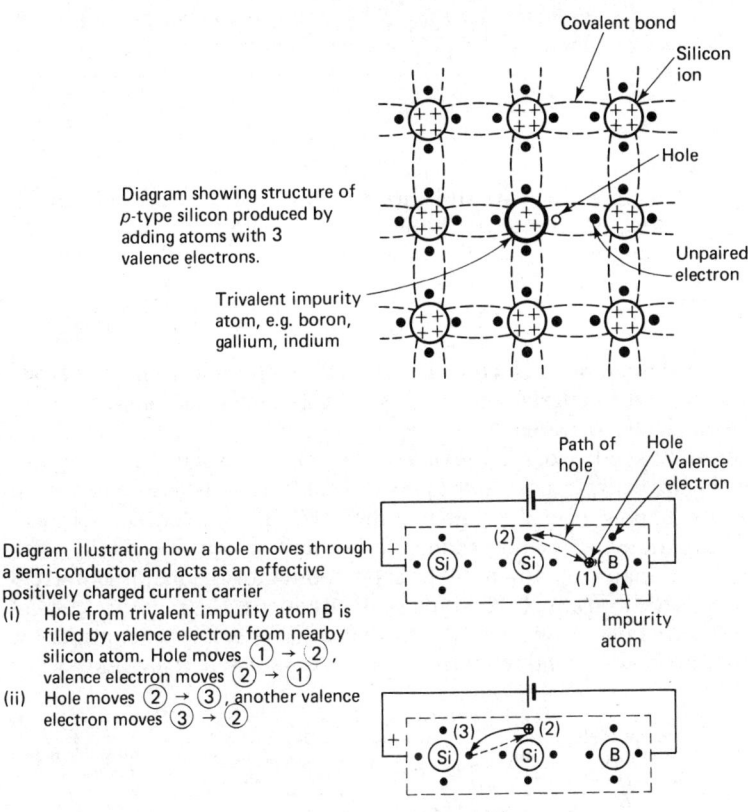

Diagram showing structure of p-type silicon produced by adding atoms with 3 valence electrons.

Covalent bond

Silicon ion

Hole

Unpaired electron

Trivalent impurity atom, e.g. boron, gallium, indium

Diagram illustrating how a hole moves through a semi-conductor and acts as an effective positively charged current carrier
(i) Hole from trivalent impurity atom B is filled by valence electron from nearby silicon atom. Hole moves ① → ②, valence electron moves ② → ①
(ii) Hole moves ② → ③, another valence electron moves ③ → ②

Path of hole

Hole Valence electron

Impurity atom

Fig. 1.5.

become established practice to regard holes as positive charges in their own right and regard conduction in p-type semi-conductors as due to positive hole current.

Note that although the charge carriers in p-type semi-conductors are predominantly holes there are always some free electrons present as well. Likewise in an n-type semi-conductor although the charge carriers are predominantly free electrons there will also be some holes present. The predominant charge carriers are known as majority carriers and are generated by the addition of impurity atoms. The minority carriers (i.e. electrons in p-type and holes in n-type) are generated by thermal energy and in some cases by light energy or a small trace of the wrong type of impurity being present. Table 1.3 gives the number of majority and minority carriers present in silicon and germanium at 27°C when doped 1 impurity atom to 10^8.

9

Table 1.3   *Approximate densities of majority and minority carriers in silicon and germanium* n *and* p-*type semi-conductors at* 27°C

	Majority carriers	Minority carriers
n-type silicon	$\sim 5 \times 10^{20}$ electrons/m^3	$\sim 5 \times 10^{11}$ holes/m^3
n-type germanium	$\sim 4{\cdot}4 \times 10^{20}$ electrons/m^3	$\sim 1{\cdot}4 \times 10^{18}$ holes/m^3
p-type silicon	$\sim 5 \times 10^{20}$ holes/m^3	$\sim 5 \times 10^{11}$ electrons/m^3
p-type germanium	$\sim 4{\cdot}4 \times 10^{20}$ holes/m^3	$\sim 1{\cdot}4 \times 10^{18}$ holes/m^3

1.4
The explanation of conduction as the movement of electrons in n-type semi-conductor material and the apparent movement of holes in p-type semi-conductor material

Most of the spade-work in the way of explanation to achieve this specific learning objective is given in the previous sub-section where we described how by doping silicon or germanium with pentavalent or trivalent impurity atoms we could produce respectively n-type and p-type semi-conductor materials. In an n-type semi-conductor the majority current carriers are electrons. In a p-type semi-conductor the majority current carriers are holes which act as positive charges. Let us now consider the actual process of conduction in n- and p-type semiconductors.

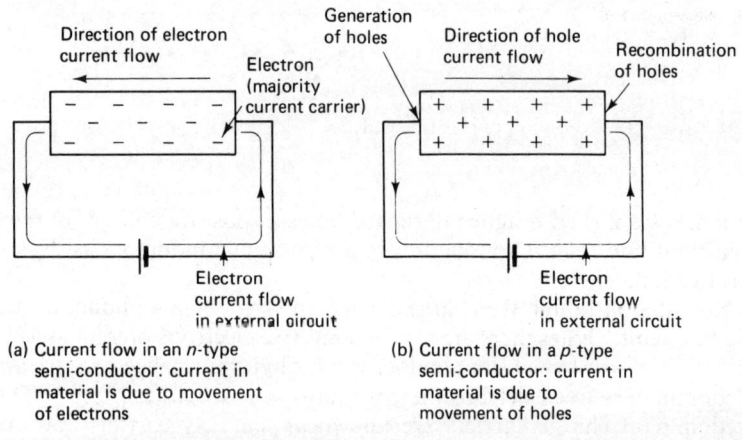

Fig. 1.6.

Fig. 1.6(a) shows a diagram of a battery connected across an n-type semi-conductor. We have simplified the model of the n-type semi-conductor by omitting details of the crystal lattice structure and showing only the n-type majority current carriers, i.e. free or conduction band electrons donated by pentavalent impurity atoms. The voltage applied across the material causes these electrons to drift from right to left towards the positive voltage terminal side of the battery. The conduction in the n-type material is exactly similar to free electron flow in a metal. For every electron leaving the left-hand side of the semi-conductor and flowing in the wire to the positive terminal of the battery, another electron leaves the negative terminal of the battery and flows in the wire to enter the right-hand side of the semi-conductor. In this way the semi-conductor always remains neutral, so at all times the number of conduction electrons within the n-type semi-conductor equals the number of donor atoms in the material. A continuous electron current is established by the battery whereby electrons flow from the negative terminal through the semi-conductor and out at the other side to enter the positive terminal.

In the case of p-type semi-conductors conduction through the material is predominantly due to holes. Thus in fig. 1.6(b) holes (and remember these act as apparent positive charges) are attracted and drift towards the side connected to the negative battery terminal. Electrons from this terminal flow through the wire and enter the p-type material where they recombine with holes, i.e. they fill the electron vacancies created by holes. For every recombination of an electron and a hole at the right-hand side another hole-electron pair is generated at the left-hand side by an electron breaking from a covalent bond, entering the wire, and then flowing into the positive terminal of the battery. Thus at all times the recombination of holes and electrons is balanced by the generation of an equal number of hole-electrons pairs. Recombination occurs at or close to the junction of the external metal conductor connected to the negative battery terminal; generation occurs at or close to the positive battery terminal side.

It should be remembered that minority carriers are also present in both n- and p-type materials. Thus they will make some contribution, although normally very small, to current flow due to holes in n-type and electrons in p-type materials. Conduction in n and p-type semi-conductors occurs principally owing to the presence of controlled amounts of donors and acceptor impurities, which set the number of majority current carriers, and is usually referred to as extrinsic conduction. In pure silicon and germanium conduction, although on a very much smaller scale at least at normal temperatures, occurs owing to the thermal generation of electron-hole pairs. In a pure semi-conductor there are thus equal numbers of electrons and holes. Conduction, where the current carriers are holes and electrons generated thermally, is known as intrinsic conduction.

Example 1.1
Calculate the resistance at 27°C of a wafer of pure silicon of thickness

1 mm and cross-sectional area 2 mm². The conductivity (σ) of silicon at 27°C may be calculated from

$$\sigma = (n\mu_n + p\mu_p)e \text{ S/m}$$

where n = number of conduction-band electrons per cubic metre = 1.6×10^{16},

p = number of holes per cubic metre = 1.6×10^{16},

μ_n = mobility of a conduction-band electron = 0.13 m/s per V/m,

μ_p = mobility of a hole = 0.05 m/s per V/m,

e = magnitude of the electronic charge = 1.6×10^{-19} C.

If donor impurities are added to silicon such that n and p are changed to $n = 2.56 \times 10^{19}$, $p = 10^{13}$, calculate the conductivity of the n-type material so produced and the resistance of a wafer 1 mm thick and area 2 mm². It may be assumed that the mobilities μ_n and μ_p are unchanged by doping.

Calculate also the resistance of such a wafer if it is made of (i) copper, of conductivity = 59×10^6 S/m, (ii) mica, of conductivity = 10^{-11} S/m.

Solution
Let us first calculate the conductivity of pure silicon using the given formula. (Note that this formula is not to be 'learnt' but shows some of the factors on which the conductivity of a semi-conductor such as silicon depends.)

$$\sigma = (1.6 \times 10^{16} \times 0.13 + 1.6 \times 10^{16} \times 0.05) \times 1.6 \times 10^{-19}$$
$$= 1.6 \times 10^{16}(0.13 + 0.05) \times 1.6 \times 10^{-19} = 0.46 \times 10^{-3} \text{ S/m}.$$

So the resistivity,

$$\rho = \frac{1}{\sigma} = \frac{1}{0.46 \times 10^{-3}} = \frac{1000}{0.46} = 2.17 \times 10^3 \ \Omega\text{m}.$$

Thus the resistance of the wafer,

$$R = \rho\frac{l}{A} \text{ where } l = 1\text{mm} = 10^{-3}\text{m}, \ A = 2\text{mm}^2 = 2 \times 10^{-6}\text{m}^2.$$

So $\quad R = 2.17 \times 10^3 \times \dfrac{10^{-3}}{2 \times 10^{-6}} = 1.08 \times 10^6 \Omega = 1.08 \text{ M}\Omega.$

When donor impurities are added n is increased, p is decreased, and the conductivity,

$$\sigma = (2.56 \times 10^{19} \times 0.13 + 10^{13} \times 0.05) \times 1.6 \times 10^{-19}$$
$$= 2.56 \times 10^{19} \times 0.13 \times 1.6 \times 10^{-19}$$

(since we can neglect the second term in the brackets—it is very small compared to the first; this also shows that conduction is due almost entirely to electrons, the holes contribution is negligible because $n \gg p$).

12

So $\qquad \sigma = 2.56 \times 0.13 \times 1.6 = 0.53$ S/m.

and the resistivity, $\rho = \dfrac{1}{\sigma} = 1.88$ Ωm.

The resistance of wafer when $\rho = 1.88$ Ωm is

$$R = \rho \frac{l}{A} = 1.88 \times \frac{10^{-3}}{2 \times 10^{-6}} = 0.94 \times 10^3 = 940 \ \Omega.$$

(i) For the case of copper:

$$\sigma = 59 \times 10^6 \text{ S/m}, \ \rho = \frac{1}{\sigma} = 1.7 \times 10^{-8} \ \Omega\text{m}.$$

So $\qquad R = 1.7 \times 10^{-8} \times \dfrac{10^{-3}}{2 \times 10^{-6}} = 0.85 \times 10^{-5} \ \Omega.$

(ii) For the case of mica (mica is a high grade insulator used in good quality capacitors):

$$\sigma = 10^{-11} \text{ S/m}, \ \rho = 10^{11} \ \Omega\text{m}.$$

So $\qquad R = 10^{11} \times \dfrac{10^{-3}}{2 \times 10^{-6}} = 5 \times 10^{13} \ \Omega.$

1.5
How a change in temperature affects the intrinsic conduction in a semi-conductor

We stated in Sub-section 1.3 that pure silicon and germanium acted as insulators at temperatures close to absolute zero, since all valence electrons were held firmly in covalent bonds. However, at room temperatures the materials gain sufficient heat energy so that the order of 1 in 10^9 covalent bonds are broken. For each bond broken a free electron and a hole are available for conduction. A semi-conductor with no donor or acceptor impurities is known as an intrinsic semi-conductor and conduction in an intrinsic conductor occurs due to thermally generated electrons and holes. These are always generated in pairs, so the number of conduction electrons equals the numbers of holes in an intrinsic semi-conductor. The number increases rapidly with temperature and thus the conductivity of an intrinsic semi-conductor rises with temperature, or equivalently the resistivity falls with rise in temperature. For example at 27°C a 1°C rise in temperature produces approximately an 8% increase in the conductivity of silicon and a 6% rise in the conductivity of germanium, corresponding to an increase in electron and hole current carriers of about 7% for silicon and 5% for germanium.

In extrinsic (doped) semi-conductors the majority current carriers are determined almost entirely by the number of impurity atoms present and obviously these cannot change with temperature. However, the minority carriers increase rapidly with temperature and can even exceed the number of majority carriers. This effect seriously limits the operation of semi-conductor diodes and transistors, and normally germanium devices are not used much above 40°C whilst silicon may be used at higher operating temperatures up to 150°C for some specialized applications.

2.1
A *p-n* junction connected in the reverse-bias mode showing current flow in the diode and the external circuit

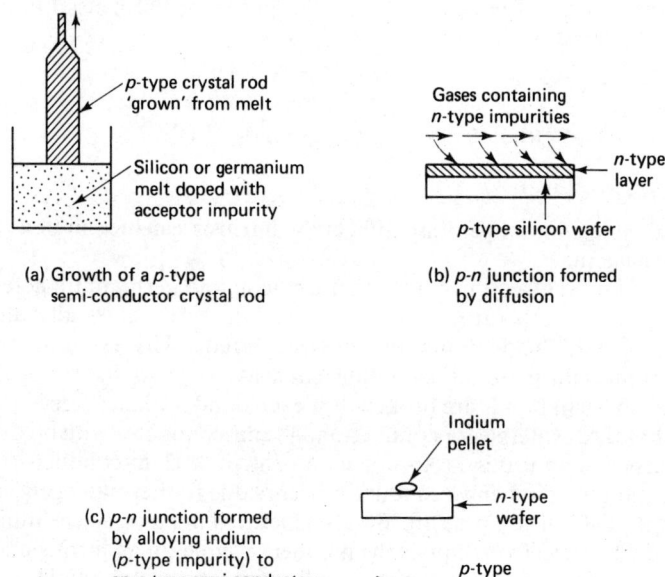

(a) Growth of a *p*-type
semi-conductor crystal rod

p-type crystal rod 'grown' from melt

Silicon or germanium melt doped with acceptor impurity

Gases containing *n*-type impurities

n-type layer

p-type silicon wafer

(b) *p-n* junction formed by diffusion

(c) *p-n* junction formed by alloying indium (*p*-type impurity) to an *n*-type semi-conductor

Indium pellet

n-type wafer

p-type

n-type wafer

Fig. 2.1.

In this section, we will consider how a *p-n* junction is made and then explain the action of a *p-n* junction when reverse-biased. A *p-n* junction is simply a single crystal of very pure semi-conductor in which the junction is defined by an interface between a *p* and an *n*-type region. The junction may be formed by separate doping processes on either side of the interface or by the 'over-doping' of one type of impurity by another in a specified region.

Fig. 2.1(a) shows a diagram illustrating how a semi-conductor crystal may be grown from a melt of semi-conductor material. The melt is doped with a controlled amount of acceptor atom impurity for the production of a p-type crystal or donor atom impurity if an n-type crystal is required. A seed crystal is dipped into the melt and slowly withdrawn. This crystal acts as a nucleus for the growth of a much larger crystal. Under carefully controlled conditions, rods of semi-conductor p-type and n-type crystals can be grown. A typical size of rod may have a diameter of 25–50 mm and be 250 mm in length. The rod is then sliced into wafers of approximately 0·2 mm thickness. These wafers serve as a substrate on which may be built semi-conductor electronic devices: diodes, transistors, or integrated circuits. Such are the advances in semi-conductor technology, that in a typical integrated circuit consisting of a small crystal of silicon of dimensions 1 mm × 1 mm × 0·1 mm whole circuits may be built consisting, for example, of resistors, capacitors, diodes, and transistors, together with their interconnections.

Fig. 2.1(b) shows a simple diagram illustrating the formation of a p-n junction by diffusing n-type impurities into a silicon wafer. By heating the wafer to about 1200°C and passing gases containing n-type impurities over the wafer, the n-type impurities diffuse into the wafer surface, producing a thin layer of n-type semi-conductor. The wafer may then be cut into literally hundreds of p-n junctions and each of these tiny pieces may be used to fabricate a semi-conductor diode by bonding metallic connection leads to each side.

Fig. 2.1(c) illustrates the formation of a p-n junction by an alloying process. A small pellet of p-type impurity, such as indium or gallium, is placed on an n-type chip of germanium or silicon (a chip is a small n or p-type crystal cut from a wafer of semi-conductor). The pellet and chip are then heated so that the pellet melts and alloys itself to the n-type chip. On solidification, the p-type impurity re-crystallizes and forms a p-layer and a p-n junction within the chip.

In both the diffusion and alloy processes, p and n-type semi conductors have been brought together to form a crystalline structure, with a p-n junction an integral part of the material. It is essential that there is a continuous crystalline structure for good p-n junction action. It is not possible to obtain good diode or transistor action merely by pressing a p and an n-type material together.

A diagrammatic representation of a p-n junction device is shown in fig. 2.2(a). We will soon see that this device in fact constitutes a diode rectifier. The current carriers in the p-type material are predominantly holes (positive charge carriers) created by the addition of trivalent impurities. Holes are the majority carriers in a p-type material. However, it must also be remembered that a number of hole-electron pairs are generated throughout the device as a result of some electrons gaining sufficient thermal energy to break their bonds. Thus, there will be a

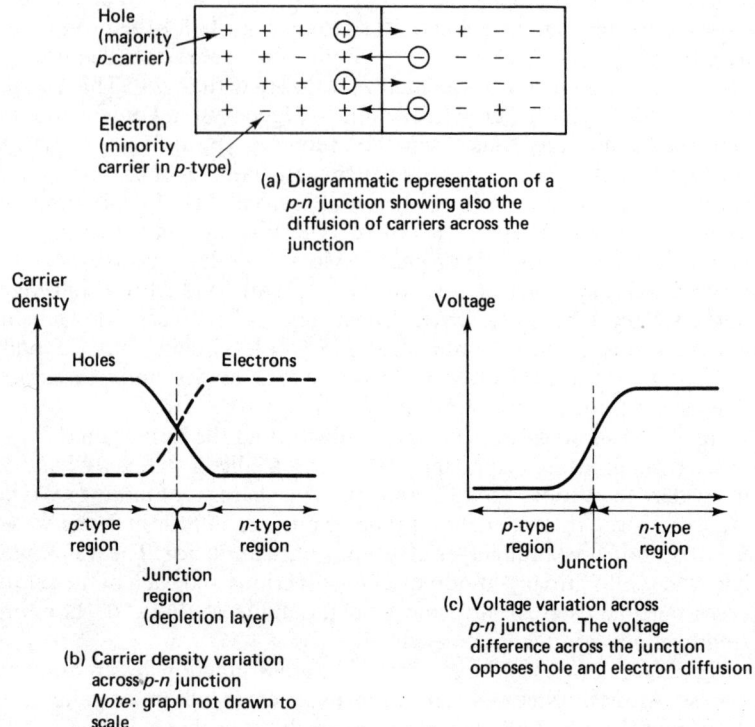

(a) Diagrammatic representation of a
p-n junction showing also the
diffusion of carriers across the
junction

(b) Carrier density variation
across p-n junction
Note: graph not drawn to
scale

(c) Voltage variation across
p-n junction. The voltage
difference across the junction
opposes hole and electron diffusion

Fig. 2.2.

number of electrons in the *p*-type region. Free electrons in a *p*-type material are minority carriers. The ratio of majority to minority carriers at room temperatures is normally in excess of 10 000 to 1. The majority current carriers in the *n*-type material are free electrons donated by the addition of pentavalent impurities. The minority current carriers in the *n*-type material are holes.

In fig. 2.2(a) the + represents holes and the − free electrons. A few minority carriers are also included in each region. Holes diffuse through the *p*-region into the *n*-region, and electrons diffuse through the *n*-region into the *p*-region, this process being indicated by the arrows in the diagram. Both processes result in a net transfer of positive charge from the *p*-type to the *n*-type side of the junction. This results in the *n*-type side acquiring a positive voltage with respect to the *p*-type side. This voltage opposes both hole and electron diffusion across the junction. Eventually, equilibrium is reached in which there is no net flow of either holes or electrons from their respective regions. Graphs showing the variation of the density of holes and electrons across the device and the variation of the voltage set up across the junction are illustrated in fig. 2.2(b) and (c) respectively. The region in the neighbourhood of the junction is often called the depletion layer region. This is because the density of the electron and hole current carriers is a minimum in this region. The region is relatively narrow, being the order of 1 μm (1×10^{-6} m) wide.

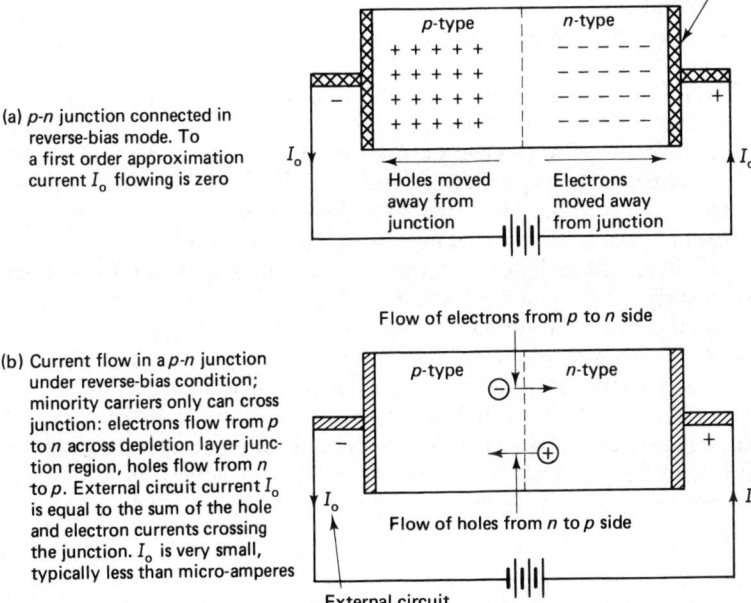

(a) *p-n* junction connected in reverse-bias mode. To a first order approximation current I_o flowing is zero

(b) Current flow in a *p-n* junction under reverse-bias condition; minority carriers only can cross junction: electrons flow from *p* to *n* across depletion layer junction region, holes flow from *n* to *p*. External circuit current I_o is equal to the sum of the hole and electron currents crossing the junction. I_o is very small, typically less than micro-amperes

Fig. 2.3.

Let us now consider the effect of applying a reverse-bias voltage across the *p-n* junction device. In the reverse-bias mode, the negative terminal of the battery is connected to the *p*-side of the junction and the positive terminal to the *n*-side. This situation is shown in fig. 2.3(a). The figure also shows metal conductors bonded into each side of the device. These bonds must produce an excellent contact with the semi-conductor material. The term bias is used to describe a fixed operating condition for a voltage across or for a current flowing through the device. For example, if a reverse-bias of 5 V is required, it means a d.c. source, such as a battery, should impress a voltage of 5 V across the device with its positive terminal connected to *n*-type side and its negative terminal to *p*-type side.

The following is a simple explanation of what happens to the charge carriers in the *p-n* device in the reverse-bias mode:

1. The reverse-bias voltage causes majority carriers to diffuse away from the junction, i.e. the holes in the *p*-type region move to the left and the electrons in the *n*-type move to the right. This is the situation one might expect, positive carriers being attracted to the negative battery side, negative charge carriers being attracted to the positive battery side. This movement of majority carriers does not give rise to any current flow in the external circuit. The holes and electrons are only moved away from the junction. Continuous current flow can only occur when holes and electrons cross the junction.

2. The minority carriers, however, do cross the junction and do provide current flow through the *p-n* device and an electron current flows in the external circuit. The electric field generated by the battery is in the correct direction to move these minority carriers across the junction, i.e. electrons cross from the *p* side and holes cross from the *n* side, as indicated in fig. 2.3(b). In practice, the magnitude of this current is very tiny, typically the order of 1 μA for a germanium *p-n* junction, and even less for a silicon *p-n* junction. As we shall see in the next section, when we change the battery terminals to forward-bias the *p-n* junction, we get currents flowing through the device and in the external circuit of the order of a thousand times greater than this. The tiny current in the reverse-bias condition, which we have seen is due to the presence of minority carriers, is known as the reverse saturation current. It increases with temperature and for many types of *p-n* junctions is largely independent of bias voltage values up to several tens of volts.

2.2
A *p-n* junction connected in the forward-bias mode showing current flow in the diode and the external circuit

In the forward-bias mode the positive terminal of the biasing battery or other suitable d.c. source is connected to the *p*-type side of the junction and the negative terminal to the *n*-type side as shown in fig. 2.4(a). Since the *p-n* junction diode conducts strongly when forward-biased, a variable resistor *R* has been included in the circuit shown so that the current through the diode may be controlled and thus prevent possible damage to the diode by over-heating.

The action of a forward-bias voltage is to effectively lower the voltage which exists across the junction region when the diode is unbiased. Remember that this voltage opposes the diffusion of majority carriers across the depletion layer-junction region. Provided the magnitude of the forward-bias voltage exceeds about 0·2 V for germanium and about

18

(a) *p-n* junction connected
in forward-bias mode. The
p-n diode conducts strongly and
so *R* is included as a safety
precaution to limit *I*

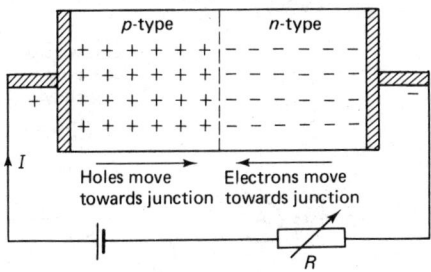

(b) Current flow in a *p-n* junction
under forward-bias conditions;
strong current flow due to
majority carriers crossing junction: holes flow from *p* to *n*,
electrons from *n* to *p*. Forward-
bias current *I* is equal to the sum
of the hole and electron currents

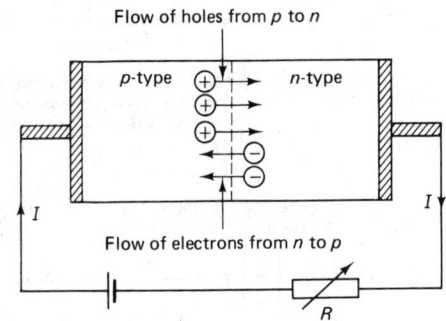

Fig. 2.4.

0·6 V for silicon diodes it will force considerable numbers of holes to
move across the junction from *p* to *n* regions and electrons from *n* to *p*
regions and establish a current of significant magnitude. For example
with a forward-bias of 0·9 V, a medium power silicon diode passes a
current of about 1 A.

Thus under the action of a forward-bias voltage there is a continual
flow of majority carriers across the junction. When holes cross the
junction ($p \rightarrow n$) and electrons cross the junction ($n \rightarrow p$) they rapidly
recombine, holes being neutralized by free electrons in the *n*-type region
and electrons being neutralized by holes in the *p*-type region. For each
recombination an electron enters the *n*-type region from the negative
battery terminal and an electron leaves the *p*-type region (a new hole is
also generated in the *p*-type region by a valence electron breaking its
covalent bond) and enters the positive battery terminal. Hence majority
current carriers are replenished and electron current flows in the external
circuit equal in magnitude to the sum of the hole and electron currents
crossing the *p-n* junction. The current flow in the diode and external
circuit is shown in the diagram of fig. 2.4.

19

2.3
The measurement of the current flow through a *p-n* junction connected in the forward-bias mode

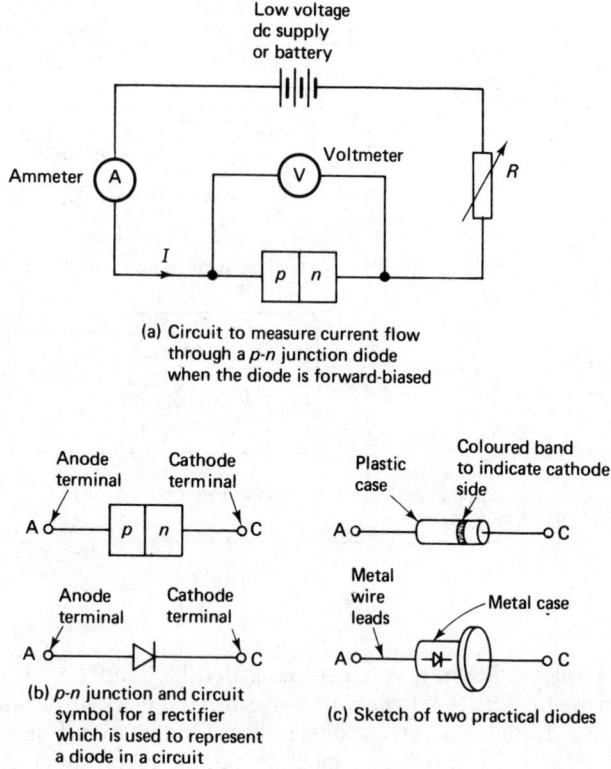

(a) Circuit to measure current flow
through a *p-n* junction diode
when the diode is forward-biased

(b) *p-n* junction and circuit
symbol for a rectifier
which is used to represent
a diode in a circuit

(c) Sketch of two practical diodes

Fig. 2.5.

The circuit shown in fig. 2.5(a) may be used to measure the current flow in a *p-n* diode in the forward-biased mode, i.e. when the *p*-side of the junction is positive with respect to the *n*-side. The variable resistance is included in the circuit to allow the forward-bias voltage across the diode to be varied. This voltage is measured by a voltmeter connected across the diode terminals. Ideally the voltmeter should have a resistance of several megohms if very low diode currents are to be measured accurately. It will then take virtually no current, or at the worst only a fraction of a microampere. The current flowing in the series circuit and therefore through the *p-n* diode is measured by an ammeter. Preferably the ammeter should be a multi-range instrument able to measure micro- and milliamperes for very low and low bias voltages. At bias voltages above the order of 0·4 V for germanium and 0·8 V or so for silicon several tens to hundreds of milliamperes and even amperes (for high-power diodes) flow through the diode and therefore may be required to be measured.

Fig. 2.5(b) shows the circuit symbol used to denote a diode in a circuit diagram. In the forward-biased mode the p-side is positive with respect to the n-side of the junction and thus, taking over the same terms used for thermionic valves and electrolytic cells, the terminal connected to the p-side is known as the anode terminal and the other lead connected to the n-type material of the junction is known as the cathode terminal. The sketches of fig. 2.5(c) show the physical appearance of a typical low-power and a medium-power semi-conductor diode.

The following measurements were obtained for two silicon diodes: Table 2.1(a) for a low-power general purpose diode, Table 2.1(b) for a medium-power diode.

Table 2.1   *Experimental results obtained for two diodes in forward-bias mode*

Forward-bias voltage, V	Diode current, I	Forward-bias voltage, V	Diode current, I
0	0	0	0
0·1 V	1·5 μA	0·1	1 μA
0·2	3·4 μA	0·2	3·5 μA
0·3	6 μA	0·3	5·5 μA
0·4	31 μA	0·4	24 μA
0·5	0·1 mA	0·5	0·3 mA
0·6	0·4 mA	0·6	2·7 mA
0·7	2 mA	0·7	26 mA
0·8	9·5 mA	0·8	280 mA
0·9	29 mA	0·85	490 mA
0·95	45 mA	0·9	910 mA

(a) *V-I* data for a low-power general purpose silicon junction diode

(b) *V-I* data for medium-power silicon junction diode

The bias voltage was first set at 0 V and it was observed that the diode current was zero. It was then increased to 0·1 V by adjustment of the variable resistor and increased in 0·1 V steps up to 0·8 V. At each setting of bias voltage the diode current I was measured. For bias voltages up to 0·6 V it is seen from the tables that I is very small. However, as V is increased beyond 0·7 V both diodes conduct strongly, giving current just below their maximum d.c. rated values at $V = 0.95$ V (*low-power diode*) *and* $V = 0.9$ V (medium-power diode). The graphs of diode current I versus bias voltage V for the two diodes are drawn in fig. 2.6. An alternative means of measuring the V-I forward-bias characteristic of a diode using a variable d.c. voltage supply is given in Problem 2.3.

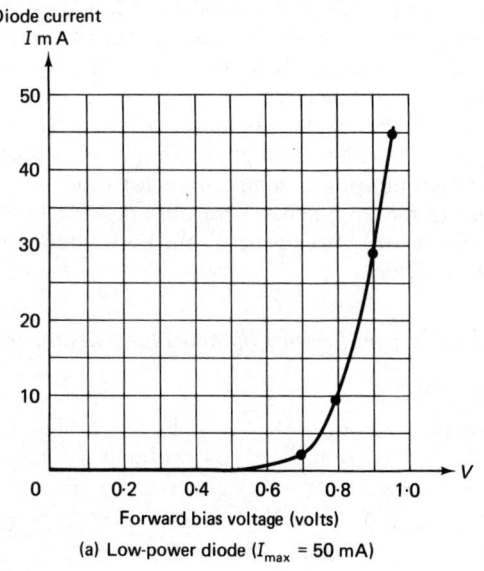

(a) Low-power diode (I_{max} = 50 mA)

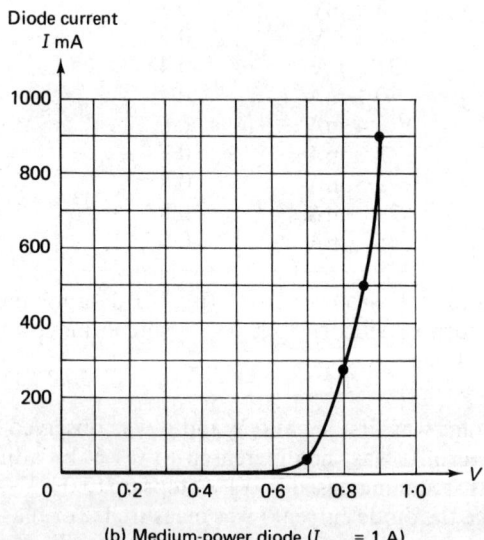

(b) Medium-power diode (I_{max} = 1 A)

Fig. 2.6. V–I forward bias static characteristics for two silicon diodes.

22

2.4

The comparison of the junction potentials of germanium and silicon diodes when connected in the forward-bias mode

The results listed in Table 2.2 were obtained using the circuit of fig. 2.5 for a germanium and a silicon *p-n* junction diode, the diodes having a comparable current *rating*, i.e. designed to be used to pass the same magnitudes of current.

Table 2.2   V-I data used for the comparison of the junction potentials of germanium and silicon diodes when forward-biased

Forward-bias voltage, V	Diode current, I	Forward-bias voltage, V	Diode current, I
0	0	0	0
0·1	5 μA	0·2 V	0·6 μA
0.2	0·25 mA	0·4 V	80 μA
0·3	2 mA	0·6 V	0·5 mA
0.4	5 mA	0·7 V	2·5 mA
0·5	13 mA	0·8 V	10 mA
0·6	33 mA	0·9 V	27 mA
0·65	50 mA	0·95 V	50 mA

(a) *V-I* data for a forward-biased germanium diode

(b) *V-I* data for a forward-biased silicon diode.

The graphs of diode current I against forward-bias voltage V are plotted in fig. 2.7 and the two curves may be used to compare the diode junction potentials—that is the voltage dropped across each *p-n* junction—for a given significant current flow. Below the forward-bias voltage of 0·2 V the current flowing through the germanium diode is very small, but as the bias voltage increases above 0·2 V the diode conducts strongly. For bias voltages up to about 0·6 V the current in the silicon diode is very small but beyond 0·6 V it conducts strongly.

We can therefore state that there exists a threshold of forward-bias voltage below which the current in germanium and silicon diodes is very small, typically less than 1% of the maximum rated current that the diodes can safely pass. Above the threshold voltage the diode currents increase very rapidly. The threshold voltage of junction potential is normally taken as about 0·2 V for germanium diodes and 0·6 V for silicon diodes. We may also state that for a given value of forward-bias current the junction potential required across a silicon diode is always higher than that across a germanium by the order of 0·3 to 0·4 V, e.g. from the diode characteristics of fig. 2.7, for a diode current $I = 30$ mA

23

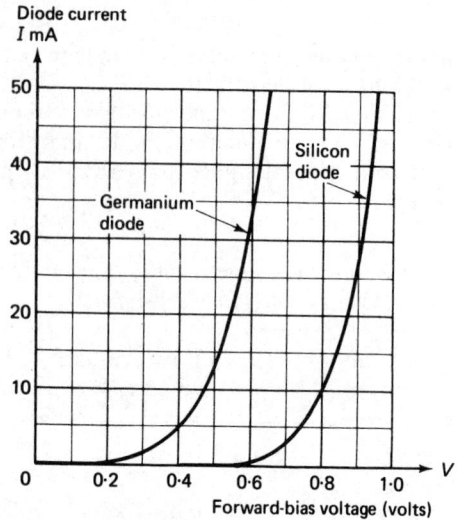

Fig. 2.7. *V–I* forward bias static characteristics for two similarly rated diodes. The characteristics are used to compare the junction potential (*V*) across germanium and silicon diodes.

the junction potential required across the germanium diode is $V \approx 0.58$ V, whilst for the silicon diode $V \approx 0.92$ V.

2.5
A sketch of the static characteristic for a diode

A plot of diode current versus voltage across the diode for both forward and reverse-bias voltages is known as the static or d.c. characteristic of the diode. Since diode characteristics vary quite considerably with temperature, the temperature at which the *V-I* values are measured is normally quoted. Typical values are at 25°C (normal ambient), 50°C to 60°C (maximum operating limit for germanium), 75°C to 150°C (for silicon, which may operate at much higher temperatures than germanium).

A typical static characteristic is sketched in fig. 2.8. Note that the scales for forward and reverse-bias currents (and voltages) are different: 1 division = 200 mA for forward-bias current, whilst 1 division = 0.01 μA (for silicon diodes) or 1 division = 1 μA (for germanium diodes). The characteristic may be divided into three regions:

1. The forward-bias range where considerable current flows in the diodes when

$$V > 0.2 \text{ V for germanium diodes,}$$
$$V > 0.6 \text{ V for silicon diodes.}$$

24

2. The reverse-bias-leakage current range.

For reverse-bias voltages up to the breakdown voltage V_B the diode current (often referred to as leakage-current) is small, typically the order of micro-amperes for germanium and nano-amperes (1 nA = 0·001 μA) for silicon diodes at 25°C. For some diodes the reverse-bias current

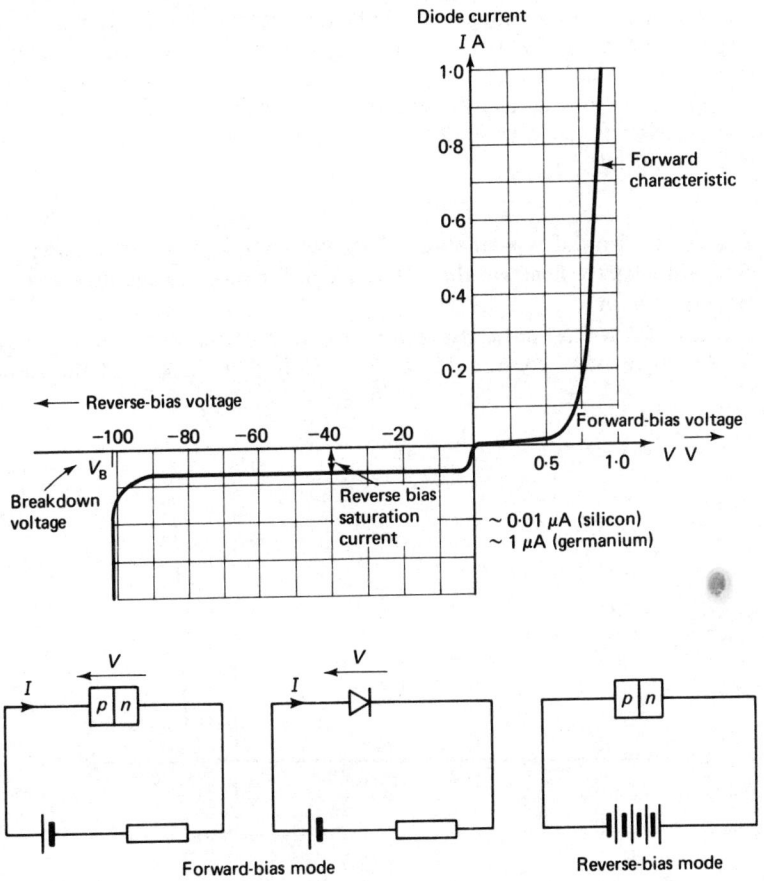

Fig. 2.8. A typical static characteristic for a diode.

remains fairly constant and largely independent of reverse-bias voltage. This situation is illustrated in our typical characteristic. The value of the current is known as the reverse-bias saturation current. In other cases (see fig. 2.10 for example) the reverse-bias current increases slowly with increase in reverse-bias voltage.

3. The breakdown region.

When a critical value of reverse-bias voltage is exceeded a relatively large reverse current flows. This critical value of voltage is known as the breakdown voltage V_B. Unless the diode is specifically designed to operate in this region it will be permanently damaged if reverse-bias voltages in excess of V_B are impressed across the diode, so in our typical case V should never be more negative than 100 V. However, diodes which are designed to dissipate safely the heat generated when the relatively large reverse currents flow, with $V = V_B$, are extensively used in voltage regulator and stabilization circuits (see Sub-sections 3.1 and 3.4). These diodes are known as Zener and avalanche diodes. A typical characteristic for a Zener diode, breakdown occurring just after $V = -6V$, is drawn in fig. 2.9.

2.6
The comparison of typical static characteristics for germanium and silicon diodes to illustrate the difference in forward voltage drop and reverse current

Fig. 2.10 shows the static characteristics for a germanium and a silicon diode which have comparable current ratings at 25°C and the same

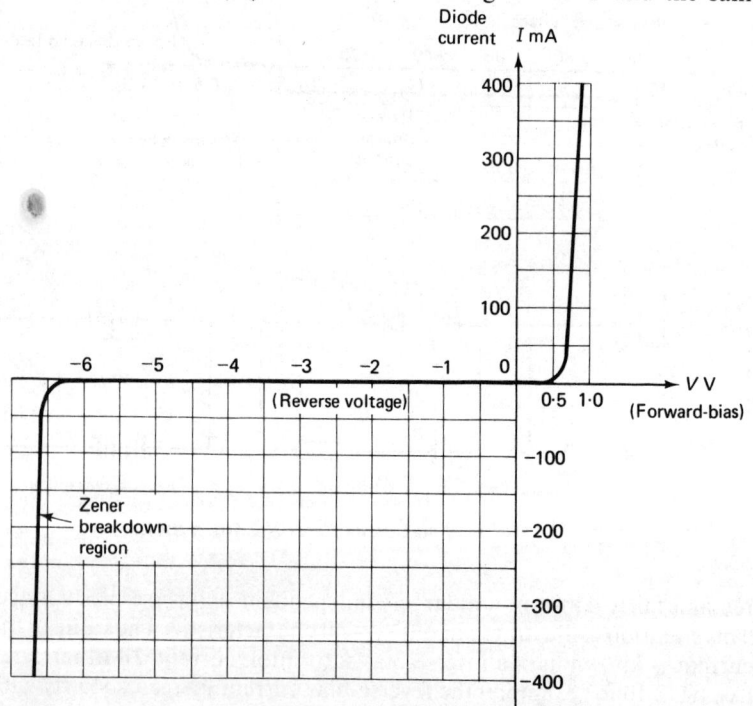

Fig. 2.9. A typical characteristic for a Zener diode.

26

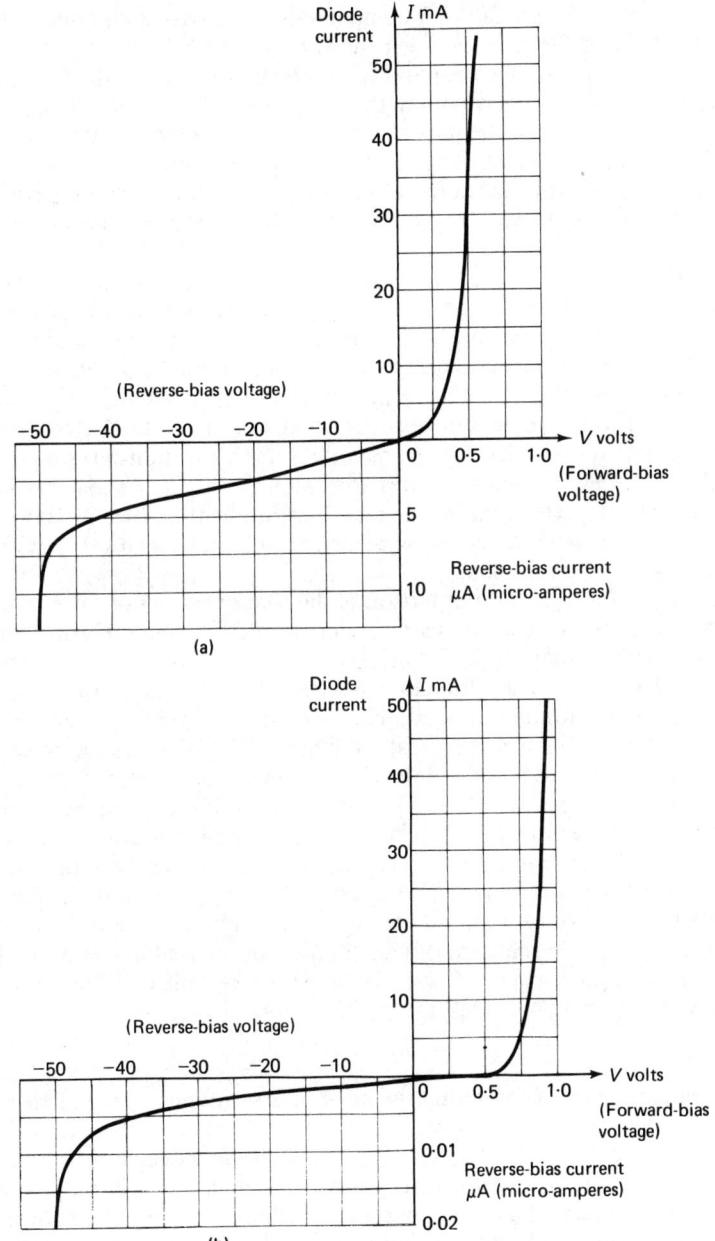

Fig. 2.10. Static characteristics of (a) a germanium, (b) a silicon diode, measured at 25°C. Current rating of both diodes is $I_{max} = 150$ mA.

breakdown voltage of 50 V. A comparison of the two curves illustrates the following differences between the two diodes:

(a) The forward voltage drop (i.e. the junction potential in the forward-bias mode) is less for germanium than for silicon. The threshold voltage for the order of 1% of the maximum rated current flow, in this case I_{max} = 150 mA so $I \approx 1.5$ mA, is about 0.2 V for germanium and about 0.6 V for silicon. For a forward current of $I = 40$ mA the forward voltage drop across the respective diode is about 0.55 V for germanium, about 0.9 V for silicon.

(b) The reverse-bias current for the germanium diode varies between 0 to 5 μA for reverse-bias voltages up to 40 V; for the same bias voltage range the reverse-bias current in the silicon diode varies between 0 and 0.005 μA. Thus the reverse-bias current in the silicon diode is of the order of 1000 times less than for the germanium diode. Remember that these values refer to an ambient or operating temperature of 25°C. (Note also that for both diodes there is no constant reverse saturation current range evident.) At higher temperatures for example 50°C, the reverse-bias current in the germanium diode may increase to the order of 100 μA, whilst the silicon diode reverse-bias current will still be relatively low and typically less than 0.1 μA.

As a general rule the comparison of the characteristics of germanium and silicon diodes enables us to conclude the following. Germanium diodes have a lower forward voltage drop than silicon, typically about 0.3 to 0.5 V less. They therefore possess this advantage over silicon diodes. Germanium diodes have relatively poor reverse-bias characteristics. They pass typically the order of 1000 times the reverse-bias current for the same reverse-bias voltage compared to silicon. Germanium diodes are also very much dependent on temperature and should not normally be used for any high temperature applications. Silicon diodes pass very low reverse-bias currents, typically of the order of nano-amperes for low-power diodes, and may be used at higher temperatures. For example silicon diodes could be used at ambients as high as 125°C. Germanium diodes should not be used above 50°C. In most applications where a forward voltage of 0.6 V plus is of no serious disadvantage silicon diodes are invariably used.

2.7
The importance of considering the peak inverse voltage of the diode

The peak inverse voltage is the maximum reverse-bias voltage to which the diode is subjected in the particular application in which the diode is being used. The maximum rating of voltage which can be impressed across the diode in the reverse-bias mode without causing breakdown is known as the peak inverse voltage rating. Thus in diode circuit applications (with the exception of voltage-regulator-Zener diode circuits where breakdown is deliberately employed) it is of utmost importance to ensure that breakdown never occurs, by selecting diodes with peak inverse voltage ratings well in excess of the largest reverse-bias voltages that might be generated across the diodes.

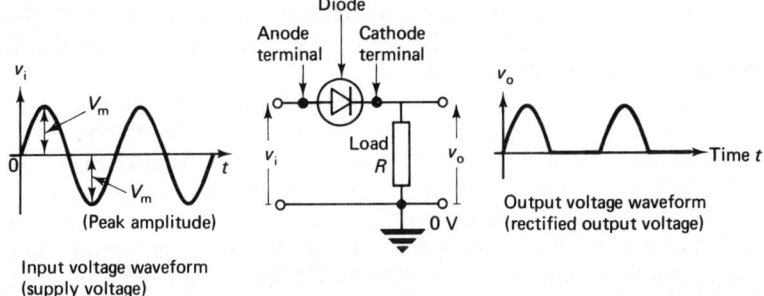

Fig. 2.11. A simple diode rectifier circuit. Peak reverse bias voltage across diode = V_m, the peak amplitude of the supply voltage.

Let us consider the circuit of fig. 2.11 to demonstrate the meaning of peak inverse voltage (i.e. the peak reverse-bias generated across the diode) in a simple rectifier application. The input voltage is an a.c. (alternating) voltage of peak amplitude V_m, i.e. its voltage varies continuously reaching a maximum of $+V_m$ on positive half-cycles and going down to a peak value of $-V_m$ on negative half-cycles. During positive half-cycles the diode is forward-biased and hence current flows through the diode and load R. The output voltage v_0 equals the input voltage less the small forward-bias potential drop across the diode. However, during negative half-cycles the diode is reverse-biased and hence virtually zero current flows in the circuit and the output voltage falls to zero. Thus with $v_0 = 0$ V the whole of the supply voltage is effectively impressed across the diode with the anode negative with respect to the cathode terminal. The peak value of the supply voltage is V_m, so in this application the diode must be able to withstand a reverse-bias voltage of V_m without breaking down, i.e. the peak inverse voltage rating of the diode must be at least V_m.

2.8
A demonstration of the breakdown effect
Under normal operating conditions only a tiny current flows in a diode in the reverse-bias mode. This current is due to the presence of thermally generated minority carriers: electrons in the p-type region, holes in the n-type region. The number of these carriers depends on temperature. Thus at a given temperature we should expect a constant reverse-bias current independent of reverse-bias voltage. This situation is indeed exhibited in

29

many diodes, and reverse-bias currents of the order of μA for germanium and nA for silicon are obtained for reverse-bias voltages from about 0·1 V to several 10's or even 100's of volts. However, when the reverse-bias voltage reaches the breakdown voltage of the diode, a sudden increase of current occurs and the reverse-bias current can rise to values comparable to forward-bias values. In normal diodes these large currents will cause permanent damage, and therefore operation under breakdown conditions must always be avoided.

The rapid increase of current at breakdown occurs owing to two effects: Zener breakdown and avalanche effects. In Zener breakdown intense electric fields set up in the semi-conductor by the application of a breakdown reverse-bias voltage 'pulls' electrons out of their covalent bonds, thereby increasing the number of electron-hole current carriers. In the avalanche effect the high electric fields accelerate the thermally generated electrons and these electrons acquire sufficient energy to 'knock-out' electrons in covalent bonds when they collide with the bound electrons in the crystal lattice. This process gives rise to more current carriers. The free electrons so generated will be similarly accelerated and produce further carries by collision, i.e. an avalanche is created.

Zener diodes and avalanche diodes are designed to operate without damage under breakdown conditions, provided an upper limit of current is not exceeded.

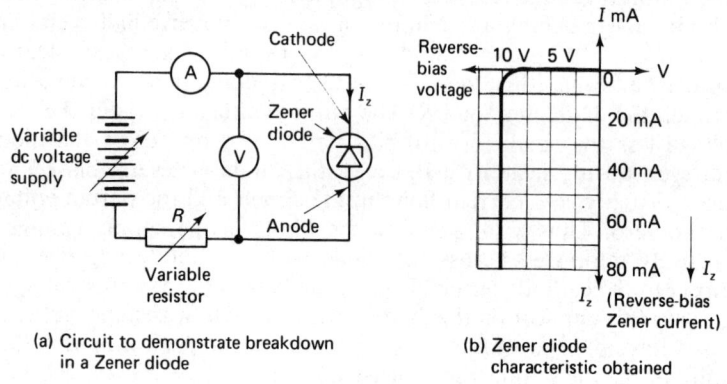

(a) Circuit to demonstrate breakdown in a Zener diode

(b) Zener diode characteristic obtained

Fig. 2.12.

The circuit shown in fig. 2.12(a) may be used to demonstrate the breakdown effect in a Zener diode. In this particular demonstration a Zener diode with a breakdown or Zener voltage of 10 V, i.e. breakdown occurs when the anode terminal is −10 V with respect to the cathode, and a maximum current rating of 100 mA, is used. The variable resistor R is included in the circuit to limit the current through the diode at breakdown. For example, with the breakdown voltage at 10 V and the bias supply set at a reverse-bias of 11 V the voltage drop across R is $11-10 = 1$ V, so the diode current $I_z = 1/R$ amperes. Thus if R were set initially at 500 Ω say, the diode current would be limited to about 2 mA when breakdown occurs.

Thus initially R is set at about 500 Ω. The reverse-bias voltage across the Zener is then increased from 0 V in say 1 V steps by increasing the output of the variable d.c. supply. For reverse-bias voltages up to 10 V the reverse-bias current is negligibly small. However, when the reverse-bias reaches 10 V a sudden increase of current is observed. Increasing the output from the d.c. supply increases the Zener current I_z but no increase in the voltage across the Zener diode is observed. The diode voltage remains constant at 10 V. Fig. 2.12(b) shows the complete reverse-bias characteristic. This may be plotted by measuring V–I values by varying both R and the supply voltage in the breakdown region, and by varying the supply voltage only in the normal reverse-bias saturation region.

Section 3: *The expected learning outcome of this section is that the student should know simple applications of semi-conductor diodes.*

3.1
Simple applications of the available range of (a) power diodes, (b) Zener diodes, (c) signal diodes

(a) *Power diodes*
The main application of power diodes is as rectifying elements to convert a.c. to d.c. Remember a.c. is the abbreviation for alternating and d.c. for direct (constant) current. Electrical power is distributed in a.c. form but many industrial processes, machines, electrical and most electronic equipments, require d.c. power supplies for their operation. The basic operation of a rectifier circuit in a d.c. power supply is illustrated in fig. 3.1(a).

Any electrical device which offers a low resistance to current flow in one direction but a high resistance to current flow in the reverse direction is called a rectifier. An ideal rectifier passes a current freely with no resistance (and therefore no voltage drop) in one direction but has infinite resistance and passes no current in the reverse direction. Silicon and germanium diodes approach more closely to the ideal rectifier than any other device. Silicon diodes are very much more widely used because of their excellent reverse-bias properties—they pass only tiny currents in the reverse direction—and their much superior temperature characteristics. Silicon power diodes may be designed to pass several tens and even hundreds of amperes in the forward-bias mode with only a volt or so potential drop across the diode. The *p-n* junction area of a diode

31

capable of handling in excess of 100 A is of the order of 100 mm², a hundred or so times greater than a low-power diode working with currents of up to a few 100 mA. A sketch of a high-power silicon diode with typical dimensions is drawn in fig. 3.1(b). In use it must be bolted down onto a large block of metal (i.e. a heat sink) to ensure that the power dissipated as heat in the diode is conducted away. For very high power applications special cooling provisions must be made.

As rectifying elements power diodes find application in d.c. power supplies of electronic and electrical equipment; in d.c. supplies for use in medium scale electrolytic plating processes; in battery-charger circuits

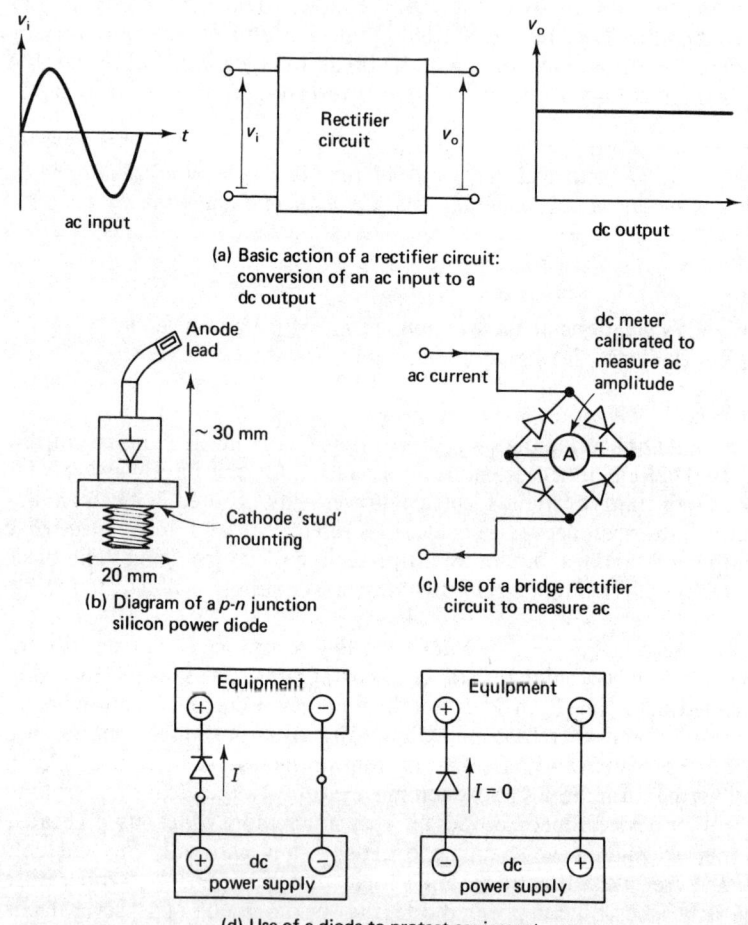

(a) Basic action of a rectifier circuit: conversion of an ac input to a dc output

(b) Diagram of a *p-n* junction silicon power diode

(c) Use of a bridge rectifier circuit to measure ac

(d) Use of a diode to protect equipment

Fig. 3.1.

to charge secondary cells; in rectifier circuits used to convert a.c. to d.c. so, with the aid of such a rectifier, a moving coil voltmeter or ammeter may be used to measure an a.c. voltage or current (fig. 3.1(c) shows a suitable circuit; the bridge circuit is considered in the next section). Power diodes may be used as protective devices to safeguard equipment. A simple example is shown in fig. 3.1(d). If the d.c. supply is connected with the right polarity the diode is forward-biased and current flows into the equipment. However, if the supply is connected the wrong way round, the diode becomes reverse-biased and hence no current flows into the equipment. Power diodes (normally medium and low-power, say diodes which can safely dissipate up to a few watts) are also widely used in electronics circuits: switching circuits, clipping circuits, clamping circuits, which have applications in control, computing, and communications.

(b) *Zener diodes*
A major application of Zener diodes is as a voltage stabilizing or regulating element. Over the reverse-bias breakdown region the voltage across the diode remains very nearly constant and independent of the diode current flowing. This constant voltage characteristic is used to stabilize the voltage output of d.c. power supplies against variations in a.c. input and changes in load current being drawn from the supply. A simple voltage stabilizer circuit is considered in Sub-sections 3.4 and 3.5. Zener diodes are available with maximum power ratings in excess of 100 W and therefore may be used in quite high power circuits.

Zener diodes also find wide application as voltage reference sources. They can be made with a wide range of breakdown voltages, typically from a few to several hundred volts, and are commonly used in circuits to set a required voltage level. They may be used also as protective devices, breaking down and conducting current when their Zener voltage is exceeded.

(c) *Signal diodes*
Signal diodes, as distinct from power diodes, handle very much smaller powers, typically of the order of milliwatts and in some cases even microwatts. They are sensitive devices with small *p-n* junctions and are used extensively to detect radio frequency signals. They are an essential element in radio, television, and communication receivers.

A simple diode detector circuit is shown in fig. 3.2. The input to the circuit is an amplitude-modulated radio frequency signal, such as is provided in many radio programme transmissions. The object of the detector circuit is to provide an output voltage which varies in accordance with the shape or 'envelope' of the input signal. The diode rectifies the radio frequency input voltage and passes a pulsating 'd.c.' current. The *R-C* network is designed to follow the positive peaks of the

33

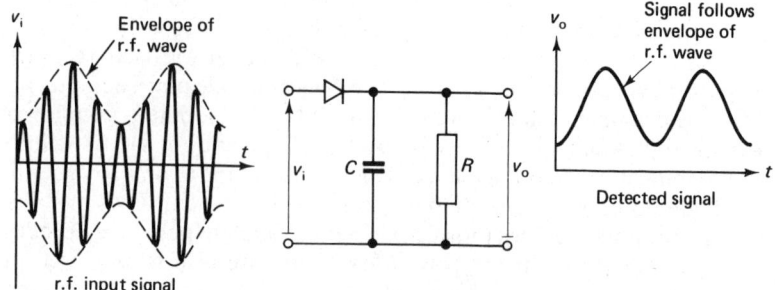

Fig. 3.2. Basic circuit for a diode detector.

radio frequency (r.f.) waveform and the voltage generated across the network reproduces the much slower variation in amplitude of the r.f. signal envelope. It is this latter variation which contains the information carried by the r.f. wave. The detector therefore serves to extract this information.

3.2
Sketches of waveforms of applied a.c. voltage and load current for diode circuits which provide half-wave and full-wave rectification into a resistive load

In many rectifier circuits it is common practice to apply the a.c. input voltage via a transformer. We shall therefore start this sub-section with a brief description of the function of a transformer. A diagram of a simple two-winding transformer and its circuit symbol is shown in fig. 3.3(a). The windings on the input side are known as the primary windings and those on the output as the secondary windings. The transformer has the property of stepping up or stepping down the amplitude of an applied a.c. voltage. If the number of primary turns is N_1, the number of secondary is N_2 and the amplitude of the input a.c. voltage is V_1, then the amplitude of the output voltage induced across the terminals of the secondary windings is

$$V_2 = \frac{N_2}{N_1} V_1$$

Fig. 3.3(b)(i) shows the case when $N_2 > N_1$, i.e. a step-up situation. Fig. 3.3(b)(ii) shows the case when $N_1 < N_2$, i.e. a step-down situation.

Thus by employing a transformer with a suitable turns ratio (N_1/N_2) we can select the value of a.c. voltage amplitude applied to a rectifier circuit, e.g. if we are operating from the mains where the peak amplitude is $\sqrt{2} \times 240 = 339$ V (240 V is the root-mean square (r.m.s.) value of the

34

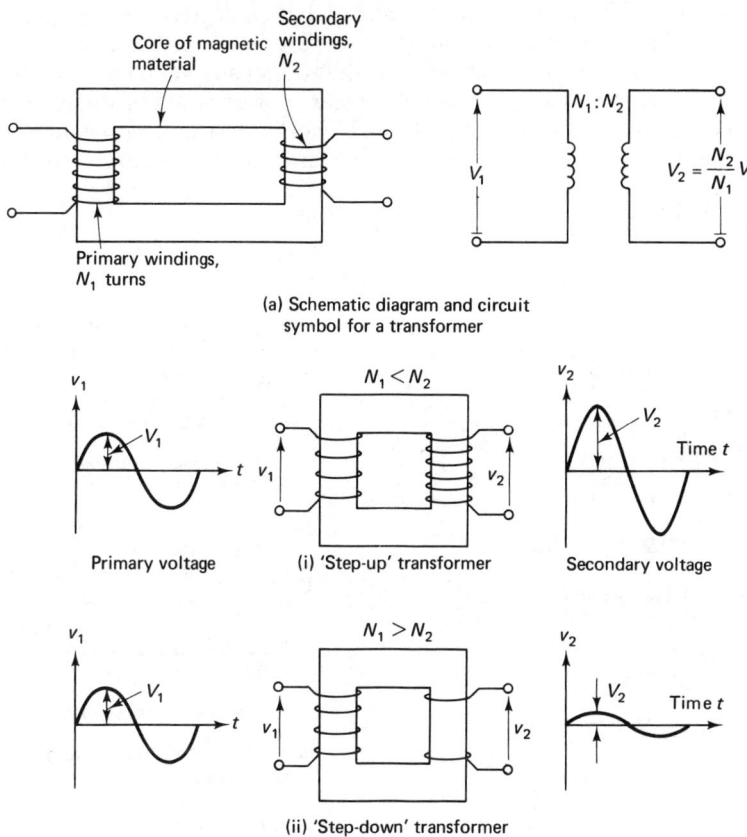

(a) Schematic diagram and circuit symbol for a transformer

Primary voltage

(i) 'Step-up' transformer

Secondary voltage

(ii) 'Step-down' transformer

(b) Schematic diagram of voltage step-up and step-down transformer

Fig. 3.3.

mains supply in the U.K.) and we wish to rectify to a maximum d.c. voltage of 20 V, which, as we will soon show, means the secondary voltage amplitude required is 20 V, we would select:

$$\frac{N_2}{N_1} = \frac{V_2}{V_1} = \frac{20}{339} = 0.06$$

that is, the secondary winding would need to have 6 turns for every 100 on the primary.

A second advantage of using a transformer is that the circuit connected to the secondary is not directly connected to the mains or a.c. input supply. Transformer action occurs by varying magnetic flux flowing through the core. There is no conductor connection between the primary and secondary turns. Thus we have a degree of isolation from the mains supply which can add to safety of operation. A third advantage is that transformer action occurs with little, ideally no, power loss. If we were to employ a resistive potentiometer, power would be dissipated in the resistances of the potentiometer network. The use of a potentiometer would also seriously impair the regulation of the rectifier (see Subsection 3.4).

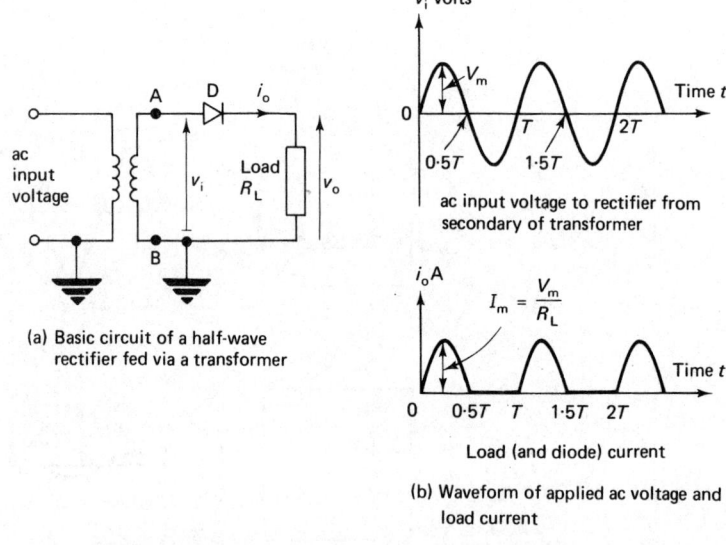

(a) Basic circuit of a half-wave rectifier fed via a transformer

ac input voltage to rectifier from secondary of transformer

Load (and diode) current

(b) Waveform of applied ac voltage and load current

Fig. 3.4. Half-wave rectifier: basic circuit and input voltage and output current waveforms.

Fig. 3.4(a) shows the basic circuit of a half-wave rectifier fed via a transformer. On positive half-cycles of supply voltage (i.e. when terminal A is positive with respect to B), the diode D is forward-biased. It therefore conducts and current flows in the secondary circuit through D and the resistive load R_L. If we neglect the voltage drop across the diode, which will only be a volt or so, the peak value of load current is

$$I_m = \frac{\text{peak value of secondary voltage across AB}}{R_L} = \frac{V_m}{R_L}.$$

On negative half cycles (i.e. A is negative) the diode D is reverse-biased and acts as a very high resistance, normally in excess of several megaohms. Thus virtually no current flows on negative half-cycles and the load current $i_0 = 0$. The waveforms of a.c. input voltage to the rectifier and of the load current are shown in fig. 3.4(b). The rectifier is called half-

36

wave for the simple reason that it only produces an output on positive half-cycle of the a.c. input voltage.

If we were to place a moving coil ammeter in series with the load the meter would measure the average value of load current taken over a complete cycle of input voltage. The average value of load current in a half-wave rectifier is

$$\bar{i}_0 = \frac{I_m}{\pi} = \frac{V_m}{\pi R_L} = 0 \cdot 318 \times \text{peak value} \left(\pi = 3 \cdot 142, \frac{1}{\pi} = 0.318 \right).$$

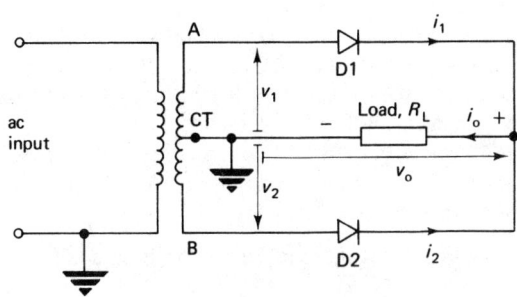

(a) Basic circuit of a full-wave rectifier

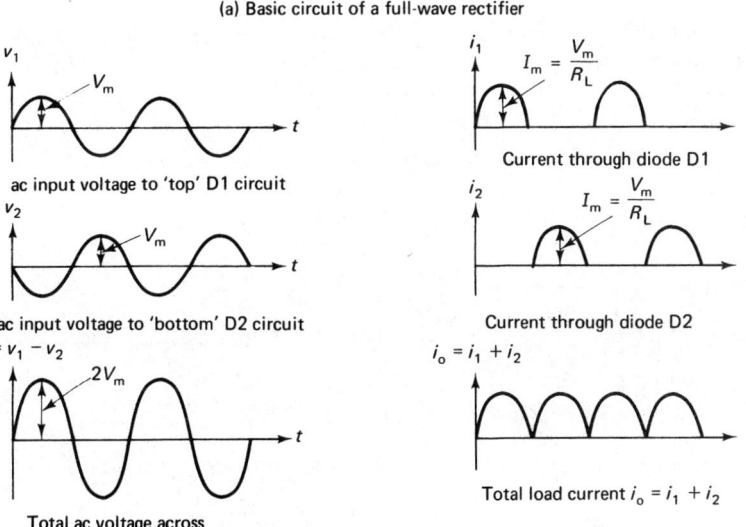

(b) Waveforms of applied ac voltage, the currents in diodes D1 and D2, and the load current

Fig. 3.5. Full-wave rectifier: basic circuit and input voltage and output current waveforms.

Fig. 3.5(a) shows the basic circuit of a full-wave rectifier, 'full-wave' because the rectifier produces an output on both positive and negative half-cycles of a.c. input voltage. The rectifier consists essentially of two half-wave circuits: the 'top' circuit containing D1 is supplied by v_1 (see fig. 3.5(b)) where v_1 is the a.c. voltage developed across terminal A and the centre-tap point CT of the secondary transformer winding; the 'bottom' circuit containing diode D2 is supplied by v_2, the voltage developed across terminal B and centre-tap point CT.

On positive half-cycles of the mains input to the primary, v_1 is positive so diode D1 is forward-biased and current i_1 flows through D1 and the load R_L. On negative half-cycles of the mains input, v_1 is negative since terminal A goes negative and hence D1 is cut off, i.e. D1 is reverse-biased. However, v_2 goes positive, i.e. B goes positive with respect to CT, and hence diode D2 is forward-biased. Current i_2 therefore flows through D2 and the load R_L, and in the same direction as i_1. Thus current flows through the load every half-cycle of a.c. input voltage. Input voltage and output current waveforms are shown in fig. 3.5(b). If we neglect the voltage drops across the diodes when they are forward-biased, the peak current flowing through the load is

$$I_m = \frac{\text{peak voltage between A or B and CT}}{R_L} = \frac{V_m}{R_L}$$

noting that in this case, $V_m = \frac{1}{2} \times$ peak voltage across secondary terminals A-B. The average value of load current for the full-wave rectifier is

$$\bar{i}_0 = \frac{2}{\pi} I_m = 0.636 \times \text{peak value}.$$

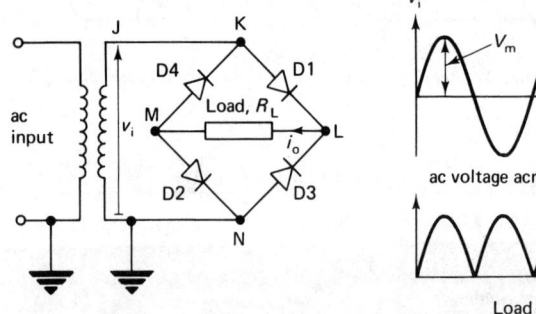

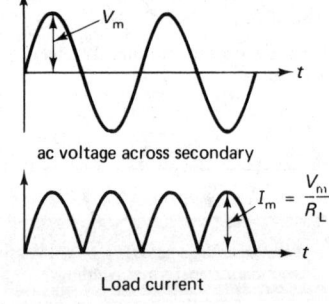

ac voltage across secondary

Load current

(a) Basic circuit of a bridge (full-wave) rectifier

(b) Waveforms of applied ac voltage and load current

Fig. 3.6. The bridge rectifier circuit.

A very popular form of full-wave rectifier is the bridge rectifier circuit shown in fig. 3.6(a). This circuit enables the full secondary voltage to be effectively impressed across the load and does away with the necessity of using a centre-tapped secondary winding. The four diodes making up the bridge may be obtained in a unit block with two pairs of connections available for connecting the load and the a.c. input.

On positive half-cycles, i.e. when terminal J is positive, diodes D1 and D2 are forward-biased. Current therefore flows through the load along the path JKLMN. On negative half-cycles, J goes negative with respect to N (N is in fact normally earthed and therefore at 0 V), which forward-biases diodes D3 and D4. Current then flows through R_L along the path NLMKJ. The a.c. input voltage and load current waveforms are sketched in fig. 3.6(b).

3.3
The effects of connecting a smoothing capacitor across the load resistor in half and full-wave rectifier circuits upon the diode current waveform, the load current waveform, the load voltage waveform, and the inverse voltage applied to the diode

Fig. 3.7(a) shows a half-wave rectifier with a capacitor C connected across the load resistor R_L. When the circuit is initially connected to an a.c. supply current will flow both to charge C and to supply the load current through R_L. This situation occurs during the first quarter period of positive supply as the a.c. input voltage rises from 0 to its peak amplitude V_m, so charging C via diode D from 0 to V_m volts. As the a.c. input reduces below V_m the diode D becomes reverse-biased, i.e. its anode voltage is falling whilst we assume C remains charged to (or almost to) V_m. However, some of the charge stored by C may now be utilized to maintain current flow through the load. Thus provided C has sufficient charge storage capacity, the load current and voltage is maintained approximately constant even though the diode is non-conducting. The load voltage which is of course equal to the capacitor voltage, since C and R_L are in parallel, decays relatively slowly, until on the next positive half-cycle of the a.c. input the diode D becomes forward-biased. This occurs when the a.c. input voltage rises to equal the voltage across C and R_L. At this point D conducts and a surge of current is passed to replenish the charge previously lost by C as well as supplying the load current.

The effect of C on the waveforms of output voltage v_0, load current i_0, and the diode current i_D is shown in fig. 3.7(b). The outputs of load voltage and current when C is absent are shown for reference as dashed curves and the actual outputs with C present by full lines. The waveforms are ideal in the sense that forward voltage drop across the diode and other practical circuit losses such as winding resistances and transformer leakage effects have been neglected. Note that the diode conducts only over a relatively short time from the cut-in point (marked X), which

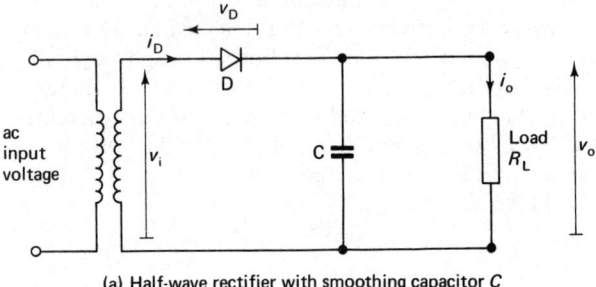

(a) Half-wave rectifier with smoothing capacitor C

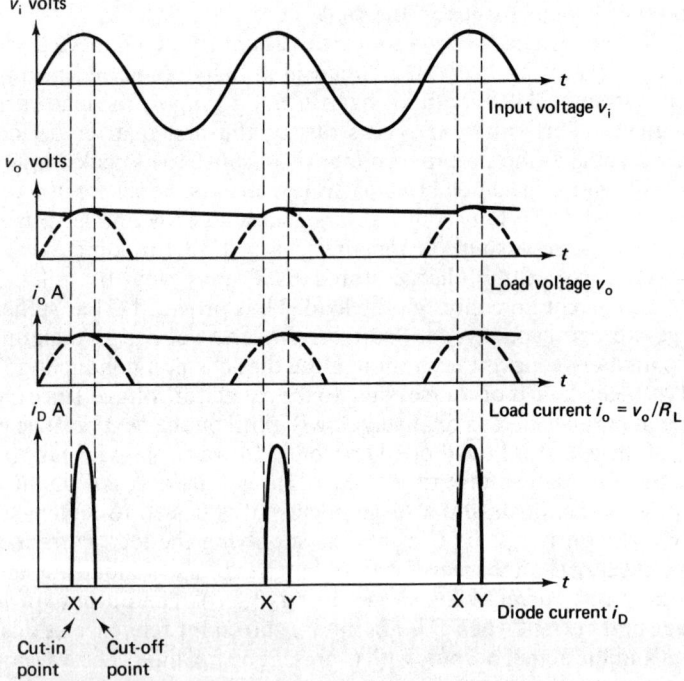

(b) Voltage and current waveforms for half-wave rectifier
with the smoothing capacitor C

Fig. 3.7. Capacitor smoothed half-wave rectifier: circuit and waveforms.

occurs when the input equals the load voltage, to the cut-off point marked Y. In practice X occurs just before the a.c. input v_i reaches its maximum value V_m and Y just after.

If the diode conduction time X–Y is short compared to the period T of the a.c. input ($T = \frac{1}{50}$ s = 20 ms for a.c. mains frequency of 50 Hz) then we can make the following estimate for the variation in load voltage:

since charge Q = current × time

the charge lost by C during its discharge time (assumed equal to T) is

$$Q = I_0 \times T$$

where I_0 = average value of load current, i.e. the d.c. value;

but as capacitance $= \dfrac{\text{charge}}{\text{voltage}}$, i.e. $C = \dfrac{Q}{V}$

the amount by which the capacitor voltage falls during its discharge is

$$V = \frac{Q}{C} = \frac{I_0 \times T}{C}$$

Thus we can state that the capacitor and therefore the load voltage varies between

$$V_m \text{ and } V_m - V = V_m - \frac{I_0 T}{C} ,$$

e.g. if $V_m = 50$ V, $T = 20$ ms, $C = 1000$ μF, $R_L = 1$ kΩ, then $I_0 \approx \dfrac{V_m}{R_L}$

$= \dfrac{50}{1000} = 0\cdot05$ A (we neglect V in this calculation) and the drop in

voltage $V = \dfrac{I_0 T}{C} = \dfrac{0\cdot05 \times 20 \times 10^{-3}}{1000 \times 10^{-6}} = 1$ V, i.e. output voltage varies

between 50 V and 49 V. V is known as the peak-to-peak ripple voltage and gives the maximum variation in output voltage.

The effect of varying R_L and C on the output voltage waveform is shown in fig. 3.8. When R_L is reduced the load current increases and so does the ripple voltage $V = I_0 \times (T/C)$; increasing C reduces the ripple voltage amplitude.

The maximum reverse-bias voltage applied across the diode occurs when the a.c. input is at $-V_m$, i.e. at its peak negative value. Thus, since the load voltage is held very nearly at V_m, the peak inverse voltage V_{PIV} of the diode is

$$V_{PIV} = (\text{cathode voltage–anode voltage})_{max} = V_m - (-V_m) = 2V_m.$$

Thus the presence of C causes the peak inverse voltage to increase from a value equal to the maximum secondary winding voltage when no capacitor is used to twice the maximum secondary voltage. The diode, of course, must be chosen with a rating to withstand such a peak inverse

41

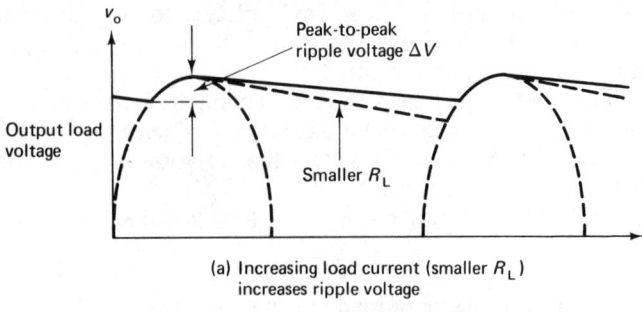

(a) Increasing load current (smaller R_L) increases ripple voltage

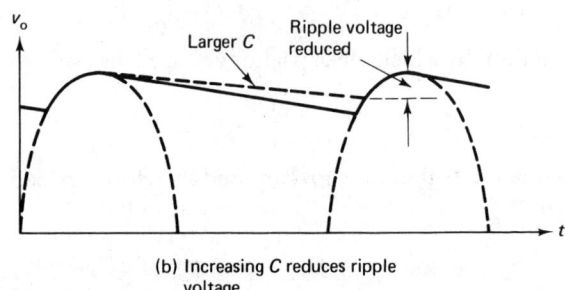

(b) Increasing C reduces ripple voltage

Fig. 3.8. The effect of increasing load current and smoothing capacitor C on output voltage of a capacitatively smoothed half-wave rectifier.

voltage. The diode must also be rated to take the heavy surge current which flows through it during its relatively short forward-biased conduction time.

Fig. 3.9 shows a capacitatively smoothed full-wave rectifier, together with waveforms of load voltage, load current, diode D1 and diode D2 conduction currents. The secondary voltages between terminal A and the centre-tap CT and terminal B and CT are shown as dashed curves for reference. The action of C is exactly similar to the half-wave case except that C is charged twice every cycle of a.c. input voltage: once via D1 and once via D2; and its discharge time is approximately $\frac{1}{2}T$ (compared to T for the half-wave case). The peak inverse voltage across each diode is approximately $2\ V_m$, where V_m = peak amplitude of secondary voltage between A (or B) and CT. Note that the load voltage waveform does not exactly follow the secondary voltage when the diodes are forward-biased and C is being charged. The waveform shown is typical of that obtained for a practical circuit and includes the effects of voltage drops across the diodes, in the secondary windings, and imperfections in the transformer. For a given load current and smoothing capacitor value the ripple voltage of a full-wave circuit is approximately half that obtained in a half-wave circuit.

42

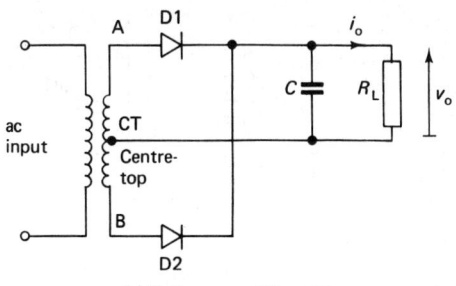

(a) Full-wave rectifier with
smoothing capacitor C

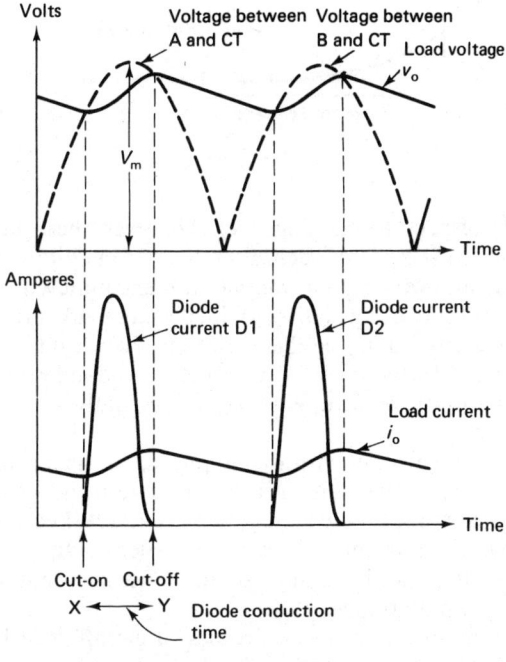

(b) Practical waveforms for full-wave rectifier
with smoothing capacitor

Fig. 3.9. Capacitor smoothed full wave rectifier: circuit and practical waveforms.

3.4
A circuit diagram of a stabilized voltage source including a Zener diode and a series resistor

The use of a smoothing capacitor across the load in a rectifier circuit produces a d.c. voltage with a low ripple content provided C is correctly

43

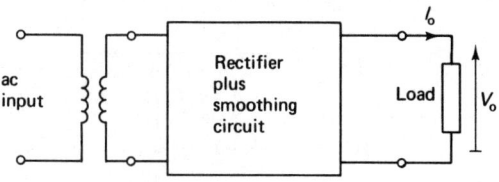

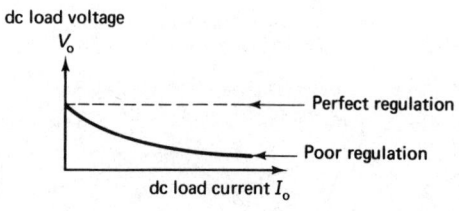

Fig. 3.10.

chosen with respect to the load R_L. However there are two main disadvantages to the use of such circuits for some applications where varying loads and varying a.c. supplies are encountered:

(a) The d.c. output voltage falls with increasing load current. The term regulation is used to describe this variation. Two regulation curves are sketched in fig. 3.10. When V_0 stays constant and independent of I_0 we say the regulation is perfect. When V_0 falls rapidly with I_0 we say the regulation is poor.

(b) The d.c. voltage varies with the amplitude of the a.c. input voltage. A.C. supply voltages vary from location to location and also owing to the varying demand made on the system by the users. When demand is very heavy the mains voltage may drop by the order of 10%. Most rectifiers are mains supplied, so if the mains voltage varies in amplitude so will the rectified d.c. output voltage.

There are many examples in electronics of circuits which need to be supplied with a constant or stabilized voltage. In such cases both of these defects must be overcome.

Fig. 3.11(a) shows a diagram of a stabilized voltage source which is designed to combat effects (a) and (b). The simple stabilizing circuit consists of two elements: a series resistor R and a Zener diode connected in parallel with the load R_L.

We may explain the action of this circuit as follows. The Zener diode is operated under reverse-bias breakdown conditions, i.e. the output voltage from the rectifier V_0 must (at least) exceed the Zener voltage V_z. We can also observe from the Zener characteristic of fig. 3.11(b) that

44

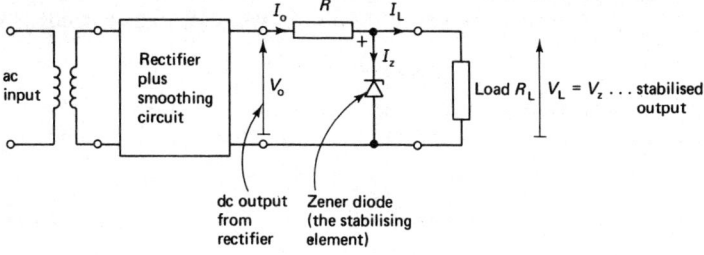

(a) A basic stabilised voltage source including a
Zener diode and a series resistor R

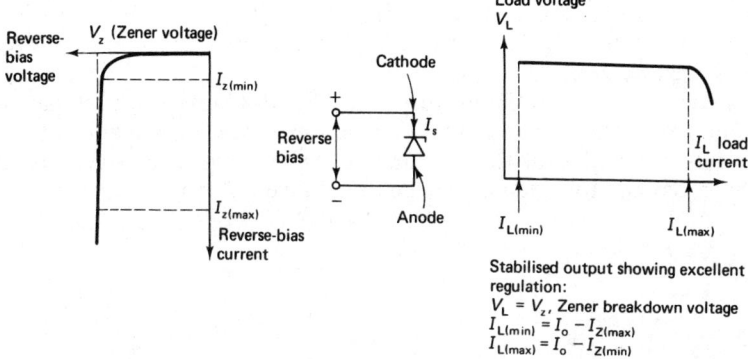

Stabilised output showing excellent
regulation:
$V_L = V_z$, Zener breakdown voltage
$I_{L(min)} = I_o - I_{Z(max)}$
$I_{L(max)} = I_o - I_{Z(min)}$

(b) Typical Zener diode characteristic
in reverse-bias mode

Fig. 3.11.

provided the Zener current I_z exceeds $I_{z(min)}$ the voltage across the Zener is virtually constant and independent of I_z. The maximum value of Zener current $I_{z(max)}$ is determined by the maximum permitted heat dissipation within the device. If the maximum device dissipation is P_{max} watts, and the Zener voltage is V_z, then $I_{z(max)}$ is found from

$$V_z I_{z(max)} = P_{max}, \text{ i.e. } I_{z(max)} = P_{max}/V_z.$$

Thus over the range I_z greater than $I_{z(min)}$ and less than $I_{z(max)}$ the voltage across the diode remains constant or very nearly so at V_z and the power dissipation of the Zener is not exceeded. Now, since the load R_L is placed in parallel with the Zener, the load voltage $V_L = V_z$ is also constant over the same range of Zener current. The rectifier output supplies both the Zener and the load, i.e.

$$I_0 = I_z + I_L.$$

At low load currents the Zener draws more current but should never exceed $I_{z(max)}$; the corresponding minimum value of load current is

$$I_{L(min)} = I_0 - I_{z(max)}$$

At high load currents the Zener draws less current but should never fall below $I_{z(min)}$; the corresponding maximum value of load current is

$$I_{L(max)} = I_0 - I_{z(min)}.$$

The rectifier output current itself is determined by V_0, V_z, and the series resistor R, i.e. the voltage drop across R is

$$RI_0 = V_0 - V_z \text{ so } I_0 = (V_0 - V_z)/R.$$

The above equations may be used to design simple stabilizer circuits and will be applied in the next sub-section. Fig. 3.11(c) shows the regulation curve of V_L versus I_L. V_L is virtually constant from $I_{L(min)}$ to $I_{L(max)}$, being stabilized by Zener diode action.

Zener diodes are available with a wide range of breakdown voltages— from a few volts to several hundred, and with power dissipations from hundreds of milliwatts to a hundred or so watts. Fig. 3.12 shows typical characteristics for a family of Zeners capable of dissipating 10 W.

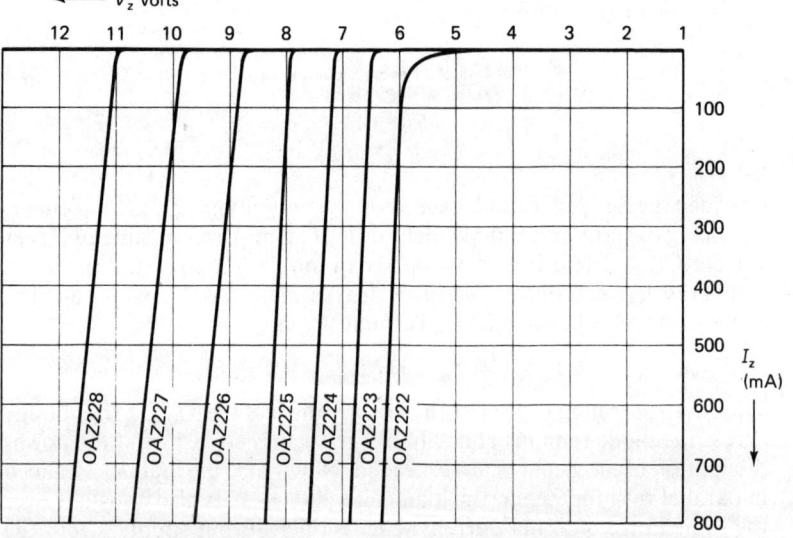

Fig. 3.12. Typical reverse-bias characteristic of a family of Zener diodes. (Courtesy of Mullard Ltd.).

3.5
The calculation of the value of the series resistor R in the simple Zener stabilizing circuit for conditions of (a) varying supply voltage, fixed load, (b) fixed supply voltage, varying load

(a) *Calculation of R for varying supply voltage, fixed load*

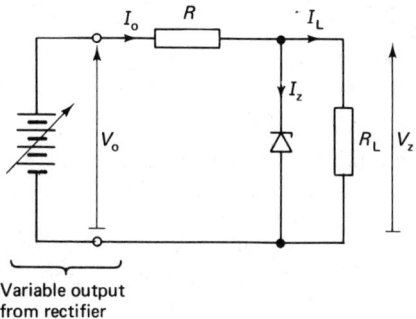

Fig. 3.13.

If the a.c. supply voltage to a rectifier varies, the d.c. output voltage V_0 will also vary. Suppose this varies between $V_{0(max)}$ and $V_{0(min)}$. We have a fixed load R_L and if the required stabilized load voltage is to be V_L we select a Zener with a breakdown voltage $V_Z = V_L$ (or as close as is available). The fixed load current is then given by

$$I_L = \frac{V_z}{R_L}.$$

Now let us refer to the basic Zener stabilizer circuit shown in fig. 3.13. We have the design criterion that the series resistor R value must be chosen to feed the load with I_L and the minimum Zener current $I_{z(min)}$ in the 'worst-case' situation when the rectifier output $V_0 = V_{0(min)}$. Thus for this case the current from the rectifier.

$$I_0 = I_L + I_{z(min)}$$

but the voltage drop across R,

$$RI_0 = V_{0(min)} - V_z$$

so our design value for R is

$$R = \frac{V_{0(min)} - V_z}{I_{0(min)}} = \frac{V_{0(min)} - V_z}{I_L + I_{z(min)}}.$$

This value of R ensures that the load voltage never drops below the 'knee' of the Zener diode characteristic and thus stabilizes V_L at V_z.

We should also check that the maximum Zener diode current is not exceeded when $V_0 = V_{0(max)}$. Now when this occurs

$$I_0 = \frac{V_{0(max)} - V_z}{R} = I_{0(max)} \text{ (say)}.$$

So $I_z = I_{0(max)} - I_L < I_{z(max)}$, the maximum Zener current rating. Finally let us consider a practical stabilizer circuit with the following specification:

Zener voltage $V_z = 20$ V
Required load current $I_L = 400$ mA
Minimum Zener current $I_{z(min)} = 20$ mA
Maximum Zener current rating $= 500$ mA

The input to the stabilizer circuit varies between $V_0 = 30$ V and 35 V. The value of series resistance R required is

$$R = \frac{V_{0(min)} - V_z}{I_L + I_{z(min)}} = \frac{30 - 20}{(400 + 20)10^{-3}} = \frac{10 \times 10^3}{420} = 23 \cdot 8 \ \Omega.$$

So we select $R = 22 \ \Omega$ (nearest preferred value of resistance to $23 \cdot 8 \ \Omega$). The maximum current supplied by the rectifier when $V_0 = 35$ V is

$$I_0 = \frac{35 - V_z}{R} = \frac{35 - 20}{22} = \frac{15}{22} = 682 \text{ mA}$$

and the corresponding maximum current drawn by the Zener,

$$I_z = I_0 - I_L = 682 - 400 = 282 \text{ mA}$$

is well below the maximum Zener current rating of 500 mA.

(b) *Calculation of R for fixed supply voltage, varying load*
In this case V_0 is fixed, so with a given value of R the supply current to the stabilizer circuit is

$$I_0 = \frac{V_0 - V_z}{R}.$$

The maximum load current that can be drawn is

$$I_{L(max)} = I_0 - I_{z(min)}$$

since we must ensure that at least $I_{z(min)}$ flows in the Zener to keep above the 'knee' of the Zener characteristic. If we wish to find the value of R for a given $I_{L(max)}$ we substitute $I_0 = I_{L(max)} + I_{z(min)}$ in the first equation, i.e.

$$I_0 = I_{L(max)} + I_{z(min)} = \frac{V_0 - V_z}{R}.$$

So
$$R = \frac{V_0 - V_z}{I_{L(max)} + I_{z(min)}}$$

The minimum value of load current corresponds to the condition that I_z equals the maximum Zener current rating, so

$$I_{L(min)} = I_0 - I_{z(max)}$$

Alternatively if we design for the output to be stabilized from zero load current, i.e. open-circuit load, then with $I_L = 0$ we have

$$I_0 = I_z \leq I_{z(max)}.$$

So designing R for the case $I_0 = I_{z(max)}$, we obtain

$$R = \frac{V_0 - V_z}{I_{z(max)}}.$$

The maximum value of load current for this value of R is

$$I_L = I_{z(max)} - I_{z(min)}.$$

Example 3.1
Fig. 3.14 shows a diagram of a simple voltage regulator circuit which employs a 20 V, 0·5 W (maximum d.c. power rating) Zener diode. The minimum operating or knee current of the Zener is 2 mA. If the input supply to the circuit is 50 V, calculate the value of R which provides a 20 V stabilized supply for a load current $I_L = 0$ to the maximum that will permit good regulation to be maintained. Calculate also this maximum value.

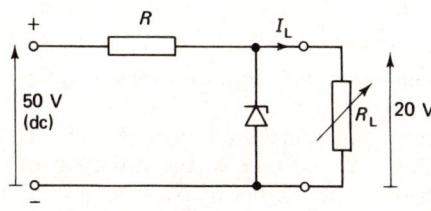

Fig. 3.14.

Solution
When the load current $I_L = 0$ all the supply current is passed through the Zener. The maximum current that the Zener can safely handle is determined from

$$I_{z(max)} \times \text{Zener voltage} = \text{maximum d.c. power rating}$$

So $I_{z(max)} \times 20 = 0·5$, $I_{z(max)} = \frac{0·5}{20} = 0·025$ A or 25 mA.

Thus R must limit the supply current to $I_{z(\max)}$, hence

$$RI_{z(\max)} = V_0 - V_z = 50 - 20 = 30$$

So
$$R = \frac{30}{I_{z(\max)}} = \frac{30}{0 \cdot 025} = 1200 \ .$$

The maximum load current at which stabilization may be maintained corresponds to the Zener taking its minimum current of 2 mA. Thus

$$I_{L(\max)} = \text{supply current} - 2 \text{ mA} = 25 - 2 = 23 \text{ mA}.$$

Section 4: *The expected learning outcome of this section is that the student should know the arrangement of transistor electrodes*

4.1
The arrangement of a bipolar transistor produced from a sandwich of semi-conductor materials

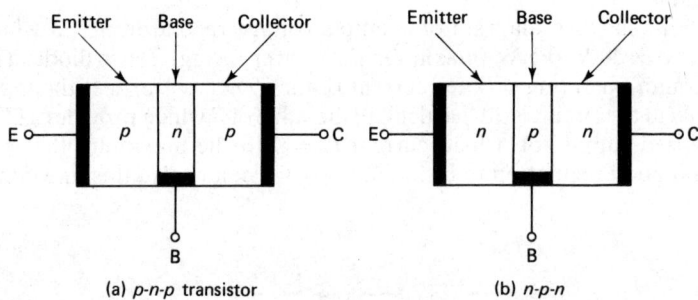

Fig. 4.1. Diagram illustrating sandwich structure of bipolar transistor.

A bipolar transistor consists of a single crystal of silicon or germanium in which either a thin layer of n-type (the order of μm) is sandwiched between two thicker p-type regions, or alternatively a thin p-type layer is sandwiched between two n-type regions. The first configuration whose physical structure is shown diagrammatically in fig. 4.1(a) is known as a *p-n-p* transistor; the second configuration shown in (b) is known as an *n-p-n* transistor. Both *p-n-p* and *n-p-n* transistors may be used as amplifying elements to produce voltage, current, and power gain (see Section 13) and also as switching elements (see Section 21).

The transistors are referred to as bipolar junction transistors for the following reasons: 'bipolar' to signify the transistor has both positive (holes) and negative (electrons) current carriers, both of which play an essential role in transistor action; 'junction' because the transistor

50

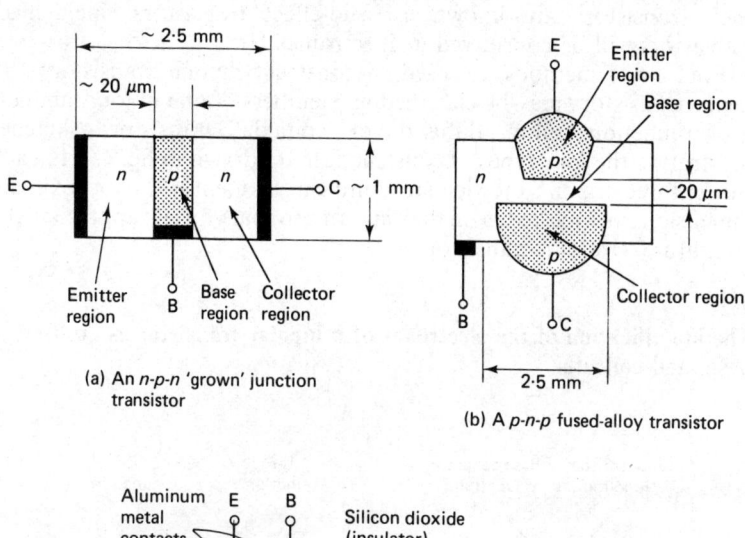

(a) An *n-p-n* 'grown' junction transistor

(b) A *p-n-p* fused-alloy transistor

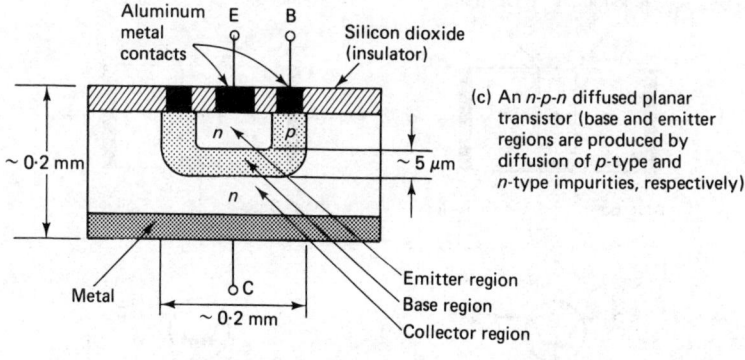

(c) An *n-p-n* diffused planar transistor (base and emitter regions are produced by diffusion of *p*-type and *n*-type impurities, respectively)

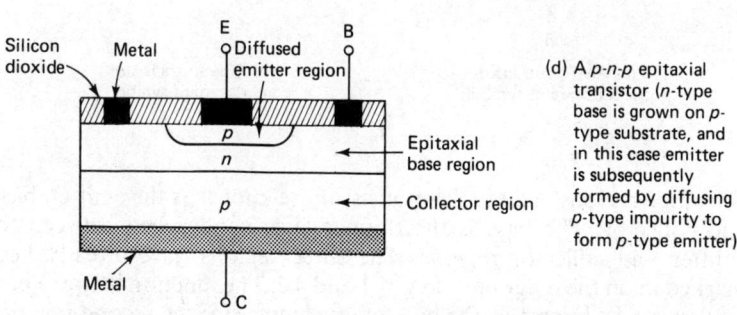

(d) A *p-n-p* epitaxial transistor (*n*-type base is grown on *p*-type substrate, and in this case emitter is subsequently formed by diffusing *p*-type impurity to form *p*-type emitter)

Fig. 4.2. Diagrams illustrating structure of the four types of bipolar transistors.

51

consists of two *p-n* junctions formed close together in a single crystal of semi-conductors. There is a second type of transistor whose operation depends on the flow of majority carriers only. It is a 'unipolar' device and such transistors are known as field-effect transistors. Field-effect transistors will be considered in Electronics III.

Four basic methods are used to construct bipolar transistors and hence transistors may be classified as members of one of four families: grown-junction, alloy, diffused, or epitaxial transistor. Sketches illustrating the four types of construction are drawn in fig. 4.2. In each case a three-region sandwich structure can be identified. Approximate dimensions are marked in on the diagrams so that you can appreciate the 'miniature' size of a transistor.

4.2
The identification of the electrodes of a bipolar transistor as emitter, base, and collector

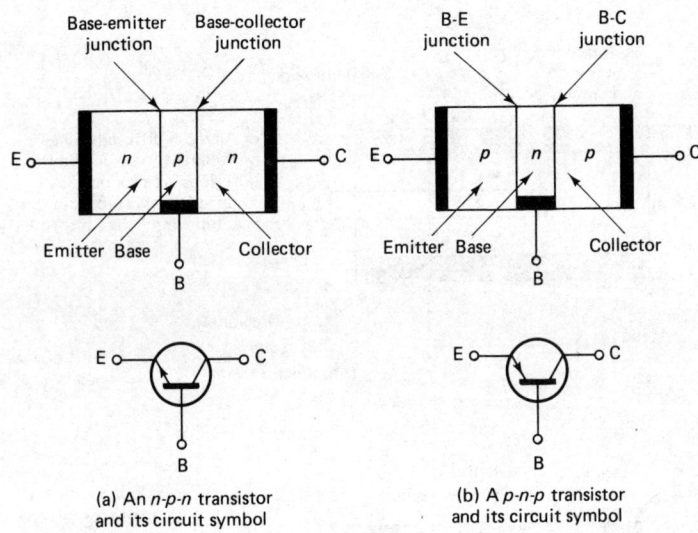

Fig. 4.3.

(a) An *n-p-n* transistor and its circuit symbol

(b) A *p-n-p* transistor and its circuit symbol

The three regions of a bipolar transistor are known as the emitter, base, and collector. The base is the thinner layer sandwiched between the emitter and collector regions. The three regions have already been marked in on the diagrams of fig. 4.1 and 4.2. The junction between base and emitter is known as the base-emitter junction; the second junction between base and collector as the base-collector junction. The three

52

regions and the two junctions are shown in fig. 4.3 together with circuit symbols used to represent *n-p-n* and *p-n-p* transistors in circuit diagrams.

Conducting leads are bonded to each of the three regions to enable external connections to be made. The lead bonded to the emitter is known as the emitter lead or electrode and the terminal marked E as the emitter terminal. Likewise the leads connected to the base and collector are known as the base and collector electrodes and their respective terminals, marked B and C, as the base and collector terminals. E (emitter), B (base) and C (collector) are used both on circuit diagrams and also as subscripts to define terminal currents and voltage differences between electrodes. For example, see fig. 4.4,

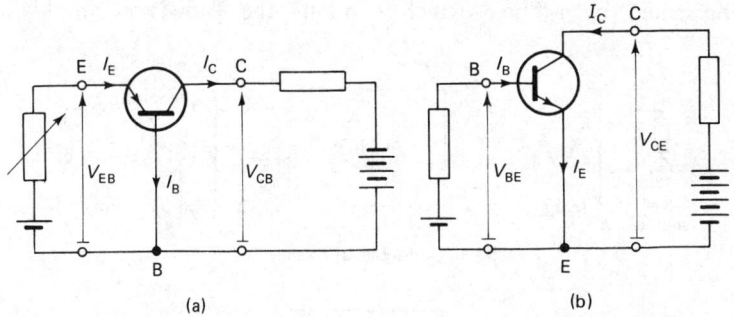

Fig. 4.4 Two circuits (a) containing a *p-n-p* (b) an *n-p-n* transistor I_E, I_B, I_C are the terminal dc currents; V_{EB}, V_{CB} and V_{BE}, V_{CE} denote the dc voltage differences between transistor terminals.

I_E = the d.c. emitter current
I_B = the d.c. base current
I_C = the d.c. collector current
V_{EB} = the d.c. voltage difference between emitter and base
V_{CB} = the d.c. voltage difference between collector and base
V_{BE} = the d.c. voltage difference between base and emitter
V_{CE} = the d.c. voltage difference between collector and emitter

We will find this notation very useful for all transistor work and especially when investigating the static characteristics of a transistor.

One final point, the direction of the arrow on the emitter electrode of a transistor circuit symbol serves to distinguish between *n-p-n* and *p-n-p* transistors. The arrow direction indicates the direction of positive current flow when the B-E junction is forward-biased (the normal mode of operation for a transistor), that is, away from the emitter for an *n-p-n*, but into the emitter for a *p-n-p* transistor.

5.1
Circuit diagrams for the common base, common emitter, and common collector modes of connection

Our learning objective in this sub-section is to know and to be able to draw circuit diagrams for the three principal ways in which a transistor is connected in a circuit. Since one of the major applications of transistors is as amplifying elements we will begin our discussion by considering a simple amplifier.

Fig. 5.1(a) shows a block diagram of an amplifier. The amplifier itself is represented by a 'box' which contains the transistor amplifying

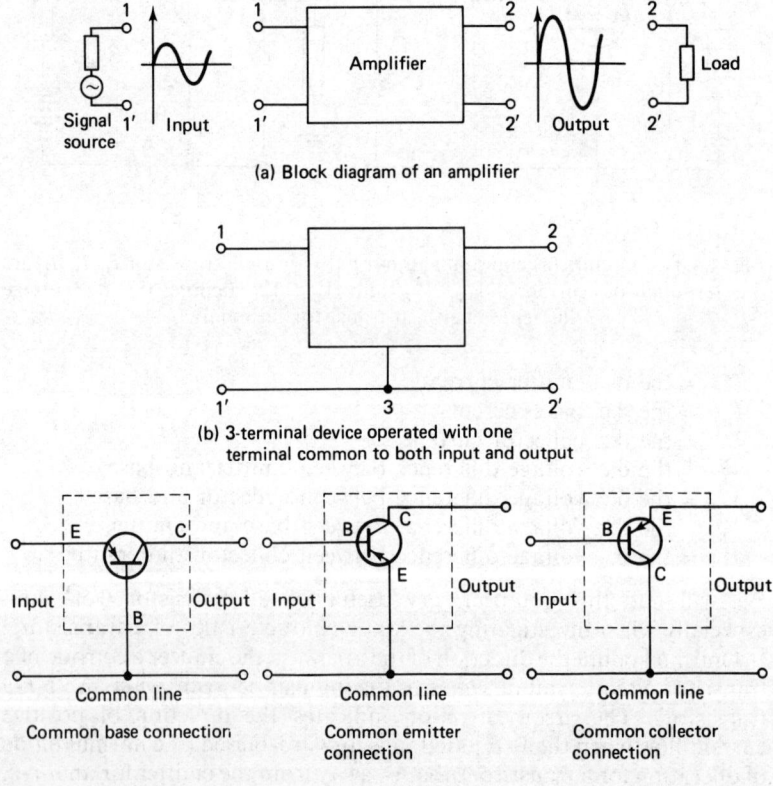

(a) Block diagram of an amplifier

(b) 3-terminal device operated with one terminal common to both input and output

Common base connection

Common emitter connection

Common collector connection

Fig. 5.1.

element(s), other circuit components such as resistors and capacitors, and the d.c. power supplies necessary to operate the transistor(s). The 'box' has a pair of input terminals at which an external signal (e.g. an a.c. voltage from a microphone) is connected; and a pair of output terminals at which a load may be connected to extract the amplified output. Now we have seen that a transistor is a three-terminal device—emitter, base, and collector electrodes being available for external connections—and so to simplify wiring, and also for many other practical reasons, it is usual to make one of these electrode terminals common to both input and output sides, as shown in fig. 5.1(b). If the base is made common, we refer to this as the common base mode of connection; if the emitter is returned to the common line between input and output, we have the common emitter connection; whilst if the collector is made common, we have the common collector mode of connection. These three modes of connection are shown in basic form in fig. 5.1(c). Each mode of connection has definite properties and is used in different amplifying (and other) applications.

However, before we can utilize a transistor, external d.c. supplies must be correctly connected to its electrodes. These are necessary to supply energy into the circuit and to ensure that the transistor is working with the correct d.c. bias currents and voltages. In the amplification process part of the energy drawn by the transistor from the d.c. supplies is converted to a.c. signal energy, the transistor being the device which makes this conversion possible. As a general rule the transistor (as an amplifying element), regardless of its mode of connection, is operated so that

(a) its base-emitter junction is forward-biased,
(b) its base-collector junction is reverse-biased.

Figs. 5.2 and 5.3 show the three modes of connection for an n-p-n and a p-n-p transistor. In each connection the B-E junction is forward-biased by a d.c. source (the resistor shown in series can be used to adjust the forward-bias current); and the B-C junction is reverse-biased by a d.c. power supply. In practice only one d.c. source is normally used to supply both forward B-E and reverse B-C bias conditions. In fig. 5.2(a) and 5.3(a) the input signal (represented by an a.c. generator of amplitude V_S in series with its source resistance R_S is introduced into the E-B circuit and the amplified output signal is developed across a resistive load connected in the C-B circuit. The base electrode is common to both input and output circuits. In (b) the signal is applied to the B-E circuit and the output taken across the C-E terminals. The emitter is common to both input and output circuits. We shall be considering the action of the common emitter amplifier circuit in detail in Chapter D. Circuits (c) are referred to as common collector amplifiers since the collector is essentially common (if we neglect the d.c. collector supply) to both input and output circuits. In fact terminals C and C' do differ in voltage but

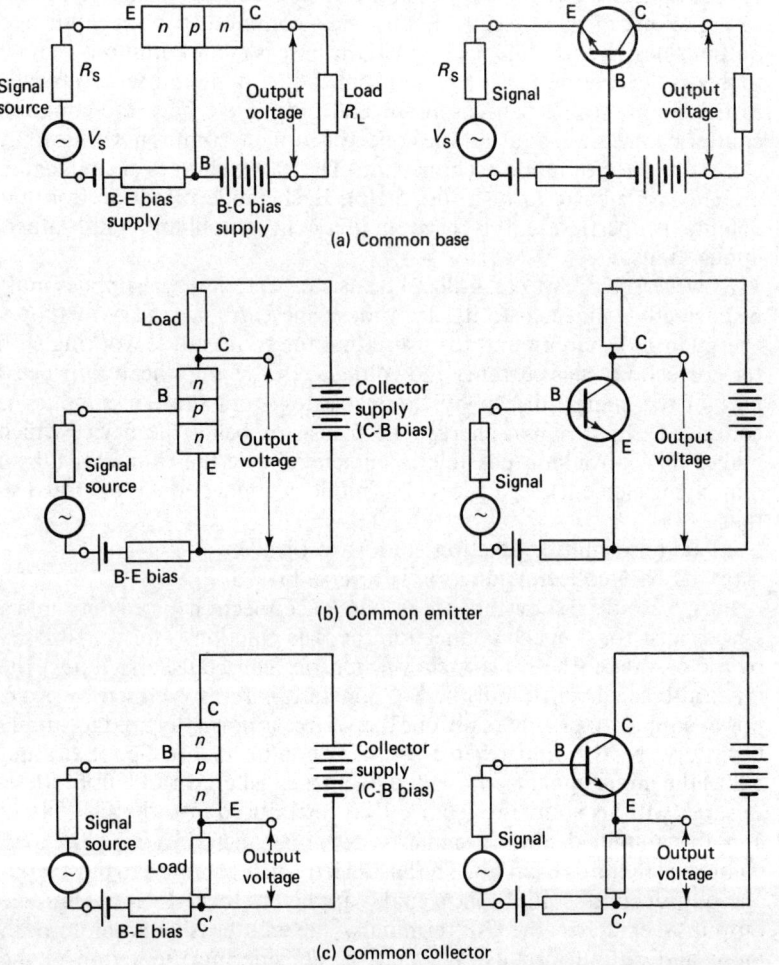

Fig. 5.2. Diagrams showing an *n-p-n* transistor connected in (a) the common base, (b) the common emitter, and (c) the common collector modes. Bias supplies, a signal source and a load resistor are also shown.

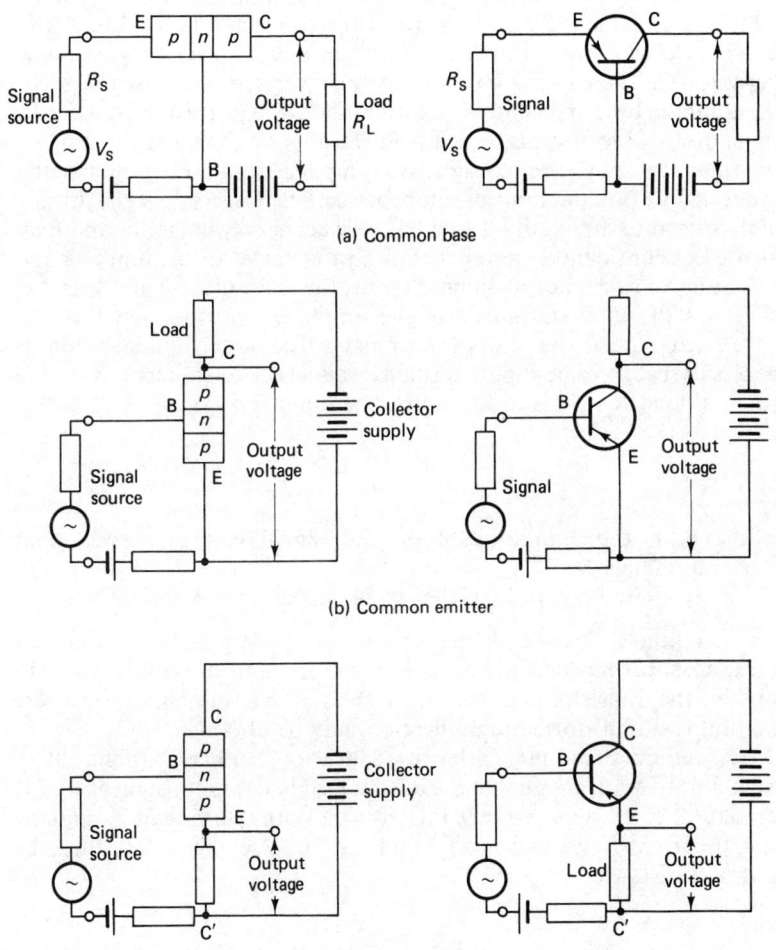

Fig. 5.3. Diagram showing a *p-n-p* transistor connected in (a) the common base, (b) the common emitter and (c) the common collector modes. Bias supplies, a signal source and a load resistor are also shown.

only by the d.c. collector voltage supply. As far as a.c. signal variations are concerned C and C' are effectively common.

5.2
The comparison of the relative values of input and output resistances for common base, common emitter, and common collector connections
Let us first consider the general definitions of input and output resistances as applied to an amplifier, before making a comparison of the relative values for the case of a transistor connected in its three basic modes. However, it is worth commenting that the word transistor is derived from the contraction of the terms 'transfer' and 'resistance'. In the common base and common emitter modes the transistor 'transfers' a signal from a low resistance to a high resistance circuit and in doing so the transistor may effect considerable amplification in the power of the input signal. The common collector connection 'transfers' a signal from a high resistance input circuit to a low resistance output circuit and may also effect considerable amplification in the power of the input signal.

The input resistance of an amplifier is the resistance 'seen' by an a.c. signal at the input terminals of the amplifier, with a given load R_L connected across the output terminals. The latter qualification is necessary because the input resistance value very often depends on the value of load R_L. Thus in fig. 5.4(b), the input resistance,

$$R_i = \frac{V_1}{I_1}$$

where V_1 is the amplitude of the a.c. signal voltage across input terminals 1-1',
I_1 is the amplitude of the signal current into 1 and out of 1'.

Note: V_1 and I_1 are a.c. signal variations only. We shall see later that these are superimposed on the d.c. bias voltages and currents required to operate the transistor and to set the transistor in a suitable region where faithful and undistorted amplification may be obtained.

The concept of output resistance is a little more difficult to define. If we were able to apply a signal to the output terminals and then measure the amplitude of the signal voltage V_2 across the output terminals 22' and the signal current I_2 flowing in at 2 and out at 2', see fig. 5.4(c), then the output resistance,

$$R_0 = \frac{V_2}{I_2} .$$

Note that the source resistance R_S (the resistance associated with the signal source which will be actually driving the amplifier in normal use) is connected across the input terminals when R_0 is measured. The reason

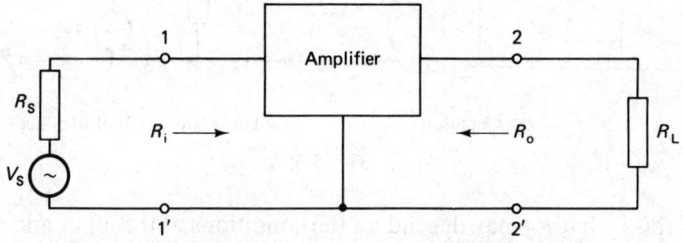

(a) General amplifier diagram:
amplifier driven by source represented by signal V_S and R_S
and terminated at output by load R_L.
R_i = effective resistance 'seen' by signal at input terminals
R_o = effective resistance 'looking back' into amplifier at output terminals

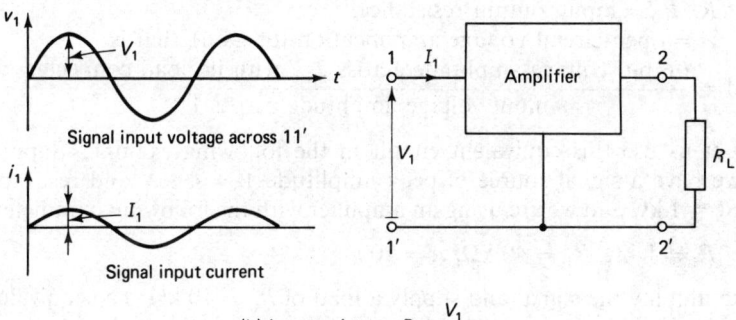

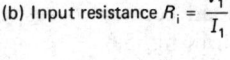

(b) Input resistance $R_i = \dfrac{V_1}{I_1}$

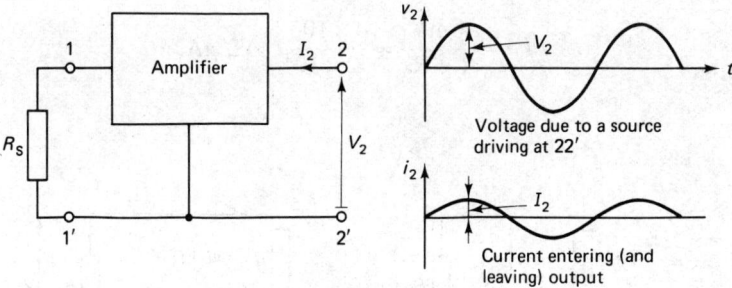

(c) Output resistance $R_o = \dfrac{V_2}{I_2}$

Fig. 5.4. The input and output resistances of an amplifier.

59

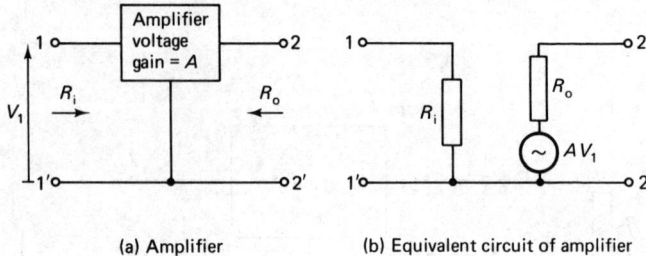

(a) Amplifier   (b) Equivalent circuit of amplifier

Fig. 5.5.

for this is that R_0 may depend on the conditions at the input and so we simulate these by connecting R_S across 1-1'.

The concept of input and output resistances is very useful in constructing a simple equivalent circuit model of an amplifier. Such a model is shown in fig. 5.5(b):

R_i, R_0 = input, output resistance,

A = open-circuit voltage amplification (or gain), that is

$$A = \frac{\text{output voltage amplitude across 2-2' with no load connected}}{\text{input voltage amplitude across 1-1'}}.$$

Let us use this equivalent circuit in the following example. Suppose we have a signal source of peak amplitude $V_S = 4$ mV and resistance $R_S = 1$ kΩ and we are using an amplifier with the following parameters:

$R_i = 1$ kΩ, $R_0 = 40$ kΩ, $A = 200$

to amplify the signal and supply a load of $R_L = 10$ kΩ. The equivalent circuit with source and load connected respectively at the input and output terminals is shown in fig. 5.6.

The input signal current,

$$I_1 = \frac{4 \text{ mV}}{R_i + R_S} = \frac{4 \times 10^{-3}}{2 \times 10^3} = 2 \ \mu A.$$

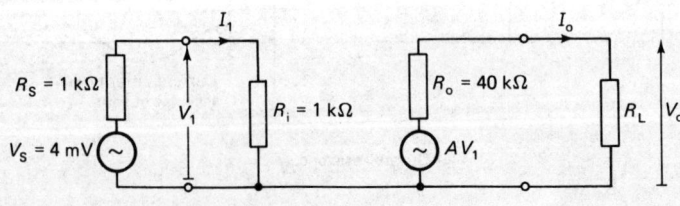

Fig. 5.6.

The input voltage,

$$V_1 = R_i I_1 = 1 \times 10^3 \times 2 \times 10^{-6} = 2 \times 10^{-3} \ V = 2 \ mV.$$

As $A = 200$, this input voltage is amplified 200 times, i.e. to 400 mV. That is, the voltage generator in the output circuit has an amplitude,

$$AV_1 = 200 \times 2 = 400 \ mV$$

and so the output current,

$$I_0 = \frac{AV_1}{R_0 + R_L} = \frac{400 \times 10^{-3}}{(40 + 10) \times 10^3} = 8 \ \mu A$$

and the output voltage,

$$V_0 = R_L I_0 = 10 \times 10^3 \times 8 \times 10^{-6} = 80 \ mV.$$

This example serves to illustrate how input and output resistances may be used in amplifier calculations and shows how they affect the input to output voltage and current gains. The values for the latter in the above case are:

$$\frac{V_0}{V_1} = \frac{\text{output voltage}}{\text{input voltage}} = \frac{80}{2} = 40$$

$$\frac{I_0}{I_1} = \frac{\text{output current}}{\text{input current}} = \frac{8}{2} = 4.$$

Finally let us return to the specific learning objective of this sub-section: the comparison of the relative values of input and output resistances for a transistor connected in the common base, common emitter, and common collector modes. This comparison is given in Table 5.1 below.

Table 5.1 *A comparison of the relative values of input and output resistances of a transistor connected in the three basic modes*

Mode	Input Resistance R_i	Output Resistance R_0
Common base	very low, $\sim$ 20 to 500 Ω	very high, $\sim$ 100 kΩ to 2 MΩ
Common emitter	low to medium, $\sim$ 500 Ω to 2 kΩ	medium to high $\sim$ 30 kΩ to 100 kΩ
Common collector	medium to very high $\sim$ 2 kΩ to 2 MΩ	very low to medium $\sim$ 20 Ω to 30 kΩ

5.3
The definition of the short-circuit current gains of a transistor connected in the common base and the common emitter modes

Let us first consider the current flow in a p-n-p transistor connected in the common base mode and working under d.c. conditions, as shown in fig. 5.7(a). The E-B junction is forward-biased by the battery V_{BB} (R limits the forward-bias current), and the C-B junction is reverse-biased by the battery V_{CC}. The following discussion is exactly similar for the case of an n-p-n transistor except that holes should be interchanged for electrons and vice versa.

Under the action of forward-bias, holes in the emitter region move across the E-B junction. Now the base region is deliberately made very thin and also since the B-C junction is forward-biased for holes by V_{CC}, most of the holes from the emitter are swept across the B-C junction into the collector region. A relatively small number of holes do recombine with electrons in the base region and this gives rise to a small base current I_B. In quantitative terms we can define the relationship between d.c. emitter current I_E and d.c. collector current I_C as

$$I_C = \alpha_{DC} I_E$$

where α_{DC} is known as the d.c. common base current gain of the transistor. In practice α_{DC} is usually between 0·95 and 0·999 and strictly speaking is not a current gain but rather a ratio, i.e.

$$\alpha_{DC} = \frac{I_C}{I_E}$$

α_{DC} depends on the reverse-bias voltage V_{CB} between C and B terminals and hence the value of V_{CB} should really be quoted in specifying the value of α_{DC}. In modern notation the symbol h_{FB} is used to denote α_{DC}.

Next let us consider the situation when an a.c. signal voltage is placed in series with the E-B bias voltage supply as shown in fig. 5.7(b). On positive half-cycles the signal source increases the forward-bias E-B voltage, i.e. it acts in the same direction as V_{BB}, and thus more emitter current will flow. On negative half-cycles the signal decreases the forward-bias voltage, i.e. it acts in opposition to V_{BB}, and less emitter current flows. Let us suppose the peak amplitude of emitter current variation is I_e, that is

I_e = maximum value of emitter current i_E − d.c. emitter current I_E

    = I_E − minimum value of emitter current i_E

Note: It is standard practice to denote a.c. amplitudes by upper case I and V with lower case subscripts e, b, c. Total currents and voltages, that is, d.c. bias values plus the signal variations are denoted by lower case i and v with upper case (capital) subscripts, e.g.

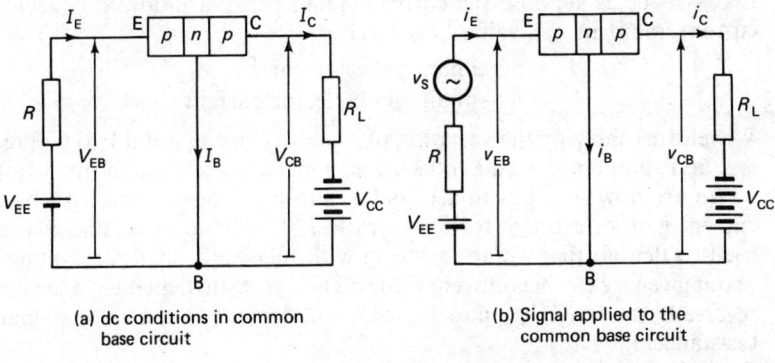

(a) dc conditions in common base circuit

(b) Signal applied to the common base circuit

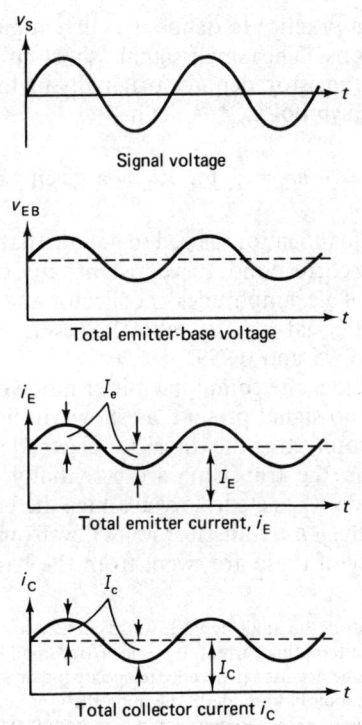

Signal voltage

Total emitter-base voltage

Total emitter current, i_E

Total collector current i_C

(c) Signal voltage v_S and transistor current and voltage waveforms for a p-n-p transistor

Fig. 5.7.

$$i_E = I_E + \text{a.c. signal current in emitter}$$
$$i_C = I_C + \text{a.c. collector current.}$$

Likewise let us suppose the corresponding peak variation in collector current about its d.c. value I_C is I_c, i.e.

$$I_c = \text{maximum collector current} - I_C$$
$$= I_C - \text{minimum collector current.}$$

Waveforms showing the variation of signal voltage v_S, total E-B voltage v_{EB}, and total emitter and collector currents are sketched in fig. 5.7(c).

We are now in a position to define the small-signal, common base current gain of a transistor as the ratio I_c/I_e. The term small-signal is used to denote that we are working with relatively small a.c. changes about given steady bias current values. The I_c/I_e ratio depends on the d.c. reverse-bias voltage V_{CB}, and V_{CB} itself will depend on the value of load resistance, i.e.

$$V_{CB} = -(V_{CC} - R_L I_C).$$

Thus it is common practice to define the current gain under given load and bias conditions. The small-signal, short-circuit, common base current gain of a transistor, denoted originally by the symbol α but now commonly by the symbol h_{fb},* is defined as

$$\alpha \equiv -h_{fb} = \frac{I_c}{I_e} \text{ for } V_{CB} = \text{a given value.}$$

The short-circuit qualification is used to denote that if $R_L = 0$ (i.e. short-circuit load) then α corresponds to a constant value of $V_{CB} = -V_{CC}$. Note that α is the ratio of a.c. amplitudes of collector and emitter current and therefore does not equal α_{DC}. Its value, however, is very similar and is typically between 0·95 and 0·999.

Let us now consider the common emitter mode of connection, firstly the d.c. case with no signal present, as shown in fig. 5.8. Although the external mode of connecting the transistor electrodes is different, the d.c. currents flowing in the transistor are essentially the same as in the common base mode. V_{BB} acts to forward-bias the base-emitter junction and in the case of the p-n-p transistor holes flow from the emitter into the base. The majority of these are swept from the base region across the

---

*h_{fb} is known formally as the small-signal, forward current, transfer ratio with the output C-B short-circuited to alternating current. It is one of four small signal parameters, known as the hybrid or h-parameters, used in practice to specify the small-signal characteristics of a transistor in the common base mode. In defining the h-parameters a slightly different current convention is used: one where both input and output currents are assumed to flow into the transistor. Thus h_{fb} is defined (using the directions shown in fig. 5.7) as $h_{fb} = \dfrac{-I_c}{I_e}$

$= -\alpha.$

64

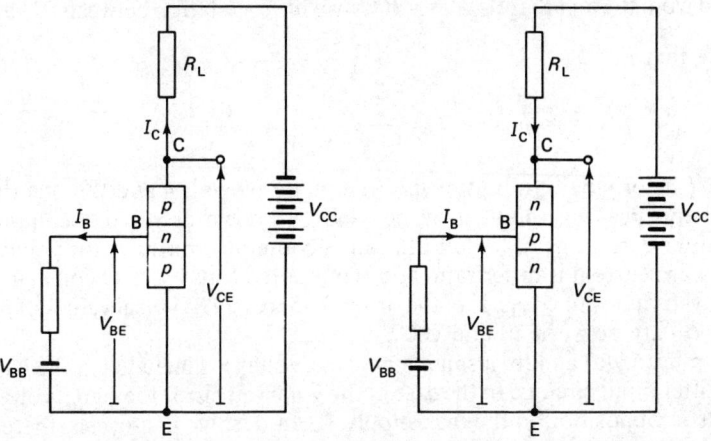

Fig. 5.8. DC conditions in a common-emitter circuit.

base-collector junction into the collector region. A small fraction of the emitter hole current recombines with the free electrons in the n-type base and this effect gives rise to the relatively small base current flowing out in the base lead. In the case of the n-p-n transistor electrons from the n-type emitter move across the emitter-base junction under the action of the forward-bias produced by V_{BB} (for n-p-n transistors V_{BB} is connected, of course, so the base is positive with respect to the emitter). A small fraction of the emitter electrons recombines in the base and this effect constitutes the base current I_B. The majority are swept across the base-collector junction into the collector region and provide the collector current I_C.

In quantitative terms we may state

$$I_C = \alpha_{DC} I_E \quad \therefore \quad I_E = I_C / \alpha_{DC}$$

and since the emitter current I_E must equal the sum of the base current I_B and the collector current I_C we have also

$$I_E = I_B + I_C$$

so the base input current,

$$I_B = I_E - I_C = \frac{I_C}{\alpha_{DC}} - I_C = I_C \left(\frac{1}{\alpha_{DC}} - 1 \right) = \frac{1 - \alpha_{DC}}{\alpha_{DC}} I_C$$

We may now define an important d.c. parameter of a transistor:

$$\beta_{DC} \equiv h_{FE} = \frac{I_C}{I_B}$$

65

and from the fact that $I_B = \dfrac{1 - \alpha_{DC}}{\alpha_{DC}} I_C$ we have a relation between β_{DC} and α_{DC}, that is

$$\beta_{DC} = \frac{I_C}{I_B} = \frac{I_C \alpha_{DC}}{I_C(1 - \alpha_{DC})} = \frac{\alpha_{DC}}{1 - \alpha_{DC}}$$

β_{DC} (the original symbol) or h_{FE} (the modern symbol and the one that strictly speaking should now be used) is known as the d.c. common emitter current gain. Its full title is the common emitter, static value of forward current transfer ratio and is measured with V_{CE} held constant at a given value. Since α_{DC} is of the order 0·95 to 0·999, typical values of β_{DC} or h_{FE} are between 20 and 1000.

In fig. 5.9(a) and (b) a small-signal a.c. voltage is included in the base-emitter input circuits. In the case of the *p-n-p* transistor the signal voltage acts in opposition to the bias supply V_B on positive half-cycles, thereby reducing the B-E forward-bias and hence the emitter, base, and collector currents. On negative half-cycles the signal increases the forward-bias voltage and hence the transistor currents increase. The resulting waveforms for total (i.e. d.c. signal) base-emitter voltage v_{BE}, base current i_B, collector current i_C, and output voltage v_C are sketched in (c).

In the case of the *n-p-n* transistor the forward-bias of the base-emitter junction is increased on positive half-cycles of the signal, thereby increasing the transistor currents. On negative half-cycles the signal acts in opposition and reduces the forward-bias and hence the base, the emitter, and the collector currents. The corresponding waveforms are sketched in (d).

Now if,

I_b = peak amplitude of signal current in base lead,
I_c = peak amplitude of signal current in collector lead,

i.e. I_b = maximum value of base current i_B – d.c. base current I_B
$= I_B$ – minimum value of base current,
I_c = maximum value of collector current i_C – d.c. collector current I_C
$= I_C$ – minimum value of collector current,

we may define the small-signal, short-circuit, common emitter current gain of the transistor as

$$\beta \equiv h_{fe} = \frac{I_c}{I_b} \text{ for a given value of bias voltage } V_{CE}.$$

h_{fe}* is the modern symbol, β is the symbol that was originally used. The

---

*h_{fe} is one of the 4 hybrid parameters used to quantify the small-signal performance of a transistor in the common emitter mode. Its full title is the small-signal current transfer ratio with the output short-circuited to alternating current. The 'f' subscript denotes forward and the 'e' that we are in the common base mode. In this case $\beta = h_{fe}$, there is no sign difference.

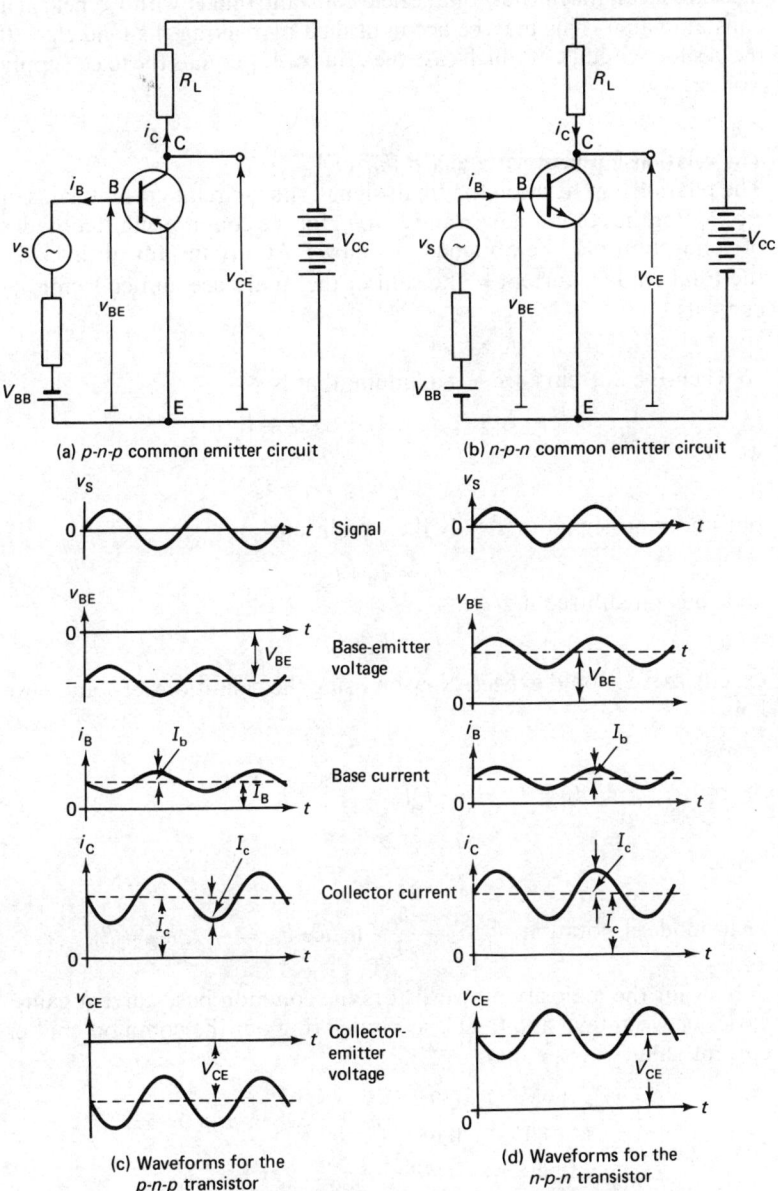

(a) *p-n-p* common emitter circuit

(b) *n-p-n* common emitter circuit

Signal

Base-emitter voltage

Base current

Collector current

Collector-emitter voltage

(c) Waveforms for the *p-n-p* transistor

(d) Waveforms for the *n-p-n* transistor

N.B. V_{BE}, I_C, I_C, V_{CE} are dc bias values; I_b, I_c are small signal amplitudes.

Fig. 5.9. Common-emitter circuits and associated waveforms.

short-circuit qualification is used to denote that I_c/I_b should be measured with the output voltage held constant, that is, with V_{CE} held at a constant value. This may be accomplished by making the load $R_L = 0$ (i.e. a short-circuit), in which case the value of V_{CE} equals the fixed supply voltage V_{CC}.

5.4
The relationship between α and β (h_{fb} and h_{fe})

The relationship between the small-signal, short-circuit current gains of α or h_{fb} for the common base and β or h_{fe} for the common emitter modes of connection may be obtained as follows. At any instant of time, the total emitter current = the sum of the total base and collector currents

i.e.
$$i_E = i_B + i_C$$

so when the currents are a maximum, that is

$$i_E = I_E + I_e, \; i_B = I_B + I_b, \; i_C = I_C + I_c$$

we have

$$I_E + I_e = I_B + I_b + I_C + I_c$$

but under no signal or steady d.c. conditions

$$I_E = I_B + I_C$$

and thus on subtracting

$$I_e = I_b + I_c$$

exactly as we should expect. Now on using the definitions of α and β we have

$$I_c = \alpha I_e, \; I_c = \beta I_b$$

so
$$I_b = I_e - I_c = I_c \left(\frac{1}{\alpha} - 1 \right) = I_c \frac{1-\alpha}{\alpha}$$

hence
$$\frac{I_c}{I_b} = \frac{\alpha}{1-\alpha} = \beta$$

or in modern notation: $h_{fe} = \dfrac{-h_{fb}}{1 + h_{fb}}$ (since $h_{fb} = -\alpha$, $h_{fe} = \beta$).

Now for the majority of transistors the common base current gain α varies between 0·95 and 0·998, so on working out the common emitter current gain β we see that:

$$\beta = \frac{0 \cdot 95}{1 - 0 \cdot 95} = \frac{0 \cdot 95}{0 \cdot 05} = 19 \quad \text{when } \alpha = 0 \cdot 95$$

$$\beta = \frac{0 \cdot 998}{1 - 0 \cdot 998} = \frac{0 \cdot 998}{0 \cdot 002} = 499 \quad \text{when } \alpha = 0 \cdot 998$$

i.e. β is very much greater than α.

Thus a transistor connected in the common emitter mode may be used to produce considerable current gain, i.e. a small a.c. signal injected into the base effects a variation in collector current approaching β times the base input current. The common base current gain is just below unity. However, the input resistance to a transistor connected in the common base mode is low, whilst the output resistance is very high. The latter fact allows us to connect a relatively large load resistor R_L in the output circuit. Thus, although the current gain is less than 1, the a.c. output voltage developed across R_L can be considerably greater than the input voltage, i.e. the common base circuit produces voltage gain. Let us illustrate this in the following example. Suppose the input signal amplitude to a common base circuit is $V_S = 1$ mV (assume all of this is developed across the emitter-base junction) and that the input resistance R_i is 50 Ω. Then the input current amplitude,

$$I_e = \frac{V_S}{R_i} = \frac{1 \text{ mV}}{50} = \frac{1}{50} \text{ mA}$$

Now the output current amplitude,

$$I_c = \alpha I_e \approx \frac{1}{50} \text{ mA if we assume } \alpha \approx 1$$

If a load $R_L = 10$ kΩ is selected, the signal voltage developed across this load has an amplitude,

$$V_{cb} = R_L I_C = 10\,000 \ \Omega \times \frac{1}{50} \text{ mA} = 200 \text{ mV}$$

$$\text{so the voltage gain} = \frac{\text{ac output voltage amplitude, } V_{cb}}{\text{signal input voltage amplitude, } V_s}$$

$$= \frac{R_L I_c}{R_i I_e} = \frac{\alpha R_L}{R_i} \approx 200$$

Section 6: *The expected learning outcome of this section is that the student should know the static behaviour of a transistor*

6.1
A test circuit diagram for determining the common base mode static characteristics
The static input characteristics of a transistor connected in a given mode are plots of d.c. input current to a transistor versus d.c. input voltage for

fixed values of d.c. output voltage. These characteristic curves enable us to determine the small-signal input resistance of the transistor at a given value of d.c. input (that is bias) current.

The static output characteristics of a transistor are plots of d.c. output current versus d.c. output voltage for fixed value of input current. These characteristics enable us to determine the small-signal output resistance and the short-circuit current gain of the transistor. The output characteristics may be used also in conjunction with the load line equation to investigate the amplifying performance of a transistor, as considered in Section 13.

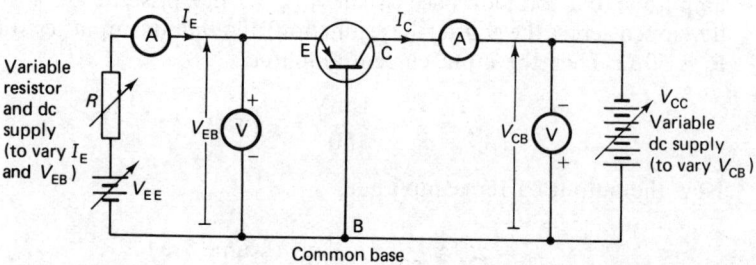

(a) Circuit to measure common base static characteristics of a *p-n-p* transistor

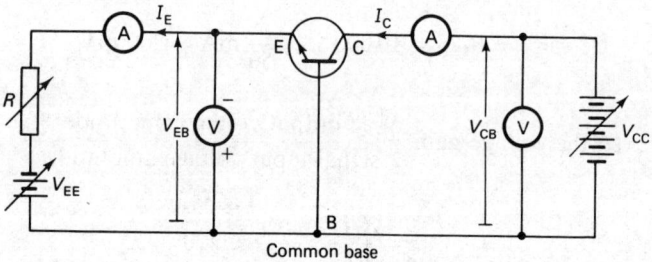

(b) Circuit to measure common base static characteristics of an *n-p-n* transistor

Note: For *p-n-p* transistors (see Fig. 6.1(a)) V_{EB} is positive. V_{CB} is negative. For *n-p-n* transistors (see Fig. 6.1(b)) V_{EB} is negative. V_{CB} is positive.

Fig. 6.1. Circuits to measure common base static characteristics.

The test circuits shown in fig. 6.1. may be used to determine the static characteristics of a transistor connected in the common base mode. Milliammeters are included in series with the emitter and collector leads to measure I_E and I_C. A voltmeter, with a typical range 0 to 1 V, is connected between the E and B terminals to measure V_{EB}. A voltmeter, with a typical range 0 to 20 V, is connected across the C and B terminals to measure V_{CB}. The variable resistance R and variable d.c. supply V_{EE} are included in the emitter-base input circuit to vary I_E and V_{BE}. A variable d.c. supply V_{CC} is used to vary V_{CB}.

An important practical point to be taken into account when you take measurements to obtain data for plotting the static characteristics is that a transistor is very sensitive to temperature changes. Thus you may find that the readings on the meters in your test circuit may drift, that is, they may tend to change with time owing to power dissipated within the transistor producing increased junction temperatures. To minimize drift your test transistor should be mounted in a heat sink. Two forms of heat sinks are shown in fig. 13.6 and the effect of heat sinks is discussed in Sub-section 13.7.

6.2
Description of the method of obtaining the common base mode static characteristics

The static output characteristics for the common base mode connection of a transistor are curves of d.c. collector current I_C versus d.c. collector-base voltage V_{CB} for fixed values of d.c. input emitter current I_E. These curves may be obtained, using the circuit of fig. 6.1, as follows:

1. Set I_E to a given value, with V_{CB} say initially at 5 V for an *n-p-n* transistor or -5 V for a *p-n-p* transistor. For example if we select $I_E = 1$ mA we then vary R until $I_E = 1$ mA. An approximate value for R may be obtained as follows. If the E-B d.c. bias supply voltage is V_{EE}, then taking $V_{BE} \approx 0.6$ V for silicon transistors and $V_{BE} \approx 0.2$ V for germanium transistors—these are typical voltage drops across the B-E junction when it is forward-biased—we have

$$V_{EE} = V_{EB} + RI_E$$

and thus if $V_{EE} = 4$ V, $I_E = 1$ mA and assuming a silicon transistor

$$R = \frac{V_{EE} - V_{EB}}{I_E} = \frac{4 - 0.6}{1 \times 10^{-3}} = 3.4 \text{ k}\Omega.$$

2. Now we can start our measurements:

Set $V_{CB} = 0$, adjust R until $I_E =$ given value, say 1 mA, measure I_C;
set $V_{CB} = 1$ V,* adjust R until $I_E = 1$ mA, measure I_C;
set $V_{CB} = 2$ V,* adjust R until $I_E = 1$ mA, measure I_C . . . . ,

and so on until the required range of V_{CB} values has been covered.

3. Repeat procedure of 2 with higher values of I_E, e.g. for $I_E = 2$ mA, 3 mA, 4 mA, 5 mA.

---

*Remember V_{CB} is positive for *n-p-n*, but negative for *p-n-p* transistors and so V_{CB} should be set at -1 V, -2 V . . . etc., for *p-n-p* transistors

4. Plot a graph of I_C versus V_{CB} for each value of I_E. A typical set of output characteristics is shown in fig. 6.2(a).

The static input characteristics for the common base mode are curves of d.c. input current I_E versus d.c. input voltage V_{EB} for given fixed values of V_{CB}. Remember V_{EB} must be positive to forward-bias a p-n-p transistor (E must be positive with respect to B) but negative for an n-p-n transistor. Actually the effect of V_{CB} on the input characteristic curve is very small. Typically a change of 1 V in V_{CB} only changes V_{EB} by a millivolt or so. The input characteristics may be obtained as follows:

1. Set V_{CB} to a given value, say 5 V (n-p-n) or -5 V (p-n-p).
2. Set $R = 1$ kΩ and increase the E-B d.c. supply V_{EE} slowly from zero volts. You will find that little or no current flows until $V_{EB} > 0.1$ V for germanium or $V_{EB} > 0.5$ V for silicon transistors. Measure I_E against V_{EB} as V_{EE} is increased.
3. Repeat procedure of 2 for $V_{CB} = 0$ and $V_{CB} = 10$ V (n-p-n) or $V_{CB} = -10$ (p-n-p).
4. Plot I_E versus V_{EB} for $V_{CB} = 0, 5$ V, 10 V. In practice there will be little observable difference, the graphs will be almost the same.

A typical input characteristic is shown in fig. 6.2(b).

6.3
A discussion of typical families of common base static characteristics:
I_C v V_{CB} **output characteristics;** I_E v V_{EB} **input characteristics;** I_C v I_E
transfer characteristics

(a) I_C *versus* V_{CB} *common base output characteristics*
The curves drawn in fig. 6.2(a) are typical output characteristics for a transistor connected in the common base mode. We may make the following observations:

(i) For each given value of input emitter current the curves of collector current I_C versus collector-base reverse-bias voltage V_{CB} are almost independent of V_{CB}. Increasing the magnitude of V_{CB} from 0 to 10 V produces only a small increase in collector current, e.g. when $I_E = 2$ mA, $I_C = 1.8$ mA for $V_{CB} = 0$ and rises to $I_C = 1.95$ mA when $V_{CB} = 10$ V.

(ii) I_C is approximately equal to but is always less than I_E. We can calculate the d.c. current gain at a given point on the characteristics. For example at $I_E = 3$ mA, $V_{CB} = 5$ V, the collector current $I_C = 2.9$ mA (point R on fig. 6.2(a)). This gives

$$\alpha_{DC} = \frac{I_C}{I_E} = \frac{2.9}{3} = 0.97.$$

Note: $\alpha_{DC} = 0.97$ is for the point $I_E = 3$ mA, $V_{CB} = 5$ V. The d.c. current gain is not constant and hence the point at which it is measured should be stated.

(iii) The slopes or gradients of the curves are very small. At $V_{CB} > 3$ V the

72

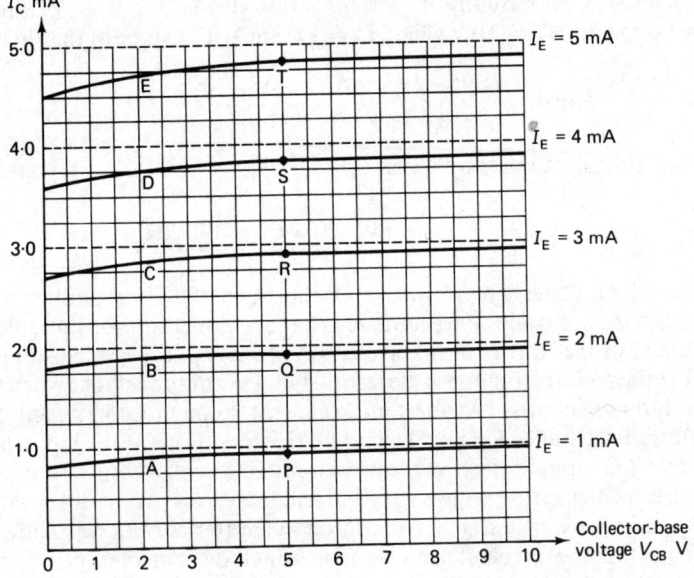

(a) Output characteristics: I_C versus V_{CB} for various I_E

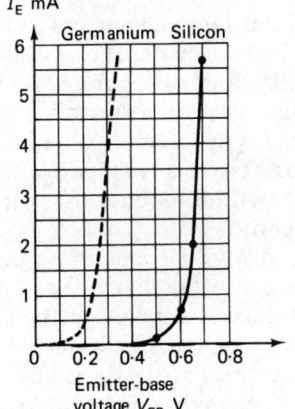

(b) Input characteristic: I_E versus V_{EB} for $V_{CB} = 5$ V

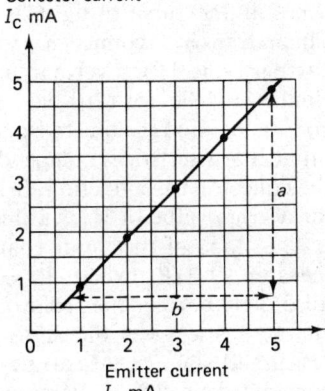

(c) Forward current transfer characteristic: I_C versus I_E for $V_{CB} = 5$ V

Fig. 6.2. Common-base static characteristics.

Note: For p-n-p transistors (see fig. 6.1(a)), V_{EB} is positive, V_{CB} is negative
For n-p-n transistors (see fig. 6.1(b)), V_{EB} is negative, V_{CB} is positive

curves approximate closely to straight lines. For example, the $I_E = 2$ mA characteristic is virtually a straight line from $V_{CB} = 3$ V where $I_C = 1.9$ mA to $V_{CB} = 10$ V where $I_C = 1.95$ mA. The slope of this 'line' is

$$\text{slope} = \frac{(1.95 - 1.9) \text{ mA}}{(10 - 3) \text{ V}} = \frac{0.05 \times 10^{-3}}{7} \text{ A/V}.$$

The reciprocal of this slope, that is,

$$\frac{7}{0.05 \times 10^{-3}} = 140\,000 \text{ V/A} = 140 \text{ k}\Omega$$

is a very high value. The reciprocal of the slope at a given point on the common base output characteristic is in fact a measure of the output resistance of the transistor in the common base mode. To be absolutely correct the reciprocal slope is the small-signal output resistance when the input is on open-circuit to alternating current. Remember we stated that the output resistance R_0 may depend on the conditions at the input. The fact that the input is effectively open-circuited to a.c. (but not to d.c.) is because in measuring a given output characteristic we maintained I_E always at a constant value. No variation in emitter current was allowed. So from the above calculation we may conclude that the open-circuit output resistance of our transistor when $I_E = 2$ mA and over the range $V_{CB} = 3$ to 10 V is approximately 140 kΩ.

(b) I_E versus V_{EB} *common base input characteristic*
The full-line curve of fig. 6.2(b) is a typical input characteristic for a silicon transistor connected in the common base mode. The dashed-line curve is typical for a germanium transistor. Both curves are in fact very similar to the forward-bias *p-n* junction characteristic of a semiconductor diode. To a reasonably good degree of approximation the input characteristic is independent of the reverse-bias voltage V_{CB} across the collector-base junction. If V_{CB} is increased the input emitter current for a given value of V_{EB} will increase slightly.

The slope of the input characteristic at a given value of I_E gives a measure of $1/R_i$ (where $R_i = $ input resistance) at that point. To be absolutely correct the reciprocal of the slope at a point on the input characteristic gives the short-circuit value of the small-signal input resistance, since in measuring the input characteristic V_{CB} is held at a constant d.c. value. Thus no a.c. variation of voltage occurs across the output, the output collector-base terminals are effectively short-circuited to a.c. (but not to d.c.). Since the input characteristic is curved, its slope and hence the input resistance varies from point to point, that is on the value of I_E. A typical value of common base short-circuit input resistance at $I_E = 1$ mA is about 25 Ω for germanium and about 38 Ω for silicon transistors, whilst at $I_E = 2$ mA, $R_i \approx 12$ Ω (Ge) and 19 Ω (Si).

(c) I_C *versus* I_E *common base transfer characteristic*

The transfer characteristic of output to input current for a given value of V_{CB} is drawn in fig. 6.2(c). The data for plotting this curve may be obtained from the output characteristics (a) by reading off the values of I_C for a given V_{CB} corresponding to $I_E = 1$ mA, 2 mA, 3 mA, 4 mA, 5 mA; i.e. when $V_{CB} = 5$ V:—

Emitter current I_E	1 mA	2 mA	3 mA	4 mA	5 mA
Collector current I_C	0·9 mA	1·9 mA	2·9 mA	3·85 mA	4·8 mA
Corresponding point marked on fig. 6.2(a)	P	Q	R	S	T

It is observed that the I_C versus I_E transfer characteristic is virtually a straight line over the plotted range. The slope of the characteristic at a given point corresponds to the small-signal, short-circuit current gain α or $-h_{fb}$ for the transistor in its common base mode. Since the characteristic is virtually a straight line

$$\text{slope} = \frac{a}{b} = \frac{4\cdot8 - 0\cdot7}{5\cdot0 - 0\cdot8} = \frac{4\cdot1}{4\cdot2} = 0.98.$$

So $\alpha = 0·98$ for $V_{CB} = 5$ V over the range $I_E = 1$ to 5 mA.

6.4
A test circuit diagram for determining the common emitter static characteristics

The test circuits shown in fig. 6.3 may be used to measure the static characteristics of a transistor connected in the common emitter mode. A milliammeter, or a microammeter for the case of a low-power transistor, is included in series with the base lead to measure I_B. A milliammeter is inserted in series in the collector circuit to measure I_C. A voltmeter with a typical range 0 to 1 V is connected across the base and emitter terminals of the transistor to measure the input base-emitter voltage V_{BE}, and the variable resistor R and d.c. supply V_{BB} is used to vary I_B and V_{BE}. A voltmeter with a typical range 0 to 20 V is connected across the collector-emitter terminals to measure the output collector-emitter voltage V_{CE}. A variable d.c. supply V_{CC} is used to vary V_{CE}. If a variable supply is not available, a fixed d.c. supply, typically 20 to 30 V, and a potential divider network may be used, as shown, for example, in fig. 6.3(b).

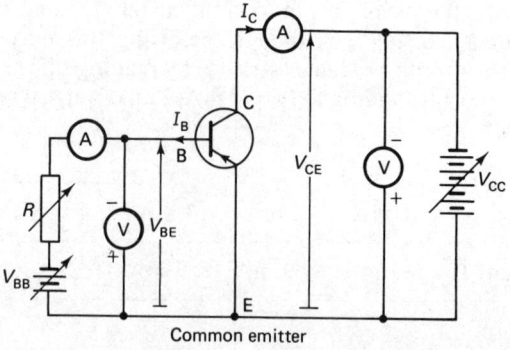

(a) Circuit to measure common emitter static characteristics of a *p-n-p* transistor

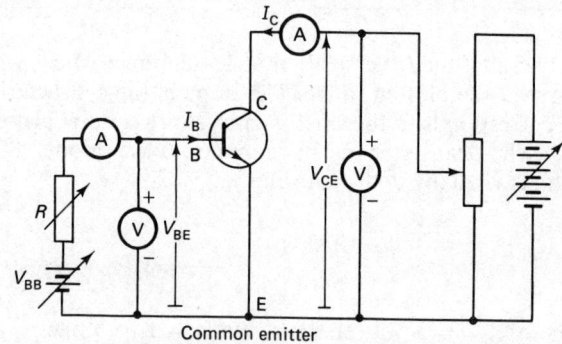

(b) Circuit to measure common emitter static characteristics of an *n-p-n* transistor

Fig. 6.3. Circuits to measure common emitter static characteristics.

6.5
The measurement of the common emitter static characteristics
The output characteristics for the common emitter mode connection of a transistor are curves of d.c. collector current I_C versus d.c. collector-emitter voltage V_{CE} for fixed values of input base current I_B. These curves may be obtained using the test circuits of fig. 6.3 as follows.

1. Set I_B to a given value with V_{CE}, say, initially at 5 V (*n-p-n*) or -5 V (*p-n-p*). I_B is adjusted to the required value by varying R and V_{BB}. For example, if $I_B = 100~\mu\text{A}$ is required and V_{BB} is set at 2 V, then as

$$V_{BB} = V_{BE} + RI_B \text{ and } V_{BE} \approx 0.6 \text{ V (Si), } 0.2 \text{ V (Ge)}$$

we have $\qquad R = \dfrac{V_B - V_{BE}}{I_B} = \dfrac{2 - 0.6}{100 \times 10^{-6}} = 14 \text{ k}\Omega \text{ (Si)}$

$$= \dfrac{2 - 0.2}{100 \times 10^{-6}} = 18 \text{ k}\Omega \text{ (Ge)}.$$

2. Set $V_{CE} = 0$, adjust R and or V_{BB} until I_B is at the required value (say 100 μA), measure I_C.

Set $V_{CE} = -1$ V (p-n-p) or $+1$ V (n-p-n), adjust R until $I_B = 100$ μA, measure I_C.

Set $V_{CE} = -2$ V (p-n-p) or $+2$ V (n-p-n), adjust R until $I_B = 100$ μA, measure I_C ... and so on until required range of V_{CE} values has been covered.

3. Repeat the procedure of 2 for other values of I_B, e.g. $I_B = 200$ μA, 300 μA, . . . etc. The characteristic for $I_B = 0$ may be obtained, if required, by open-circuiting the base circuit, i.e. by breaking or disconnecting the base bias supply circuit.

4. Plot the graph of I_C versus V_{CE} for each value of input current I_B.

The results obtained for a low to medium power n-p-n transistor are shown in Table 6.1, and the static common emitter output characteristics plotted from this data are drawn in fig. 6.4(a).

Table 6.1   *Values obtained for plotting of I_C v V_{CE} common emitter mode output characteristics*

Collector-emitter voltage V_{CE} (V)	Collector current I_C (mA) for				
	$I_B = 100$ μA	$I_B = 200$ μA	$I_B = 300$ μA	$I_B = 400$ μA	$I_B = 500$ μA
0	0	0	0	0	
1	3·0	7·0	10·0	15·0	18·0
2	4·8	10·0	12·0	19·6	23·0
3	6·0	12·0	16·1	22·0	26·1
4	6·9	12·8	17·9	24·0	28·2
5	7·4	13·5	18·8	25·0	29·7
6	7·9	13·9	20·0	25·8	30·6
6	7·9	13·9	20·0	25·8	30·6
7	8·0	14·1	20·6	26·3	31·7
8	8·1	14·3	21·0	27·0	32·3
9	8·1	14·6	21·2	27·8	33·0
10	8·2	14·9	21·8	28·0	34·0
12	8·3	15·1	22·4	28·8	35·0
14	8·5	15·3	23·0	29·4	36·2
16	8·6	15·4	23·5	30·5	37·5
18	8·8	15·5	24·0	31·2	38·2
20	9·9	15·6	24·4	32·0	40·0

The static input characteristics for a transistor connected in the common emitter mode are curves of I_B versus V_{BE} for given fixed values of V_{CE}. They may be obtained, with reference to fig. 6.3, as follows.

1. Set V_{CE} to a given value, say -10 V (p-n-p) or $+10$ V (n-p-n).

2. Set R to a suitable value, e.g. the order of kΩ if I_B values of the order of mA are anticipated, or the order of 100 kΩ to MΩ if I_B is to be smaller.

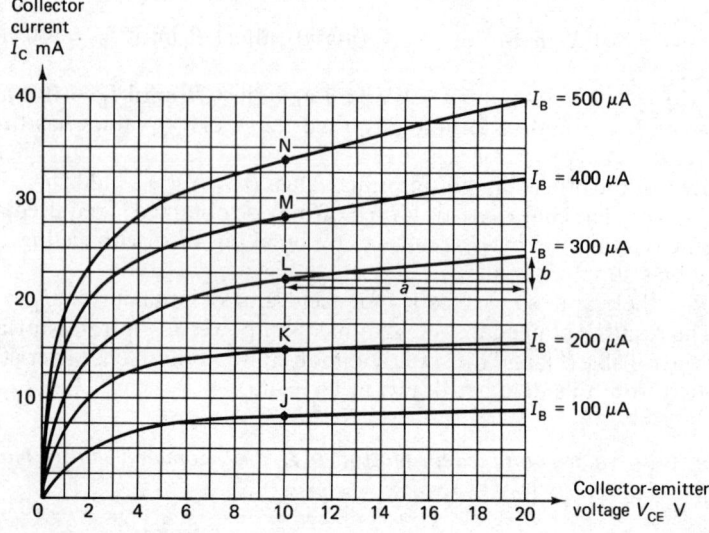

(a) Output characteristics: I_C versus V_{CE} for various I_B

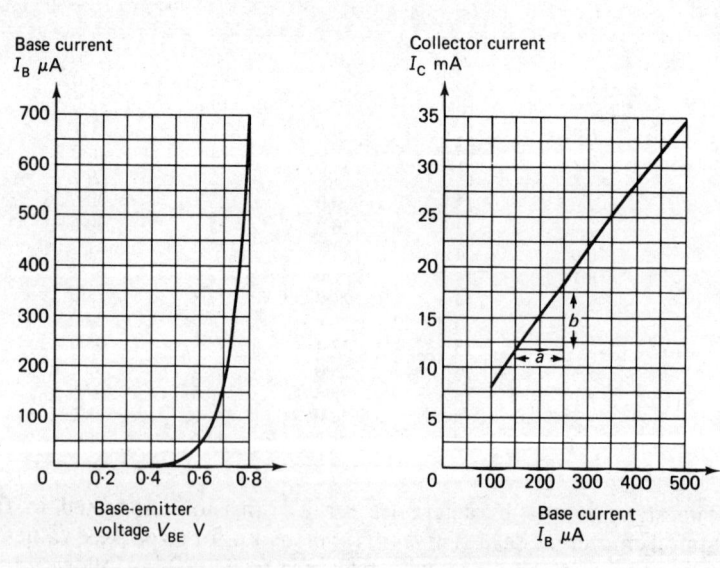

(b) Input characteristic I_B versus V_{BE} for V_{CE} = 10 V

(c) Transfer current characteristic I_C versus I_B for V_{CE} = 10 V

Fig. 6.4 Common emitter static characteristics

Increase V_{BB} gradually from 0 V. Negligible base current flows until $V_{BE} > 0\cdot1$ V for germanium and $V_{BE} > 0\cdot5$ V for silicon transistors. Measure V_{BE} for increasing values of I_B, for example by increasing V_{BB} from zero with R fixed, or alternatively by setting V_{BB} to a fixed value, say 5 V, and using a precision resistance box for R, setting R initially to the MΩ range.

3. Repeat the procedure of 2 whilst maintaining V_{CE} constant at, for example, 0 V and 20 V.

4. Plot the input characteristic curves of I_B versus V_{BE} for $V_{CE} = 0, 10, 20$ V. In practice the input characteristics will be largely independent of the V_{CE} value and the curves will therefore be very similar.

The results obtained for a low to medium power n-p-n transistor, with V_{CE} held constant at 10 V are shown in Table 6.2, and the static input characteristic plotted from this data is drawn in fig. 6.4(b).

Table 6.2   Values obtained for plotting I_B v V_{BE} common emitter input characteristic for $V_{CE} = 10$ V

Base current I_B μA		10	40	100	200	300	400	500	600	700 mV
Base-emitter voltage V_{BE} mV			500	600	660	718	740	770	781	795	800 μA

6.6
The plotting and description of typical families of common emitter static characteristics: I_C v V_{CE} output characteristics; I_B v V_{BE} input characteristics; I_C v I_B transfer characteristics

(a) I_C *versus* V_{CE} *common emitter output characteristics*
A typical set of output characteristics for a low to medium power transistor connected in the common emitter mode are drawn in fig. 6.4(a). The data for plotting these curves are recorded in the previous sub-section. When $V_{CE} = 0, I_C = 0$ for all values of input base current. As the collector-emitter voltage is increased the collector current, for each given value of input base current, rises rapidly at first and then gradually and approximately in a straight line. The slopes of these 'lines' are greater than the corresponding common base output characteristics, indicating that the output resistance of a transistor in the common emitter mode is lower than when the transistor is connected with its base common. We can use the curves to make the following calculations:

(i) *D.C. common emitter current gain, β_{DC} or h_{FE}*
E.g. when $V_{CE} = 10$ V, $I_B = 400$ μA and $I_C = 28$ mA (point M on Fig. 6.4(a)).

So $\quad \beta_{DC} \equiv h_{FE} = \dfrac{I_C}{I_B} = \dfrac{28 \text{ mA}}{400 \text{ μA}} = \dfrac{28 \text{ mA}}{0\cdot4 \text{ mA}} = 70$ at this point.

(ii) *The small-signal output resistance with input open-circuited to alternating current, $R_{0(OC)}$*

The slope at a given point on an output characteristic equals $1/R_{0(OC)}$, e.g. the $I_B = 300\ \mu A$ characteristic is very closely linear over the range $V_{CE} = 10$ to 20 V, the slope of this 'line' is

$$\frac{b}{a} = \frac{(24\cdot4 - 21\cdot8)\ \text{mA}}{(20 - 10)\ \text{V}} = \frac{2\cdot6}{10}\ \text{mA/V} = 2\cdot6 \times 10^{-4}\ \text{A/V}.$$

So $R_{0(OC)} = \dfrac{1}{\text{slope}} = \dfrac{10\,000}{2\cdot6} = 3850\ \Omega$

which is the open-circuit output resistance of the transistor when $I_B = 300\ \mu A$ over the range $V_{CE} = 10$ to 20 V.

(iii) *To determine the I_C versus I_B transfer characteristics and hence calculate the small signal current gain β or h_{fe}*

A typical common emitter transfer characteristic is drawn in fig. 6.4(c). The data used to plot this curve was obtained from the output characteristics of fig. 6.4(a), i.e. at $V_{CE} = 10$ V:

Base current I_B (μA)	100	200	300	400	500
Collector current I_C (mA)	8·2	14·9	21·8	28	34
Corresponding point marked on fig. 6.4(a)	J	K	L	M	N

The slope at a given point on the current transfer characteristic equals the common emitter small-signal short-circuit current gain at that point, e.g. the β or h_{fe} value at $I_C = 14$ mA is

$$\beta = (\text{slope at } I_C = 15\ \text{mA}) = \frac{b}{a} = \frac{(18 - 11\cdot5)\text{mA}}{(250 - 150)\mu A} = \frac{6\cdot5\ \text{mA}}{0\cdot1\ \text{mA}} = 65.$$

Note that in calculating the slope we have assumed that the characteristic about the point $I_C = 15$ mA, $I_B = 2$ is a straight line. The value of β_{DC} at the same point is $\beta_{DC} = I_C/I_B = 15\ \text{mA}/200\ \mu A = 15/0\cdot2 = 75$.

(b) *I_B versus V_{BE} common emitter input characteristic*

A typical common emitter input characteristic for a low to medium power transistor is drawn in fig. 6.4(b). For a power transistor the input base current values would be the order of milliamperes whilst for a low-

80

power transistor maximum base current employed may be only up to a few tens of microamperes. The input characteristic is similar in basic shape to the forward-bias characteristic of a semi-conductor diode except that the base current is very much smaller in magnitude, in fact smaller by a factor of β_{DC}, the d.c. common emitter current gain.

The slope of the input characteristic at a given value of I_B gives the reciprocal of the common emitter small-signal short-circuit input resistance of the transistor at that point; the short-circuit qualification again being used to denote that the output voltage V_{CE} is held constant, thus simulating an effective shoft-circuit to alternating current. The input resistance of a transistor in the common emitter mode is greater by the factor β (small-signal current gain) than the input resistance in the common base connection. For example, if the input resistance of a silicon transistor at $I_E = 1$ mA were 20 Ω in the common base mode and its corresponding current gain $\alpha = 0.98$ (remember $\alpha = I_c/I_e$ where I_c and I_e are small current changes and also $\beta = \alpha/(1-\alpha)$), then $\beta = 0.98/(1-0.98) = 0.98/0.02 = 49$ and so the common emitter input resistance would be $49 \times 20 \approx 1$ kΩ.

6.7
The determination of the small-signal short-circuit current gains, α and β, from given characteristics

We explained in Sub-section 6.3 (for the common base case) and 6.6 (for the common emitter case) that the small-signal current gains, α and β, could be determined from the static current transfer characteristics:

α = small-signal short-circuit common base current gain

$$= \left[\frac{\text{small change in collector current about a given value of } I_C}{\text{corresponding change in emitter current bringing about above change}} \right]$$

with collector-base voltage V_{CB} held constant

= (slope of I_C versus I_E curve) at a given V_{CB} and I_C.

Note: V_{CB} defines the static characteristic we are using (I_C v I_E characteristics are plotted for fixed values of V_{CB}). I_C defines the point on the characteristic at which we are measuring the slope.

β = small-signal, short-circuit, common emitter current gain

$$= \left[\frac{\text{small change in collector current about a given value of } I_C}{\text{corresponding change in base current to effect above change}} \right]$$

with collector-emitter voltage V_{CE} held constant

= (slope of I_C versus I_B curve) at a given V_{CE} and I_C.

V_{CE} defines the static transfer characteristic we are using and I_C defines the point on the characteristic at which we measure the slope.

The following two examples demonstrate the calculation of α and β from given characteristics.

Example 6.1

Fig. 6.5 shows a typical common emitter transfer characteristic for a low-power *n-p-n* transistor measured with V_{CE} held constant at 10 V. Calculate (a) the d.c. short-circuit current gain (β_{DC} or h_{FE}) at $I_C = 2\,mA$ (b) the small-signal short-circuit current gain (β or h_{fe}) at $I_C = 2\,mA$.

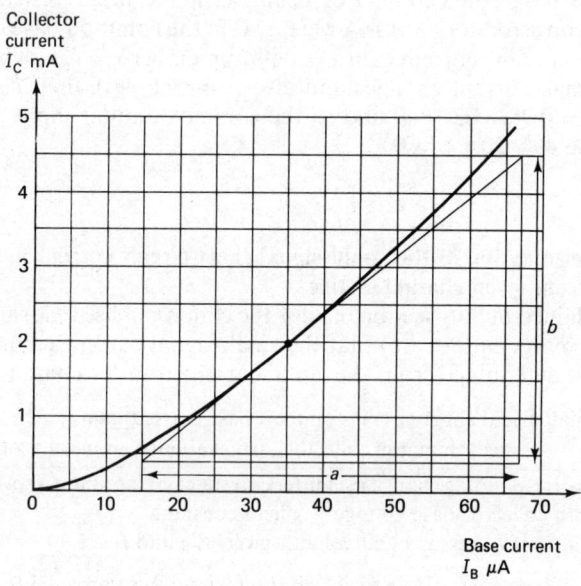

Fig. 6.5. Common-emitter current transfer characteristic for $V_{CE} = 10V$.

Solution

(a) When $I_C = 2$ mA, $I_B = 35\ \mu A$

So $\beta_{DC} \equiv h_{FE} = \dfrac{I_C}{I_B} = \dfrac{2 \times 10^{-3}}{35 \times 10^{-6}} = \dfrac{2000}{35} = 57.$

(b) The slope of the I_C v I_B characteristic at the point where $I_C = 2$ mA is the value of β or h_{fe} at that point,

$$\beta = h_{fe} = \frac{b}{a} = \frac{4\cdot1\ \text{mA}}{52\ \mu A} = \frac{4100}{52} = 79.$$

Example 6.2
Using the common base output characteristics of fig. 6.2(a), construct the transfer characteristic I_C v I_E for the case $V_{CB} = 2$ V. Hence show that the small-signal current gain α is very nearly constant over the range $I_C = 1$ mA to $4 \cdot 7$ mA and calculate its value. Calculate also the value of the d.c. current gain α_{DC} at $I_C = 3$ mA.

Solution

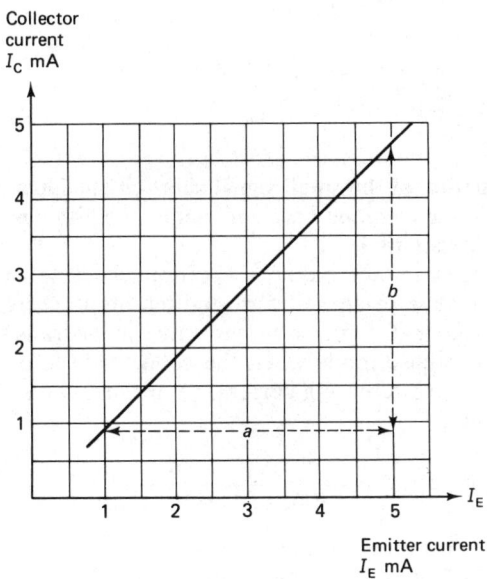

Fig. 6.6. Common-base current transfer characteristic for $V_{CB} = 2$V derived from Fig. 6.2(a).

The following table of I_C v I_E values was obtained from the output characteristics of fig. 6.2(a) when $V_{CB} = 2$ V:

Emitter current, I_E (mA)	1	2	3	4	5
Collector current, I_C (mA)	0·9	1·9	2·85	3.75	4.7
Point on fig. 6.2(a)	A	B	C	D	E

and the common base current transfer characteristic for $V_{CB} = 2$ V is plotted from this data in fig. 6.6. Over the plotted range $I_C = 0.9$ mA to 4.5 mA the points lie very nearly in the same straight line. In fig. 6.6 a

83

straight line has been drawn through the plotted points. The slope of this line gives the small-signal current gain α,

$$\alpha = \frac{b}{a} = \frac{3 \cdot 8 \text{ mA}}{4 \cdot 0 \text{ mA}} = 0 \cdot 95$$

over the range of I_C values 0·9 to 4·7 mA.

Since the characteristic is a straight line (the slope of a straight line is, of course, constant) α is constant over the range 1 mA to 4·5 mA.

When $I_C = 3$ mA the corresponding value of emitter current read-off the characteristic is $I_E = 3 \cdot 2$ mA.

So $\alpha_{DC} = \dfrac{I_C}{I_E} = \dfrac{3}{3 \cdot 2} = 0 \cdot 94$.

6.8
The determination of the small-signal short-circuit input resistances of a transistor in the common base and common emitter modes from given input characteristics

The slope of the common base I_E v V_{EB} input characteristic for V_{CB} held constant and at a given value of emitter input current equals the reciprocal of the small-signal input resistance to the transistor connected in the common base mode when the collector-base output circuit is short-circuited to alternating current, i.e. the small-signal short-circuit input resistance,

$$R_i \equiv h_{ib}{}^* = \frac{1}{(\text{slope of } I_C \text{ v } V_{EB} \text{ characteristic})}$$

with V_{CB} held constant and I_E = given value.

Similarly the slope of the common emitter I_B v V_{BE} input characteristic equals the reciprocal of the small-signal input resistance to the transistor connected in the common emitter mode when the collector-emitter output circuit is short-circuited to alternating current, i.e.

$$R_i \equiv h_{ie}{}^* = \frac{1}{(\text{slope of } I_B \text{ v } V_{BE} \text{ characteristic})}$$

with V_{CE} held constant and I_B = given value.

---

*The common base small-signal short-circuit input resistance is denoted by h_{ib}, the subscript i denotes input, b common base. h_{ib} is one of the four hybrid parameters used to define the small-signal characteristics of a transistor connected in the common base mode. h_{fb} ($h_{fb} = -\alpha$) is another (see Sub-section 5.3).

The common emitter small-signal short-circuit input resistance is denoted by h_{ie}, i denotes input, e common base. It is one of the four hybrid parameters used to define the small-signal characteristics of a transistor connected in the common emitter mode. $h_{fe} = \beta$ is another.

The calculation of input resistances from given input characteristics is undertaken in the following two examples.

Example 6.3
Fig. 6.7 shows an 'expanded' input characteristic for an n-p-n transistor connected in the common base mode, with the output collector-base voltage held constant at $V_{CB} = 5$ V. Calculate the small-signal short-circuit input resistance of the transistor when $I_E = 1$ mA.

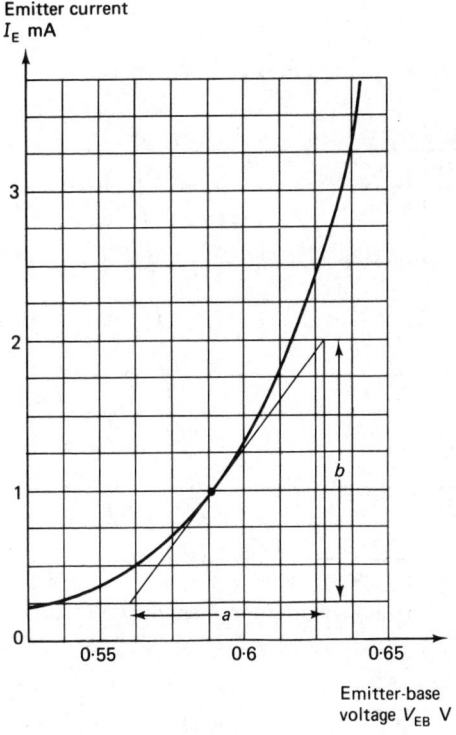

Fig. 6.7. Common-base input characteristic for $V_{CB} = 5$V.

Solution
The slope of the input characteristic at the point where $I_E = 1$ mA is

$$\frac{b}{a} = \frac{1 \cdot 75 \text{ mA}}{0 \cdot 0675 \text{ V}} = \frac{1 \cdot 75 \times 10^{-3} \text{ A}}{0 \cdot 0675 \text{ V}} = 26 \times 10^{-3} \text{ A/V}$$

The reciprocal of the slope is the small-signal, short circuit input resistance when $I_E = 1$ mA, $V_{CB} = 5$ V, so

$$h_{ib} = \frac{1}{26 \times 10^{-3}} = \frac{1000}{26} = 38 \ \Omega$$

Example 6.4

Fig. 6.8 shows the common emitter input characteristic for an *n-p-n* transistor, for the collector-emitter voltage held constant at $V_{CE} = 10$ V. Calculate the small-signal short-circuit input resistance (h_{ie}) when the input base current $I_B = 100 \ \mu A$.

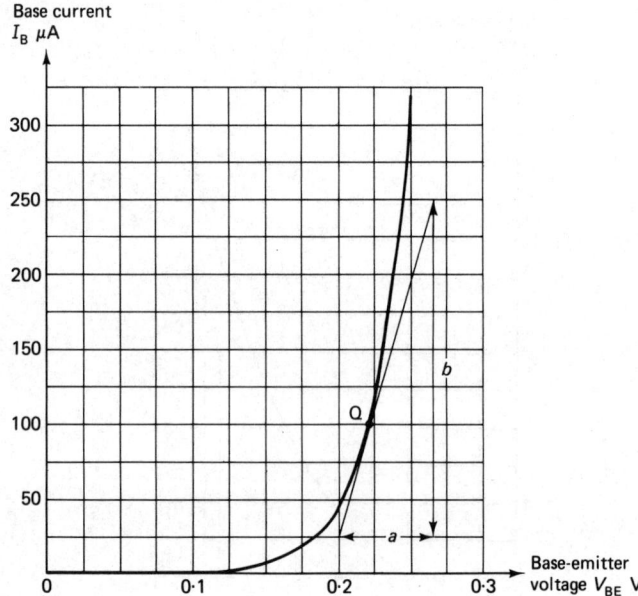

Fig. 6.8. Common-emitter input characteristic for $V_{CE} = 10$V.

Solution

At the point $I_B = 100 \ \mu A$,

$$h_{ie} = \frac{1}{\text{slope at point Q on the input characteristic}}$$

$$= \frac{1}{\frac{b}{a}} = \frac{a}{b} = \frac{0.065 \text{ V}}{225 \ \mu A}$$

$$= \frac{0.065 \text{ V}}{225 \times 10^{-6} \text{ A}} = 289 \ \Omega$$

i.e. the small-signal input resistance of the transistor in the common emitter mode with the output collector-emitter terminals held constant at $V_{CE} = 10$ V and with an input current $I_B = 100$ μA is 289 Ω.

6.9
The determination of the small-signal open-circuit output resistances of a transistor in the common base and common emitter modes from given output characteristics

The slope at a given point on the static output characteristic of a transistor equals the small-signal output conductance of the transistor with the input terminals of the transistor open-circuited to alternating current. The reciprocal of the slope equals the small-signal open-circuited output resistance. The symbols used to denote output conductance are h_{ob} for the common base and h_{oe} for the common emitter connection, where the subscript o denotes output, b common base, and e common emitter. The subscripts are lower-case ones to indicate that we are considering small-signal parameters. h_{ob} is one of the four common base hybrid parameters and h_{oe} one of the four common emitter hybrid parameters used to define the small-signal characteristic of a transistor in its given mode of connection.

For the common base connection:
the small-signal open-circuit output conductance at a given point (defined by I_E = fixed value, V_{CB} (or I_C) = a given value) is

$$h_{ob} = \frac{\text{small change in collector current}}{\text{corresponding change in collector-base voltage about given } V_{CB}}$$

with input emitter current held constant,

$$= (\text{slope of } I_C \text{ v } V_{CB} \text{ curve}) \text{ at given } I_E, V_{CB}.$$

The small-signal open-circuit output resistance,

$$R_{0(oc)} = \frac{1}{h_{ob}} \, .$$

For the common emitter connection:
at a given point [defined by I_B = fixed value, V_{CE} (or I_C) = a given value],

$$h_{oe} = \frac{\text{small change in collector current}}{\text{corresponding change in collector emitter voltage about given } V_{CE}}$$

with input base current held constant,

$$= (\text{slope of } I_C \text{ v } V_{CE} \text{ curve}) \text{ at given } I_B, V_{CE}.$$

And the small-signal open-circuit output resistance,

$$R_{0(oc)} = \frac{1}{h_{oe}} \, .$$

Calculations of output resistance for the common base and common emitter connections were made in Sub-sections 6.3 and 6.6 respectively. The following example demonstrates the calculation for a transistor in the common emitter mode.

Example 6.5
Fig. 6.9 shows the common emitter output characteristics of a transistor, for $I_B = 20, 40$, and $60\ \mu A$. The latter characteristic is expanded for $I_C > 4$ mA. Calculate the small-signal open-circuit output conductance and resistance of the transistor when $I_B = 60\ \mu A$ and $V_{CE} = 5$ V.

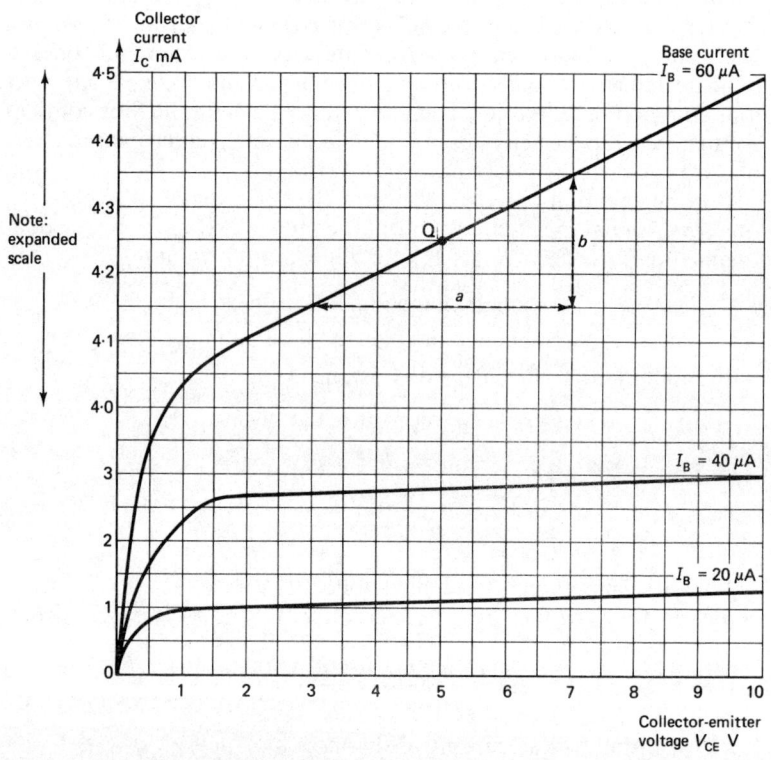

Note: expanded scale

Fig. 6.9.

Solution
At the point $V_{CE} = 5$ V on the static output $I_B = 60\ \mu A$ characteristic (marked Q on fig. 6.9), the small-signal output conductance,

88

$$h_{oe} = \text{(slope of curve at Q)}$$

$$= \frac{b}{a} = \frac{0.2 \text{ mA}}{4 \text{ V}} = 0.05 \text{ ma/V or } 50 \text{ } \mu\text{A/V}.$$

Note: the SI units of conductance are siemens, abbreviation S. However, mA/V and μA/V are still frequently used in Electronics.

The small-signal output resistance,

$$R_{0(oc)} = \frac{1}{h_{oe}} = \frac{1}{50 \times 10^{-6}} = 20\,000 \text{ } \Omega = 20 \text{ k}\Omega.$$

Problems: A *Elementary Theory of Semi-conductors*

1.1 State two common types of semi-conductor materials and compare their properties in relation to a metallic conductor and a good insulating material.

1.2 Explain with the aid of labelled diagrams the structure of
(a) an intrinsic (that is, pure) semi-conductor,
(b) a p-type semi-conductor,
(c) an n-type semi-conductor.

1.3 Explain with the aid of suitable diagrams the mechanism of electrical conduction in (a) a metal, (b) an n-type semi-conductor, (c) a p-type semi-conductor.

1.4 Complete the following statements:
(a) the majority current carriers in an n-type semi-conductor are . . .
(b) the majority current carriers in a p-type semi-conductor are . . .
(c) the addition of a relatively small number of pentavalent impurity atoms to silicon would produce . . .
(d) the addition of a relatively small number of trivalent impurity atoms to germanium would produce . . .
(e) atoms of phosphorus, arsenic, and antimony are known as donor impurities because . . .
(f) atoms of boron, aluminum, gallium, and indium are known as acceptor impurities because . . .
(g) an increase in temperature affects the intrinsic conduction in a semi-conductor because . . .

1.5 State whether the following statements are either *true* or *false* giving a brief reason for your answer.
(a) Pure germanium at 0°C acts as a very good insulator.
(b) The conductivity of intrinsic silicon increases with temperature.
(c) A p-type semi-conductor contains a greater number of holes than free electron current carriers.
(d) The addition of impurity atoms in the ratio 1 to 10^8 silicon atoms produces a higher conductivity than in the ratio 1 to 10^6.

2.1 Describe the basic structure of a p-n junction and draw labelled diagrams showing a p-n junction connected in both forward and reverse-bias modes. Explain the mechanism of current flow in the forward and in the reverse-bias mode.

2.2. Draw a labelled diagram of the circuit symbol used to represent a semi-conductor diode. Describe an experiment you would undertake to measure the static $V-I$ characteristic of a p-n junction diode and sketch a typical characteristic for (a) a germanium, (b) a silicon diode. Compare these characteristics as regards forward voltage drop and reverse-bias current.

2.3 The circuit shown in fig. P2.3 was used to measure the forward-bias I-V characteristic of a silicon diode. Results of diode current I and voltage V dropped across the diode as the d.c. supply voltage V_{dc} was varied are tabulated below. Plot the forward-bias characteristic

d.c. voltage V_{dc} (volts)	diode current I (mA)	diode voltage V (volts)
0	0	0
2	0·2	0·56
4	0·56	0·61
6	0·89	0·63
8	1·30	0·65
10	1·65	0·66
12	2·0	0·67
14	2·35	0·68
16	2·75	0·69
18	3·1	0·70

V and I are related by the equation,

$$V_{dc} = V + RI$$

Use this equation to estimate the value of the resistor R used in the measurement circuit of fig. P2.3.

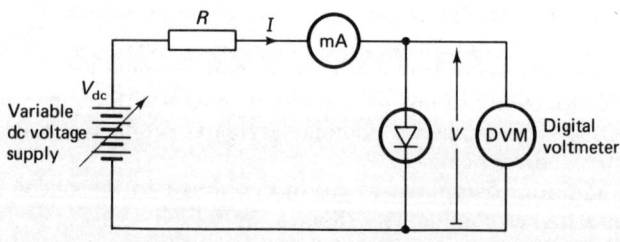

Fig. P2.3 for Problem 2.3.

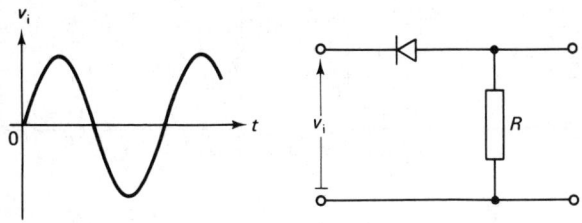

(a) rms amplitude of input voltage is 250 V

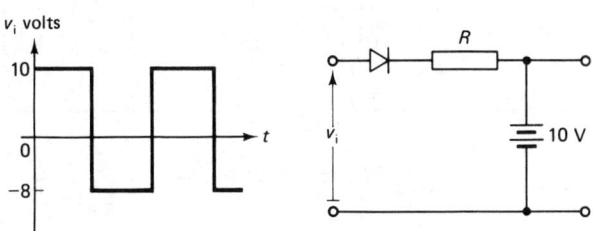

(b) Input square wave voltage varies between 10 V and −8 V; battery voltage = 10 V

Fig. P2.4 for Problem 2.4.

2.4 Define (a) the peak inverse voltage of a diode in a given circuit and (b) the peak inverse voltage rating of a diode. Explain the importance of considering the peak inverse voltage in diode circuit applications.

Determine the peak inverse voltage of the diodes shown in the circuits of fig. P2.4(a) and (b).

2.5 Sketch a typical $V-I$ characteristic of a Zener diode and describe an experiment which demonstrates the breakdown effect in such a diode.

2.6 The diodes used in the circuits shown in fig. P2.6 are silicon. Calculate, stating any assumptions made, the following:

(a) The current I in the circuit of fig. P2.6(a).

(b) The voltage V across the 1-kΩ resistor in the circuit of fig. P2.6(b).

(c) The current I in the 100 Ω-resistor connected across the Zener diode in the current of fig. P2.6(c). The breakdown voltage of the Zener diode is 20 V.

(d) The currents I_1 and I in the circuit of fig. P2.6(d).

3.1 Describe with the aid of suitable diagrams one application of (a) a power diode, (b) a Zener diode, and (c) a signal diode.

3.2 Describe with the aid of a circuit diagram the operation of a half-wave rectifier fed via a mains transformer and working into a resistive load. Draw also sketches of the applied a.c. voltage and load current waveforms.

Calculate the peak value of load current if the load resistor is 100 Ω

91

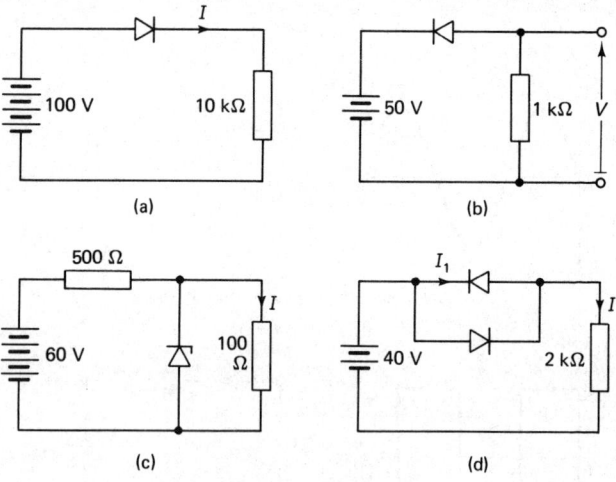

Fig. P2.6 for Problem 2.6.

and the peak amplitude of the a.c. voltage across the transformer
secondary terminals is 200 V. If a moving coil ammeter, which registers
average current, is connected in series with the load resistor, calculate its
expected reading.

3.3 The full-wave rectifier shown in fig. P3.3 is supplied by an a.c. voltage
source of 240 V r.m.s. amplitude and 50 Hz frequency via a step-down
transformer of turns ratio 4:1, that is the secondary winding has 1 turn
for every 4 on the primary winding. Calculate:

(a) The peak value of a.c. voltage across the primary winding.

(b) The peak a.c. voltage amplitude between A (or B) and the centre-tap
CT of the secondary winding.

(c) The peak and average values of the current flowing through the load
resistor $R_L = 20\ \Omega$ (neglect any forward voltage drop across the diodes
and any losses in the transformer).

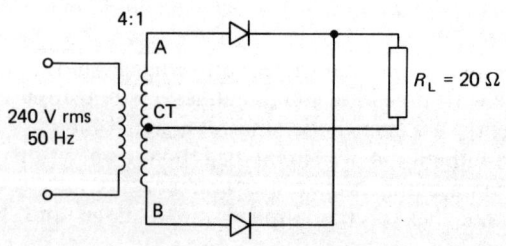

Fig. P3.3 for Problem 3.3.

Sketch also the waveforms of the applied voltage across the primary windings and the load current in R_L.

3.4 Calculate the peak and average values of load current in the bridge rectifier circuit of fig. P3.4 assuming that the forward voltage drop across the diodes may be neglected. The transformer turns ratio is unity, the a.c. supply voltage has an r.m.s. amplitude of 120 V, and the load resistor R_L = 1 kΩ. Sketch also the waveforms of applied voltage across A-B and the load current through R_L.

If a capacitor of large value (say C greater than 500 μF) is connected across R_L, calculate the new value of load current in R_L. Sketch also the waveforms of load current and voltage.

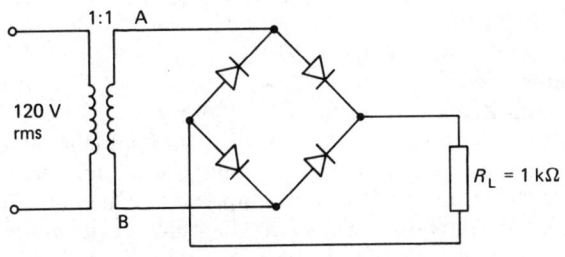

Fig. P3.4. for Problem 3.4.

3.5 Fig. P3.5 shows the circuit diagram of a half-wave rectifier with a capacitor C connected in parallel with the load. Explain the purpose of this capacitor and draw waveform sketches of the applied a.c. voltage, the load current, the load voltage, and the diode current.

If the a.c. input is at the mains frequency of 50 Hz and the peak a.c. voltage amplitude across the secondary winding is 100 V, and the load resistor R_L = 2·5 kΩ and the capacitor C = 200 μF, calculate the following:
(a) The variation in the output voltage across the load between its maximum and minimum values.
(b) The d.c. load current.
(c) The peak inverse voltage experienced by the diode.

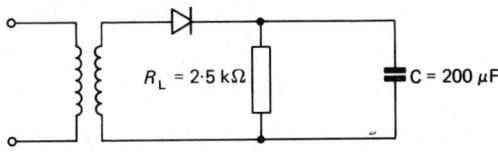

Fig. P3.5. for Problem 3.5.

3.6 Draw the circuit diagram of a simple voltage stabilizer which employs a Zener diode and a series resistor and explain its action.

In such a circuit a Zener diode with a breakdown voltage of 25 V is used. The minimum operating or knee current of the Zener is 1 mA and the maximum rated power dissipation is 0·5 W. If the input supply voltage to the circuit is 40 V d.c., calculate the value of the series resistor R which provides a 25 V stabilized supply for a load current range from zero to the maximum which will permit good regulation to be maintained. Calculate also this maximum value of load current.

3.7 A voltage stabilizer circuit employing a Zener diode and a series resistor R is required to produce a constant voltage output of 50 V at a fixed load current of 250 mA when the d.c. input to the circuit may vary between 100 V to 120 V. The Zener diode used in the circuit has the following specification:

Zener breakdown voltage = 50 V,
minimum Zener current = 5 mA,
maximum Zener current rating = 200 mA.

Calculate a suitable value of the series resistor R and the value of Zener current when the supply is at its maximum and its minimum values.

4.1 Explain the meaning of the terms 'bipolar' and 'junction' when used to describe a transistor. Draw a diagram of a p-n-p and an n-p-n transistor and label their respective regions. Draw also the circuit symbols used to represent p-n-p and n-p-n transistors and label their respective electrodes.

5.1 (a) identify the mode of connection and type of transistor in the diagrams shown in fig. P5.1.

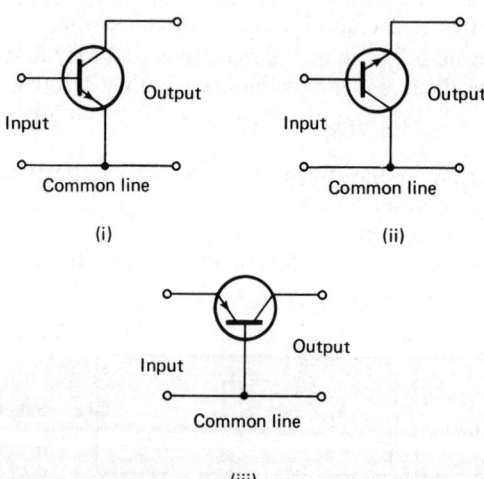

Fig. P5.1. for Problem 5.1(a).

(b) Draw labelled circuit diagrams including bias supplies showing a *p-n-p* transistor connected in the common base and common collector modes.

5.2 (a) Define the input and output resistances of a transistor amplifier and compare their relative values for the common base, common emitter, and common collector modes of connection.

(b) Determine the input resistance and voltage gain for the amplifier stage shown in fig. P5.2, given that the ammeter and the two voltmeters read the following a.c. values:

$$I_1 = 100 \ \mu A, \ V_1 = 150 \ mV, \ V_2 = 3 \ V.$$

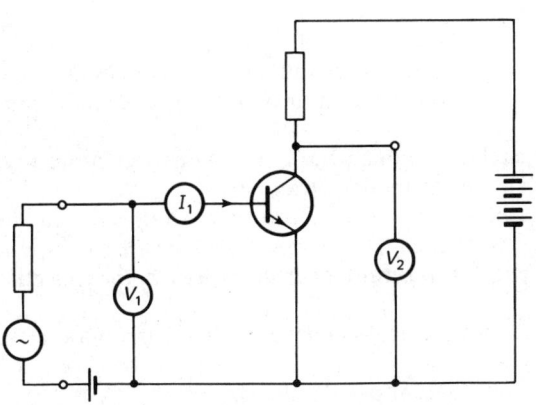

Fig. P5.2. for Problem 5.2(b).

5.3 (a) State which mode of connection you would use for the following amplifier applications:

    (i)   high input resistance, low output resistance,
    (ii) medium high input and output resistances,
    (iii) low input resistance, very high output resistance.

(b) Fig. P5.3 shows an a.c. equivalent circuit of an amplifier. The input

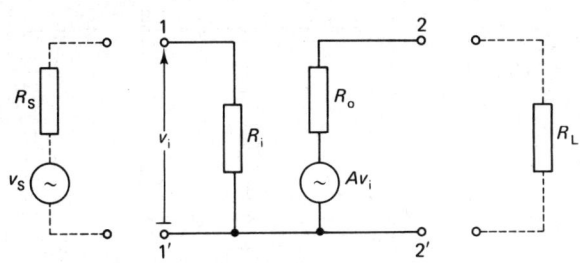

Fig. P5.3. for Problem 5.3(b).

and output resistances of the amplifier have the values
$$R_i = 5 \text{ k}\Omega, \; R_0 = 50 \text{ k}\Omega.$$
The open-circuit voltage amplification of the amplifier, $A = 250$.

If a signal generator of peak voltage amplitude $V_S = 10$ mV and internal resistance $R_S = 600 \, \Omega$ is connected across the input terminals 1-1' and a load resistor $R_L = 10$ kΩ is connected across the output terminals 2-2', use the equivalent circuit to determine:

(i) the peak value of input signal voltage across 1-1',
(ii) the peak values of the signal output current and the signal voltage developed across 2-2',
(iii) the voltage amplification of the stage,
(iv) the current amplification of the stage.

5.4 (a) Define the small-signal and the d.c. short-circuit current gains of a transistor when connected in the common base and common emitter modes.

(b) State the relationship between these gains and quote a typical range of values for each for modern transistors.

(c) The d.c. current gain of a transistor connected in the common emitter mode is 50 when $I_C = 10$ mA. Calculate the corresponding values of base current (I_B) and emitter current (I_E) when the collector current $I_C = 10$ mA.

5.5 Fig. P5.5 shows a small-signal equivalent circuit for a common base amplifier:

R_i = common base input resistance,
α = common base short-circuit current gain,
R_0 = common base output resistance.

If $R_i = 20 \, \Omega$, $\alpha = 0.98$, and R_0 is several megaohms, calculate stating any assumptions made, the following when a signal generator of peak voltage amplitude 20 mV and internal resistance 5 Ω is connected across the input terminals 1-1' and a load resistance of 2 kΩ is connected across the output terminals 2-2':

(a) The peak amplitudes of input emitter current and the input voltage across 1-1'.

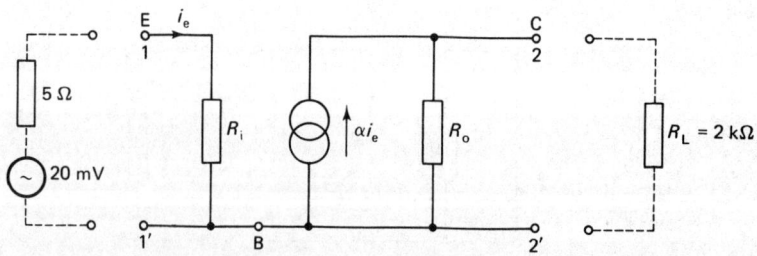

Fig. P5.5. for Problem 5.5.

(b) The peak amplitudes of collector current and the output voltage developed across 2-2'.

(c) The voltage amplification of the stage.

6.1 Draw the circuit diagram of a test circuit for measuring the common base static characteristics of a transistor and describe how you would obtain data to plot (a) the input $V_{EB} - I_E$, (b) the output $V_{CB} - I_C$, and (c) the transfer $I_E - I_C$ characteristics.

6.2 Draw the circuit diagram of a test circuit for measuring the common emitter static characteristics. Sketch a typical family of output characteristics and describe how you would use these to determine (a) the small-signal output resistance, (b) the small-signal current gain at a given value of collector current.

6.3 Fig. P6.3 shows the common emitter output characteristics for a commercial n-p-n power transistor. Using these characteristics draw up a table of I_B and I_C values when $V_{CE} = 4$ V, and hence plot the current transfer of the transistor. Use the latter to determine:

(a) The base current required to produce a collector current of 200 mA at $V_{CE} = 4$ V.

(b) The d.c. and small-signal current gains at $V_{CE} = 4$ V when (i) $I_B = 4$ mA, (ii) $I_C = 300$ mA.

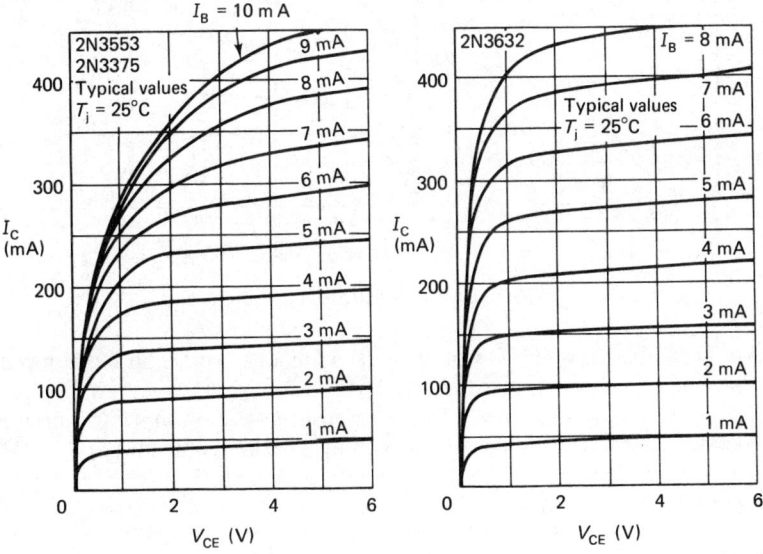

Fig. P6.3. Characteristics for Problem 6.3.

Fig. P6.4. Characteristics for Problem 6.4.

6.4 Explain how the small-signal input and output resistances of a transistor may be determined from its static characteristics.

Fig. P6.4 shows the output characteristics of an *n-p-n* transistor when connected in the common emitter mode. Use the characteristic to determine:
(a) The collector-emitter voltage when $I_B = 4$ mA and $I_C = 210$ mA.
(b) An estimation of the output resistance when $I_B = 4$ mA and over the range $V_{CE} > 1.0$ V.
6.5 Fig. P6.5 show the common emitter input and output characteristics of an *n-p-n* power transistor. Use the characteristics to determine the following:
(a) The base current when $V_{BE} = 0.8$ V (with $V_{CE} = 10$ V).
(b) The base-emitter voltage when $I_B = 3$ mA ($V_{CE} = 10$ V).
(c) The values of I_B and I_C corresponding to $V_{BE} = 0.9$ V, $V_{CE} = 10$ V.
(d) The d.c. current gain when $I_B = 6$ mA, $V_{CE} = 15$ V.

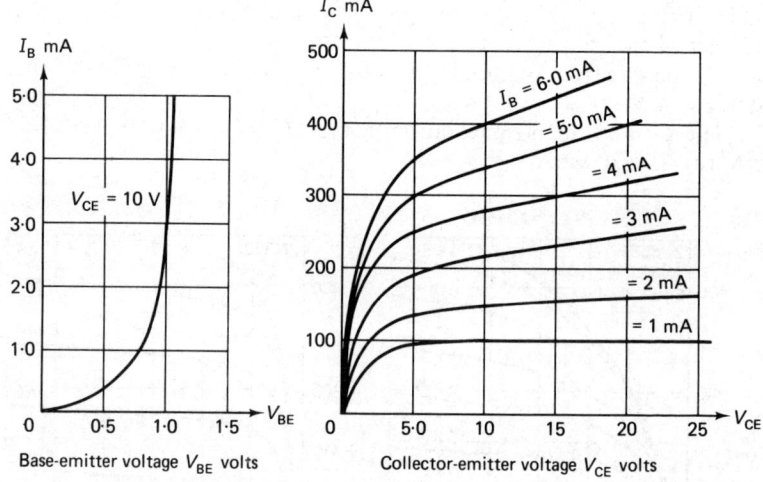

Fig. P6.5. Characteristics for Problem 6.5.

6.6 Describe how you would determine the common-base input characteristics of a transistor.

In such an experiment the following results were obtained for a *p-n-p* transistor when the collector-base voltage was held constant at -10 V.

Emitter current (mA)
    0·2 0·35 0·5   0·75   1·0   1·25 1·5   1·75
Emitter-base voltage (V)
    0·5 0·55 0·562 0·577 0·588 0·597 0·605 1·61

Plot the input characteristic and use it to determine the common base small-signal input resistance when $I_E = 1$ mA.

6.7 Describe how you would use the test circuit shown in fig. P6.7(a) to determine the common base output characteristics of a transistor.

Fig. P6.7(b) shows the common base output characteristics of an n-p-n transistor. Use these to plot the $I_E - I_C$ transfer characteristics for

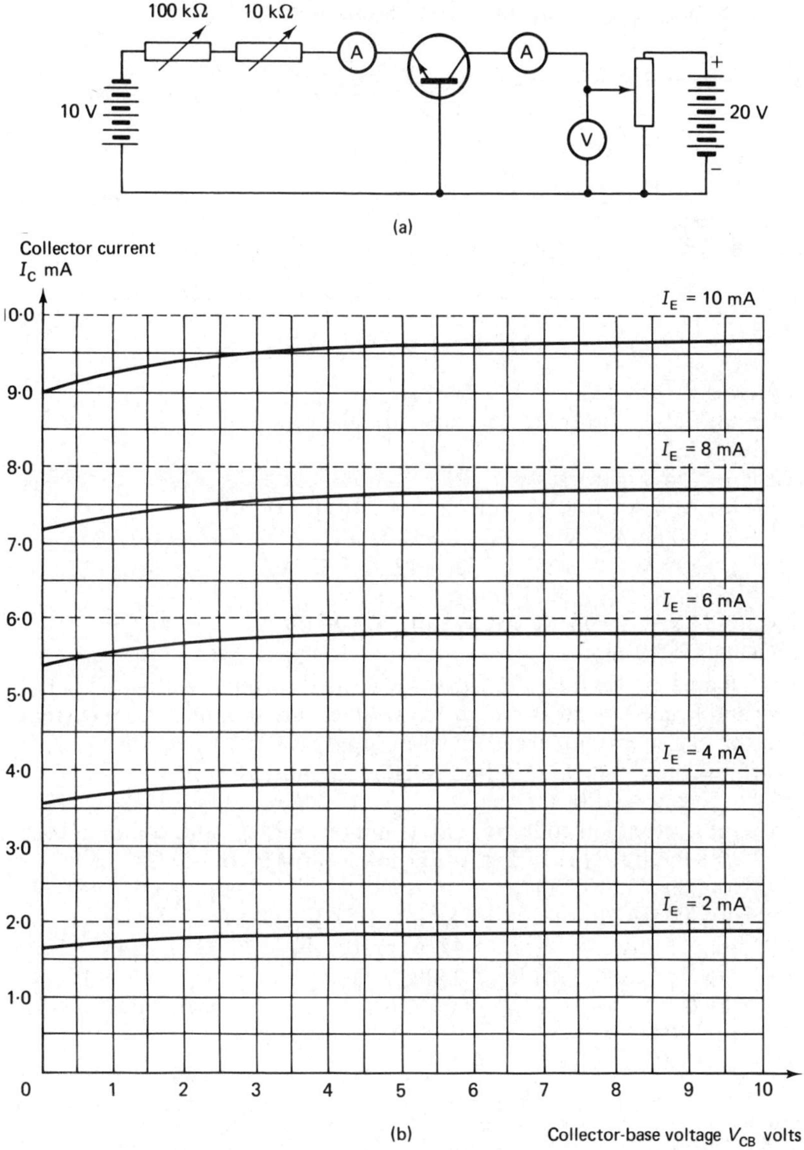

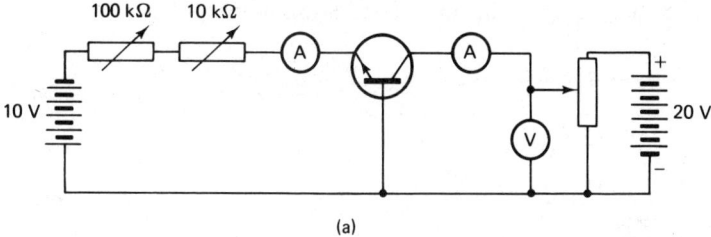

(a)

Collector current
I_C mA

$I_E = 10$ mA

10·0

9·0

8·0
$I_E = 8$ mA

7·0

6·0
$I_E = 6$ mA

5·0

4·0
$I_E = 4$ mA

3·0

2·0
$I_E = 2$ mA

1·0

0   1   2   3   4   5   6   7   8   9   10

(b)   Collector-base voltage V_{CB} volts

Fig. P6.7. for Problem 6.7.

$V_{CB} = 5$ V and determine the d.c. and small-signal current gains of the transistor when $I_E = 6$ mA, $V_{CB} = 5$ V.

6.8 Describe how you would use the test circuit of fig. P6.8 to determine the common emitter small-signal current gain and output resistance of a transistor at a given quiescent operating point.

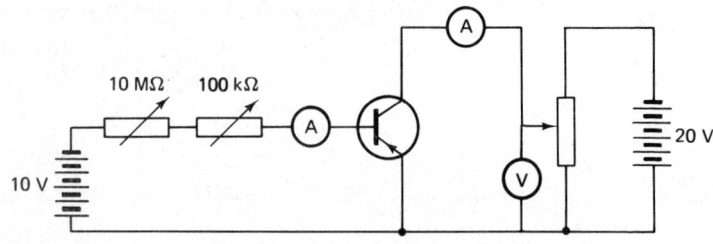

Fig. P6.8. for Problem 6.8.

Answers

1.5 (a) False   (b) true   (c) true   (d) false

2.3 5·6 kΩ

2.4 (a) 354 V   (b) 18 V

2.6 (a) 10 mA   (b)0 V   (c) 200 mA   (d) 0, 20 mA

3.2 2 A, 0·64 A

3.3 (a) 339 V   (b) 42·4 V   (c) 2·12 A, 1·35 A

3.4 170 mA, 108 mA; 170 mA

3.5 (a) 100 V to 96 V   (b) 39 mA   (c) 200 V

3.6 750 Ω, 19 mA

3.7 R must be less than 196 Ω (to provide load current when supply is 100 V) and R must be greater than 155 Ω (to ensure that maximum current rating of Zener is not exceeded when supply is 120 V). Select $R = 180$ Ω (nearest preferred value within range); 139 mA, 28 mA

5.2   (b) 1·5 kΩ, 20

5.3 (a) (i) common collector   (ii) common emitter   (iii) common base
(b) (i) 8·93 mV   (ii) 37·2 mA, 372 mV   (iii) 42   (iv) 21

5.4 (c) 200 μA, 10·2 mA

5.5 (a) 800 μA, 16 mV   (b) 784 μA, 1·57 V   (c) 98

6.3 (a) 4·2 mA   (b) (i) $h_{FE} \approx 47$, $h_{fe} \approx 50$ (ii) $h_{FE} \approx 48$, $h_{fe} \approx 45$

6.4 (a) $V_{CE} = 3$ V   (b) $R_0 \approx 250$ Ω

6.5 (a) 1 mA   (b) 1 V   (c) $I_B = 1·6$ mA, $I_C = 130$ mA   (d) $h_{FE} = 73$

6.6 $R_i \approx 41$ Ω

6.7 $\alpha_{DC} = 0·97$, α (or h_{fb}) $\approx 0·96$

B   Thermionic Valves

The thermionic diode was discovered in 1902 and first amplifying valve, the triode, was discovered in 1906, and from that time up to the late fifties, thermionic valves formed the basic active elements in electronic circuits, and in fact were responsible for the development of electronics and communications into major industries. However, for all but very specialist applications, the thermionic valve has now been virtually superseded by the transistor.

The following paragraph is included in the Technician Education Council's standard unit syllabus for Electronics II:

Note: College staff are advised that the treatment of thermionic devices and applications should depend upon industrial requirements.

Section 7: *The expected learning outcome of this section is that the student should understand the concept of thermionic emission.*

7.1
A description of the effect known as thermionic emission
Present in a metal are free electrons which may travel more or less freely through the crystal lattice of metal ions making up the body of the metal. In Sub-section 1.3, we used this simple model to explain the conduction of current in a metal. Current flow in a metal is the drift of these free electrons under the action of an applied voltage source. However, we did not consider what happened when free electrons approached the surface of a metal. When this occurs, forces are exerted from within the metal which tend to prevent the free electrons from escaping. At normal temperatures, only a relatively few free electrons possess sufficient thermal energy to overcome these forces and escape from the metal. As the temperature of the metal is raised by heating, increased thermal energy is supplied from the lattice of metal ions to the free electrons. By gaining extra thermal energy, more free electrons will have sufficient energy to overcome the surface forces and escape. The effect produced when electrons are emitted from a metal by heating the metal (normally to a high temperature) is known as thermionic emission.

The three most important thermionic emitters used in practice are pure tungsten metal, thoriated-tungsten, and oxide-coated metals. These materials are used to construct the cathodes—that is the source of electrons—in thermionic devices. Tungsten cathodes are normally operated at about 2200°C and are used for very high power applications, for example, high power transmitter valves used in radio broadcasting. Thoriated-tungsten cathodes consist of tungsten filaments (a filament is a relatively fine wire or ribbon) coated with a thin layer of thorium. They are normally operated at about 1600°C and are also used in high power applications. Oxide-coated cathodes are the most efficient type of emitters and are those most commonly used in thermionic valves. They consist of a metal base coated with oxides of barium and strontium and are normally operated at about 700°C.

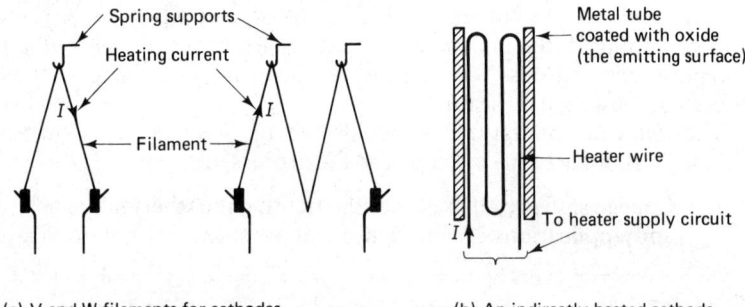

(a) V and W filaments for cathodes of directly heated valves

(b) An indirectly heated cathode

Fig. 7.1.

Cathodes used in thermionic valves may be directly heated or indirectly heated. In the directly heated case, heating current is passed directly through the cathode material. Directly heated cathodes are constructed from tungsten and are normally made in the shape of V or W, as shown in fig. 7.1(a). In the case of an indirectly heated cathode, the emitting surface is heated by a separate heating coil, as shown in fig. 7.1(b). The emitting surface consists of an oxide-coated metal. Indirect heating has the advantage that an a.c. heating current may be used without causing undesirable effects, such as hum. Typical heater supplies for an indirectly heated cathode are 0·3 A/6·3 V and are usually obtained from a mains transformer with 6·3 V tapping points.

Section 8: *The expected learning outcome of this section is that the student should know the static behaviour of a thermionic diode.*

8.1
A test circuit diagram for determining the anode characteristic of a thermionic diode
The thermionic diode is the simplest thermionic valve. It contains two electrodes: the cathode, which emits electrons by thermionic emission,

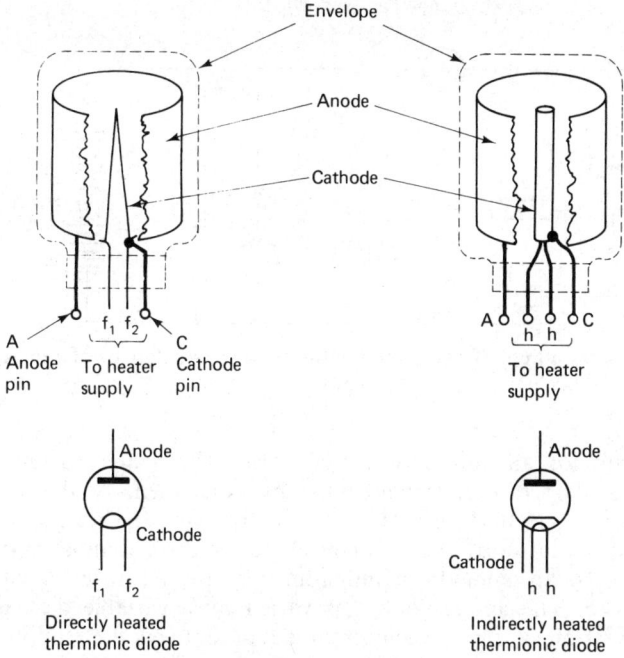

Fig. 8.1. Diagrams showing structure and circuit symbols for a directly and an indirectly heated thermionic diode.

and the anode (normally consisting of a metal cylinder surrounding the cathode), which collects these electrons when at a positive voltage with respect to the cathode.

The anode and cathode are housed in a glass, ceramic, or metal envelope, the last sometimes being called a can. In a vacuum diode, the envelope is evacuated. In a gas-filled diode, the envelope contains a gas at very low pressure, typically 5×10^{-2} mm of mercury. Gases used are mercury vapour, helium, argon, and neon. Diagrams illustrating the basic structure and showing the circuit symbols used for directly and indirectly heated thermionic diodes are given in Fig. 8.1.

The circuit shown in fig. 8.2 may be used to determine the anode characteristic of a thermionic diode. The anode characteristic is a plot of the d.c. current flowing through the diode—normally referred to as the anode current and denoted by the symbol I_A—versus the voltage across the diode between anode and cathode—normally referred to as the anode voltage and denoted by V_A. The I_A–V_A anode characteristic depends upon temperature, and a family of characteristics may be plotted for different cathode temperatures. The cathode temperature may be controlled by varying the cathode heater current. Fig. 8.2 shows

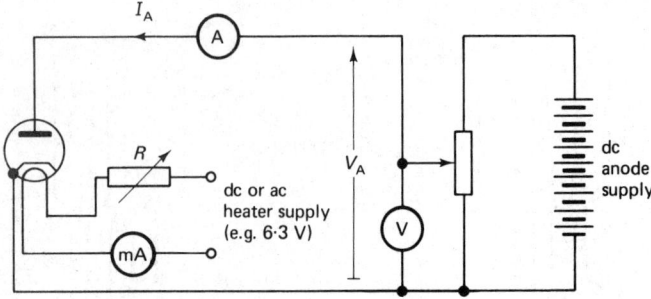

Fig. 8.2. A test-circuit for determining the anode characteristic of a thermionic diode.

the circuit for an indirectly heated valve. The heater current (and therefore temperature) is controlled by adjustment of the variable resistance R and may be monitored by observing the reading on the milliammeter included in series in the heater circuit. The anode current is measured by an ammeter or milliammeter inserted in series with the anode lead. The anode voltage is varied by a variable d.c. supply, typically 0–50 V for a vacuum diode or a fixed d.c. supply plus potentiometer as shown in the diagram.

8.2
The description of how current flow is produced in a thermionic diode by the correct application of a voltage difference between the anode and cathode

When the cathode is at a given operating temperature, typically about 700°C for an oxide-coated cathode or about 2200°C for a directly heated tungsten cathode, it emits an abundance of electrons. The rate of emission of electrons into the free space surrounding the cathode is approximately constant and depends principally on the cathode temperature. Little thermionic emission occurs below about 200°C, even for an oxide-coated cathode, the most efficient of thermionic emitters.

When the anode is positive with respect to the cathode, the emitted electrons are attracted to the anode. This flow of electrons constitutes a current through the valve and thus allows current flow in the external circuit connected across the diode. The conventional current (positive current flow), which is opposite in direction to the electron current, is from anode to cathode.

When the anode is negative with respect to the cathode, the emitted electrons are repelled by the anode back to the cathode or the region surrounding the cathode; hence zero current flows. Thus the thermionic diode acts as a rectifying element.

104

To forward-bias a thermionic diode and effect current flow, the anode must be positive with respect to the cathode. If the anode is negative with respect to the cathode, the thermionic diode is reverse-biased and no current flows.

8.3
The plotting and description of a typical anode characteristic
The plotting of the I_A–V_A anode characteristic may be obtained, with reference to the test circuit of fig. 8.2, as follows.

1. Set the variable resistance R to give a maximum value of heater current to the cathode, e.g. if a 6·3 V supply is used and the filament or heater coil resistance and the maximum permissible heater current are respectively 10 Ω and 300 mA, set R to

$$R = \frac{6\cdot3}{0\cdot3} - 10 = 21 - 10 = 11 \ \Omega$$

2. Measure the anode current I_A as the anode voltage V_A is varied over the following ranges (note the ranges given are only approximate and would obviously depend on the actual thermionic diode used):

(i)   from 0 to 10 V in 1 V steps
(ii) from 10 to 20 V in 2 V steps
(iii) subsequently in 5 V steps up to about 50 V.

3. Repeat above procedure for values of heater current 0·8, 0·6, 0·4, 0·2 times the maximum rated heater current set in 1.

4. Change over the polarity of the anode d.c. supply: anode to the negative supply terminal and cathode to the positive terminal. Note that $I_A = 0$ for all negative values (do not exceed the maximum reverse voltage rating of the valve) of anode voltage, even with maximum heater current. Actually a tiny anode current may be observed at $V_A = 0$.

5. Plot a graph of I_A (y axis) versus V_A (x axis) for each value of heater current.

Typical curves for a directly heated tungsten and an indirectly heated oxide-coated cathode thermionic diode are sketched in fig. 8.3. It is observed that at low anode voltages the anode characteristic is non-linear (curved) and I_A increases with V_A. In fact over this region $I_A = KV_A^{3/2}$, where K is a constant which depends on the size and geometry of the valve. As the anode voltage is increased a point is reached where the anode current does not increase significantly with even a large further increase in V_A. This effect, known as saturation, is quite pronounced for directly heated metal cathodes and results in the flat region of the characteristic where I_A is largely independent of V_A. In the case of valves having indirectly heated oxide-coated cathodes; saturation effects occur more gradually as indicated in fig. 8.3.

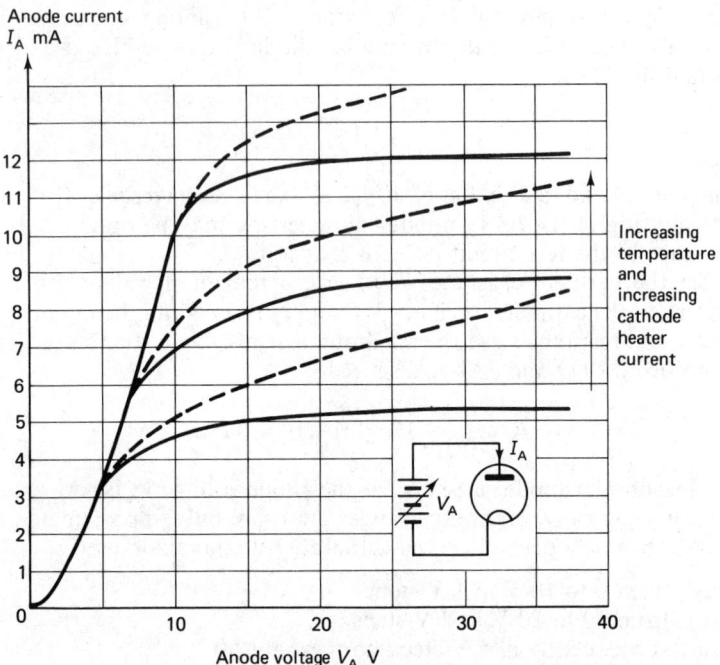

Fig. 8.3. Typical anode characteristics of a thermionic diode valve. Full line curves are typical of a directly heated tungsten cathode. Dashed-line curves are typical of an indirectly heated oxide-coated cathode.

8.4
Explanation of the saturation effect in a thermionic diode in relation to the anode characteristics

The forward-bias (V_A positive) characteristic of a diode can be divided into two regions: one where the anode current is controlled by the anode voltage, and the saturation region where increasing the anode voltage has little effect in increasing the anode current. Let us explain why these two regions occur.

When electrons are emitted from the cathode they are attracted to the anode. However, at low to moderate anode voltages, typically in the range 0–10 V, only a fraction of the emitted electrons reaches the anode and so the anode current is less than the number of electrons per second which the cathode is capable of supplying. This is because the electrons emitted from the cathode initially have low velocities and thus a high density of electrons is produced near the cathode. This region of high density is known as the space charge region and has the effect of neutralizing the attraction of the positive anode voltage on those

106

electrons just leaving the cathode. The total number of electrons moving from cathode to anode at any instant of time does not exceed the number required to provide a negative space charge just sufficient to neutralize the effect of the anode voltage at the cathode surface. All electrons in excess of this number are returned to the cathode.

However, at high anode voltages all electrons are drawn away from the cathode as soon as they are emitted and are collected by the anode. No space charge region is formed, and under these conditions the anode current equals the rate at which electrons are emitted from the cathode. This rate depends principally on cathode temperature, so for a given temperature the anode current will saturate provided the anode voltage is high enough. No increase in anode current at even higher anode voltages should occur. This explains why I_A is largely independent of V_A in the second or saturation region of the diode I_A-V_A characteristic. The limiting value of anode current above which it is not possible to increase I_A significantly by increasing V_A is known as the saturation current, and a thermionic diode operated in the saturation region is said to be working under temperature limited conditions.

8.5
The effect on the anode characteristics of a change in heater current
The effect on the anode characteristics of changing the cathode heater current is illustrated by the characteristic curves of fig. 8.3. Increasing the heater current increases the cathode temperature and therefore the electron emission. At lower anode voltages a negative space charge region is formed close to the cathode and this region acts as a buffer, neutralizing the effect of the anode voltage at the cathode surface. Thus although more electrons are emitted more electrons are returned to the cathode so, at a given anode voltage, the anode current is largely independent of cathode heater current. However, in the saturation region where the anode voltage is sufficiently high to draw off all emitted electrons, increasing cathode heater current results in increasing the anode saturation current.

Section 9: *The expected learning outcome of this section is that the student should know and be able to compare the rectifying action of diodes.*

9.1
Comparison of the merits of thermionic and semi-conductor diodes as rectifiers
In Sub-section 3.1 we defined an ideal rectifier as a device which passes current freely in one direction and presents zero resistance in that direction, whilst in the opposite direction it acts as a resistor of infinite

resistance and passes no current. The *I–V* characteristic of an ideal rectifier is drawn in fig. 9.1(a). Sketches of the *I–V* characteristic of a semi-conductor *p-n* junction diode and a thermionic vacuum diode are drawn in (b) and (c), respectively. These curves illustrate graphically the departure of the devices from the ideal rectifier characteristic.

Let us now briefly review the rectifying action and characteristics of the semi-conductor and thermionic diodes. We will then use the facts brought out to make a comparison of their relative merits as rectifying elements.

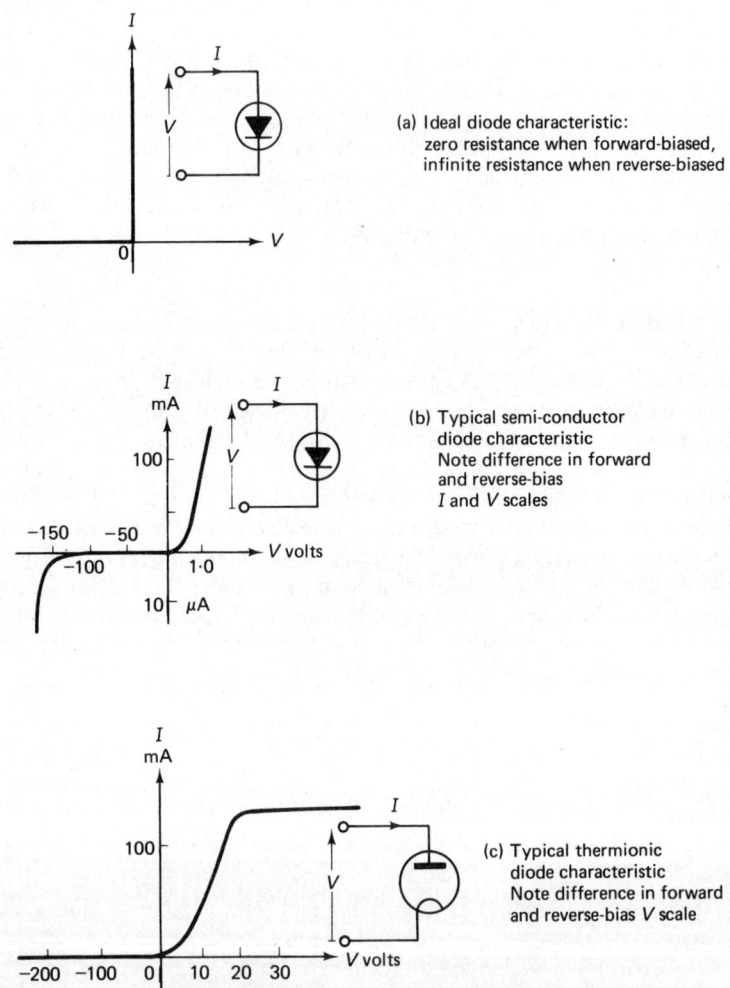

(a) Ideal diode characteristic:
zero resistance when forward-biased,
infinite resistance when reverse-biased

(b) Typical semi-conductor diode characteristic
Note difference in forward and reverse-bias *I* and *V* scales

(c) Typical thermionic diode characteristic
Note difference in forward and reverse-bias *V* scale

Fig. 9.1. Ideal and practical diode characteristics.

To forward-bias a *p-n* junction diode the applied voltage must be connected so that the *p*-type region is positive with respect to the *n*-type region side of the junction. Holes may then cross the junction from *p* to *n* and electrons from *n* to *p* regions. Under forward-bias conditions only a small voltage is dropped across the diode, e.g. for a silicon diode the voltage drop would be about 0·7 V for a current of about 100 mA, for a germanium diode with the same current about 0·3 V; whilst for power diodes several amperes and even hundreds of amperes may be conducted with only the order of a volt or so being dropped across the diode terminals. To reverse bias a *p-n* junction diode the *p*-type region must be negative with respect to the *n*-type region side of the junction. Small but finite reverse-bias currents of the order of nanoamperes for silicon and microamperes for germanium flow at normal ($\sim 25°C$) temperatures. The values of both forward and reverse-bias currents are extremely sensitive to changes in temperature, and increase with increasing temperature. Maximum junction temperatures are of the order of 50°C for germanium and 150°C for silicon. It is extremely important to provide a good heat sink when using semi-conductor diodes, so the heat dissipated in the device can be conducted away. Semi-conductor diodes may be designed to withstand reverse-bias voltages of several hundreds of volts before breaking down. Zener and avalanche diodes, which utilize the breakdown region for voltage stabilization and regulation applications, may be designed to have breakdown voltages from a few to several hundred volts. Normally, however, a diode is not operated in its breakdown region, and in rectifying applications the maximum reverse-bias that the diode can withstand should be as high as possible and certainly greater than that ever likely to be met in the given application.

A thermionic valve is forward-biased when the anode electrode is positive with respect to the cathode. Under this condition electrons emitted from the cathode are attracted to the anode. The cathode must be raised to a high temperature ($\sim 700°C$ for indirectly heated oxide-coated and $\sim 2000°C$ for directly heated tungsten cathodes) before an efficient thermionic emission of electrons is obtained. Power to heat the cathode is therefore required from an external source. Several volts or even tens of volts are dropped across a thermionic vacuum diode in the forward-bias mode. To reverse-bias a thermionic diode the anode must be negative with respect to the cathode. The *I–V* characteristic does not pass through the origin, as it does for semi-conductor diodes. At $V = 0$ a small forward current (anode to cathode) of the order of tens of microamperes flows. However, when $V = -1$ V the anode current is virtually zero. For greater values of reverse-bias voltage the thermionic diode presents an effective resistance of several 100-1000 MΩ. Thermionic diodes may be designed to withstand very high reverse voltages of up to several kilovolts.

Finally let us compare the relative merits of semi-conductor and thermionic diodes as rectifiers:

(a) *Semi-conductor diodes* are compact in size, are extremely robust, and have a long operating life. They require no cathode heater supply. They have a very much smaller voltage drop developed across their terminals in the forward-bias mode, typically less than 1 V, whereas a thermionic diode may drop $\sim 10+$ V.

Their reverse-bias characteristics at normal temperatures, although perhaps inferior to those of thermionic diodes, are satisfactory. They pass only very small reverse-bias currents and may withstand reverse voltages of several hundreds of volts. In all but very specialized applications, semi-conductor diodes have replaced thermionic valve diodes.

(b) *Thermionic valves*, by virtue of their fragile construction, are very much less robust. They require cathode heater supplies and have a much greater anode-cathode voltage drop in the forward-bias mode. Consequently they dissipate far more power than semi-conductor diodes of comparable rating. Their reverse-bias characteristics are, however, superior (at least for voltages more negative than ~ -1 V) and they may withstand reverse-bias voltages of several kilovolts. They have much better temperature characteristics than semi-conductor diodes, their $I-V$ characteristic being largely independent of ambient temperature.

Thermionic diodes are used only in specialized applications: where very large reverse voltages are encountered; where very low currents only are permitted to flow under reverse-bias conditions; in high temperature environments; in environments where the diode may be subjected to radiation (e.g. X-ray or nuclear equipment)—radiation may generate intrinsic carriers in a semi-conductor diode, which could swamp its rectification properties.

Section 10: *The expected learning outcome of this section is that the student should know the static behaviour of a thermionic triode.*

10.1
A sketch of the thermionic triode and its circuit symbol showing the anode, cathode, grid, and heater

A schematic diagram illustrating the basic construction of a thermionic triode is drawn in fig. 10.1(a). A triode has three electrodes: an anode, a cathode, and a control grid. The triode can be thought of as a diode with a third electrode—the control grid—located between the cathode (electron emitter) and the anode (electron collector). The function of the grid is to control the flow of electrons from the cathode to the anode. The grid is an open-mesh structure and does not physically block the flow of

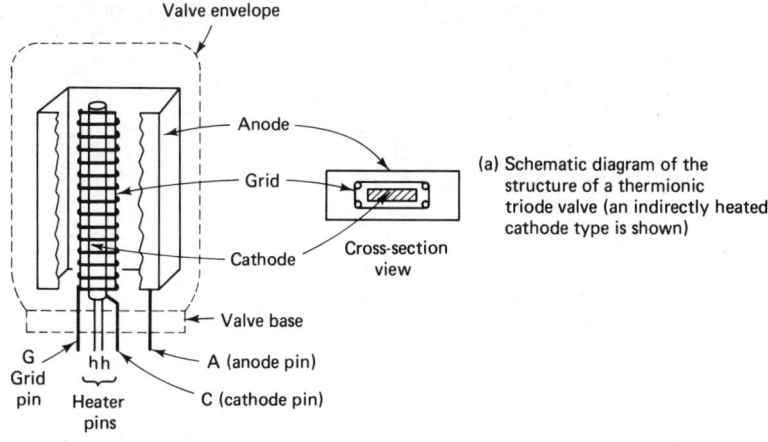

(a) Schematic diagram of the structure of a thermionic triode valve (an indirectly heated cathode type is shown)

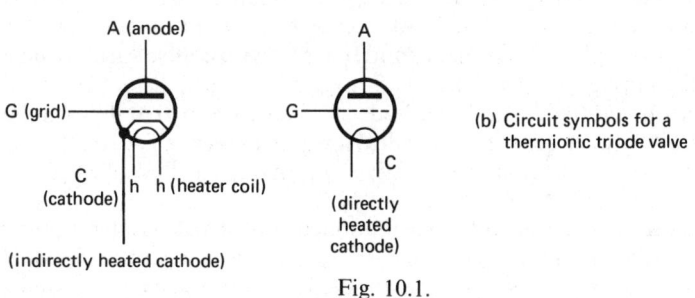

(b) Circuit symbols for a thermionic triode valve

Fig. 10.1.

electrons. It is normally held at a negative voltage with respect to the cathode and thus takes no electron current. The grid shown in fig. 10.1(a) consists of a number of turns of fine wire wound on a supporting frame.

The circuit symbols used to represent a thermionic triode in a circuit diagram are shown in fig. 10.1(b).

10.2
The effect of control grid voltage on anode current

For most applications the grid electrode of a triode is operated at a negative voltage with respect to the cathode. The grid-cathode voltage is known as the grid bias or control-grid voltage and will be denoted by V_G.

The effect of varying the control-grid voltage V_G on the anode current I_A flowing through a triode may be investigated using the simple circuit shown in fig. 10.2(a). The anode-cathode voltage V_A is set at a fixed value, say 200 V, and the control-grid voltage is varied from 0 V to -15 V (say).

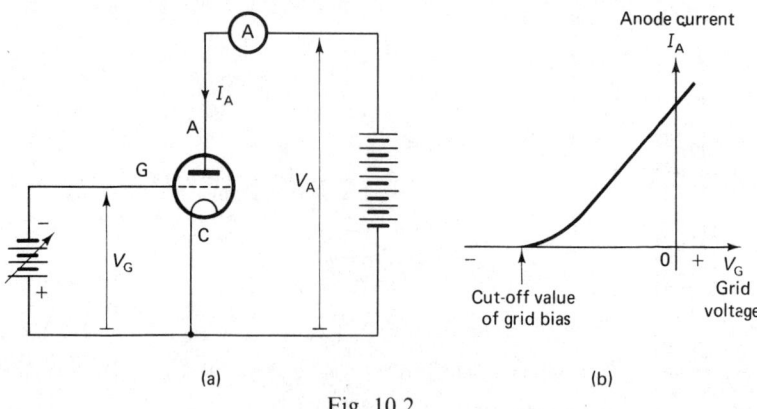

Fig. 10.2.

As V_G is made more negative the anode current falls and, in fact, if V_G is made sufficiently negative I_A is reduced to zero. The triode is then said to be in its cut-off condition and the corresponding value of grid bias voltages which first cuts off the valve is known as the cut-off grid bias (for the set V_A). A typical variation of anode current with control-grid voltage is sketched in fig. 10.2(b). When V_G is negative the grid attracts none of the electrons emitted from the cathode and hence no current flows in the grid circuit between grid and cathode. However, as seen in fig. 10.2(b), the magnitude of V_G controls very effectively I_A, that is, the flow of electrons between cathode and anode.

Let us now give a simple explanation of this effect. Under normal operating conditions a space charge region consisting of a cloud of electrons is formed close to the cathode, just as in the case of a diode operating below saturation. This cloud of electrons tends to shield the attractive effect of the positive anode voltage and thus not all the electrons emitted from the cathode contribute directly to the anode current. The actual number reaching the anode per unit time is determined mainly by the electrostatic field close to the cathode. Thus by inserting the grid structure near the cathode this electrostatic field (and hence the flow of electrons away from the cathode) can be controlled by the grid bias voltage V_G. The more negative the grid voltage, the greater the shielding effect and the smaller is I_A. Once electrons have passed the grid they travel rapidly to the anode. For negative grid voltages and for a symmetrical grid structure it can be shown that the anode current, from low to medium values, is given by

$$I_A = K \left[V_G + \frac{V_A}{\mu} \right]^{3/2} \text{ amperes}$$

where K and μ are constants determined by the valve geometry.

112

Remember for the case of a diode we stated $I_A = V_A^{3/2}$, so we may regard $[V_G + V_A/\mu]$ as an effective 'anode' voltage of a triode. The constant μ is a measure of the relative effectiveness of control-grid and anode voltages in controlling anode current.

Example 10.1

The anode current in a triode valve is given by

$$I_A = 10 \left[V_G + V_A/30 \right]^{3/2} \text{ mA.}$$

If the anode voltage $V_A = 300$ V, determine the anode current when the control-grid voltage is $V_G = -5$ V. Calculate also the minimum negative value of grid bias voltage required to cut off the valve, V_A being maintained at 300 V.

Solution

On substituting $V_G = -5$ V, $V_A = 300$ into the formula, we have

$$I_A = 10[-5 + 300/30]^{\frac{3}{2}}$$
$$= 10[-5 + 10]^{\frac{3}{2}} = 10(5)^{\frac{3}{2}}$$
$$= 10 \times 5 \times \sqrt{5} = 112 \text{ mA.}$$

In the cut-off condition $I_A = 0$ and so the minimum negative value of V_G to effect this condition when $V_A = 300$ V is found from

$$I_A = 10 \left[V_G + 300/30 \right]^{3/2} = 0$$

i.e. $I_A = 0$ when $(V_G + 300/30) = 0$

hence $\qquad\qquad V_G + 10 = 0$, and $V_G = -10$ V.

10.3
A test circuit diagram for determining the static characteristics of a thermionic triode.

The static output characteristics (also known as anode characteristics) of a triode are curves of d.c. anode current I_A versus d.c. anode-cathode voltage V_A for fixed values of grid-cathode voltage V_G. The static transfer characteristics (commonly known as mutual characteristics) are curves of I_A versus V_G for fixed values of V_A. These two families of characteristic define circuit behaviour of a triode and may be used to investigate the performance of a triode in practical applications, e.g. as an amplifying or switching element.

The test circuit drawn in fig. 10.3 may be used to determine the static

113

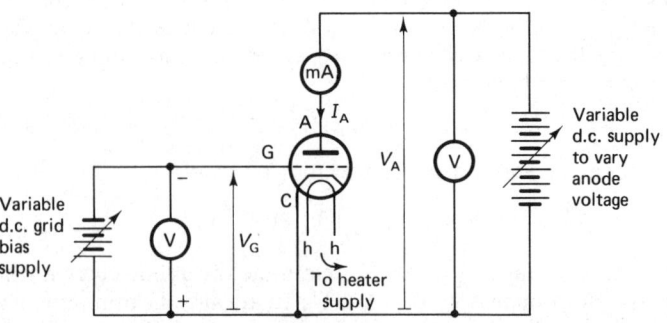

Fig. 10.3. Test-circuit for measuring the static characteristics of a thermionic triode.

characteristics of a triode. A milliammeter is included in series with the anode lead to measure I_A. A voltmeter, with a typical range 0–20 V, is connected between grid and cathode to measure the grid bias voltage V_G. A voltmeter, with a typical range 0–300 V, is connected between anode and cathode to measure V_A. A low-power variable d.c. supply, e.g. a grid bias battery, is used to vary V_G. A variable H.T. (high tension—tension being an old word used to denote voltage) d.c. supply, capable of supplying up to ~ 300 V and current up to ~ 100 mA or even higher for power triodes, is used to vary V_A. Alternatively a fixed H.T. supply plus a potentiometer may be used.

10.4
The determination of the static characteristics
The following procedures may be used with reference to the test circuit diagram of fig. 10.3. The maximum and step interval values of grid bias voltage V_G and anode voltage V_A given below are typical and can of course be varied to give more or finer detail. The maximum rated value of V_A as specified by the manufacturer should, however, not be exceeded.

(a) *The output or anode characteristics:* I_A *versus* V_A *for fixed values of* V_G
1. Set the grid bias voltage to zero, i.e. $V_G = 0$, and increase the anode voltage from zero in 20 V steps up to 300 V. Record the value of anode current I_A at each value of anode voltage V_A. Note that the correct cathode heater supply as specified by the manufacturer must be used.
2. Set the grid bias voltage $V_G = -1$ V and measure I_A as V_A is increased from 0 V in 20 V steps up to 300 V.
3. Repeat 2 for $V_G = -2$ V, -3 V $\cdots$ up to say -10 V.
4. Plot curves of I_A versus V_A (I_A along the y or vertical axis, V_A along the x or horizontal axis) for each value of grid bias voltage V_G.

114

(b) *The transfer or mutual characteristics:* I_A *versus* V_G *for fixed values of* V_A
1. Set $V_A = 50$ V and measure I_A as V_G is varied from zero volts in the negative direction, say for example, in 1 V steps, i.e. record I_A for $V_G = 0$, -1 V, -2 V, -3 V $\cdots$ up to cut-off where $I_A = 0$.
2. Repeat 1 for V_A held constant at 100 V, 150 V, 200 V, 250 V.
3. Plot curves of I_A versus V_G (I_A along the y axis, V_G along the x axis) for each value of anode voltage.

10.5
Sketches of typical families of I_A-V_A output and I_A-V_G transfer characteristics
The results shown in Table 10.1 were obtained for a medium-power triode:

Table 10.1   *Data for plotting output characteristics of fig.* 10.4(a): *anode current and anode voltage values for five fixed values of grid bias voltage*

Anode voltage V_A (V)	Anode current I_A (mA) for				
	$V_G = 0$	$V_G = -2.5$ V	$V_G = -5$V	$V_G = -7.5$ V	$V_G = -10$ V
0	0	0	0	0	0
20	1·6	0·2	0	0	0
40	3·5	0·5	0	0	0
60	6·0	1·2	0·1	0	0
80	8·6	2·8	0·6	0	0
100	11·5	4·6	1·3	0·2	0
120	14·8	7·0	2·7	0·8	0
140	18	9·6	4·5	1·5	0·2
160	21·6	12·7	6·5	2·8	0·8
180		16·0	9·1	4·4	1·6
200		19·5	11·9	6·6	2·8
220			15·0	8·7	4·4
240			18·5	11·5	6·5
260				14·3	8·6
280				17·5	11·0
300					14·0

The data of Tables 10.1 and 10.2 were used to plot the families of static output and transfer characteristics drawn in fig. 10.4(a) and (b), respectively. These curves illustrate the typical features of triode static characteristics. Also shown in fig. 10.4(c) are plots of V_A versus V_G for I_A held constant at 6 mA and 12 mA. The data for these curves were

115

Table 10.2 *Data for plotting transfer characteristic of fig.* 10.4(*b*): *anode current and grid bias voltage values for four fixed values of anode voltage*

Grid bias	Anode current I_A (mA) for anode voltage held constant at			
V_G (V)	$V_A = 50$ V	$V_A = 100$ V	$V_A = 150$ V	$V_A = 200$ V
0	4·7	12·0	20·0	28·0
−1	2·5	8·7	16·5	25·0
−2	1·0	6·0	13·1	21·6
−3	0·5	3·9	10·0	18·1
−4	0·1	2·4	7·6	14·9
−5	0	1·3	5·5	12·0
−6	0	0·8	4·0	9·6
−7	0	0·4	2·7	7·5
−8	0	0·1	1·7	5·5
−9	0	0	1·0	4·0
−10	0	0	0·5	2·8
−11	0	0	0·2	1·9
−12	0	0	0	1·1
−14	0	0	0	0·4
−16	0	0	0	0

Table 10.3 *Data for plotting* V_A-V_G *transfer characteristic: anode and grid bias voltage values for two fixed values of anode current*

Grid-bias voltage V_G (V)	0	−2·5 V	−5 V	−7·5 V	−10 V
Anode voltage V_A (V) for $I_A = 6$ mA	60	112	155	196	235
Corresponding point on fig. 10.4 (a)	a	b	c	d	e
Anode voltage V_A (V) for $I_A = 12$ mA	102	156	201	244	286
Corresponding point on fig. 10.4(a)	f	g	h	j	k

obtained by reading off the respective values of V_A and V_G from the output characteristics of (a) and are summarized in Table 10.3. The curves of (c) show that a relatively small change in grid bias requires a relatively large change in anode voltage to hold I_A constant, and thus demonstrate the greater effect of the grid voltage in controlling the anode

116

current. For example, to hold $I_A = 6$ mA when V_A is increased from 155 V to 196 V (21 V change) requires a change in V_G from -5 V to $-7\cdot5$ V ($2\cdot5$ V change).

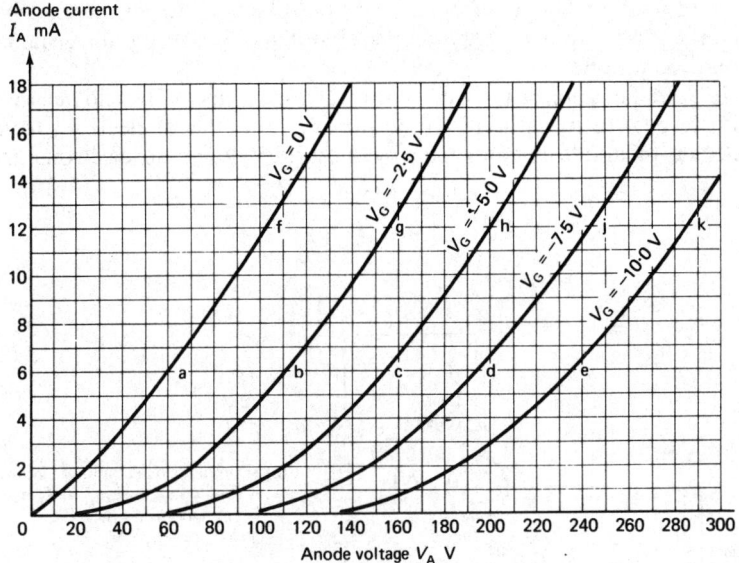

(a) Static output characteristics: I_A versus V_A for fixed values of V_G

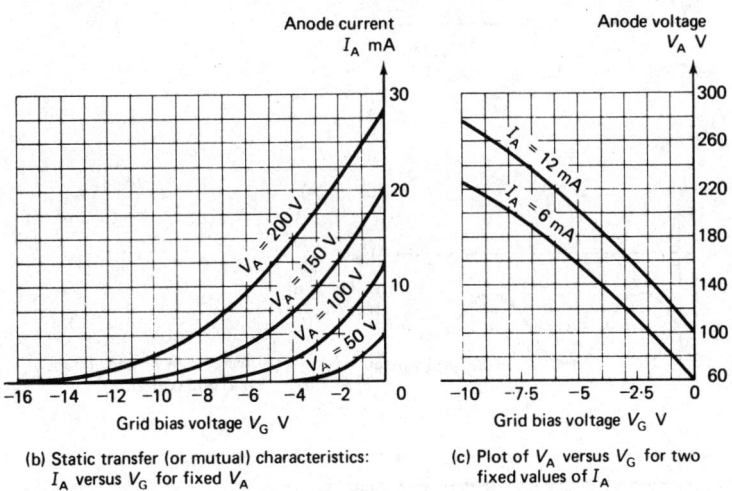

(b) Static transfer (or mutual) characteristics: I_A versus V_G for fixed V_A

(c) Plot of V_A versus V_G for two fixed values of I_A

Fig. 10.4. Typical static characteristics for a thermionic triode.

117

10.6
The input resistance of a thermionic triode is high since the grid current is normally negligible

The input resistance of a triode is the effective resistance to alternating current seen looking in at the grid-cathode terminals. Thus, if no current were to flow into these terminals, we would be left with the conclusion that the triode input acted as an open circuit and thus its input resistance would be infinite.

In amplifier applications, the grid of a triode is normally held negative with respect to its cathode. Thus, although the flow of electrons from cathode to anode is controlled by the grid voltage, no electrons are

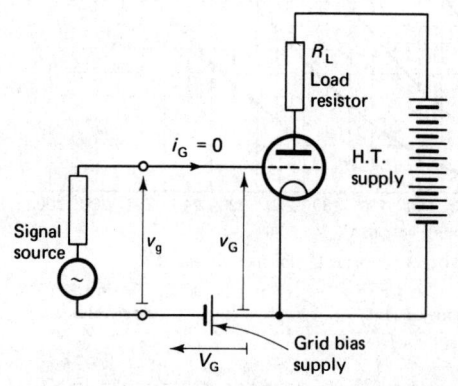

(a) A simple triode amplifier circuit: if total grid voltage $v_G = V_G + v_g$ is negative at all times, the grid current $i_G = 0$

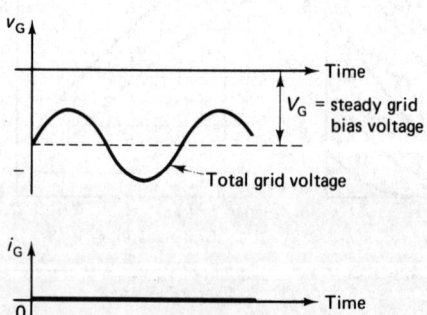

(b) Waveforms of grid voltage v_G and grid current i_G

Fig. 10.5.

actually collected by the negative grid and hence there is no current flowing in the grid input circuit. Under these conditions of zero grid current flow, the input resistance looking into the triode is infinite. For example, in the simple amplifier circuit of fig. 10.5(a), the grid bias voltage V_G maintains the grid negative with respect to the cathode at all times. The amplitude of the a.c. signal in series with V_G would be restricted to be less than the magnitude of V_G so that the resultant grid-cathode voltage v_G, even on positive peaks of the input signal, is negative. This situation is shown in Fig. 10.5(b). Thus, the grid current $i_G = 0$ at all times and the input resistance is infinite. In practice, some a.c. signal current may flow in the input grid circuit of the amplifier especially for high-frequency signal inputs. The reason for this is that there is a capacitive effect between the cathode and grid structures, i.e. the grid-cathode structures are separated by a vacuum (insulator) and act as a capacitor. The values of this capacitance are normally very low. Typical values may vary from a few to several picofarads, and with signal frequencies below a megahertz, the capacitive effect may normally be neglected.

To summarize: since the grid is normally maintained at negative voltage with respect to the cathode, the grid circuit input current is normally negligible and the input resistance of the triode is of very high value.

10.7
The definition of anode slope resistance r_a, mutual conductance g_m, and amplification factor μ

There are three important parameters which quantify the circuit behaviour of a thermionic triode. They are essentially small-signal parameters since they are defined as ratios of small voltage and current changes about a selected operating point on the static characteristics. Their definitions are as follows:

1. *The anode slope resistance r_A*

$$r_a = \left(\frac{\text{small change in anode voltage, } \delta V_A}{\text{corresponding change in anode current, } \delta I_A} \right)$$

with the grid-cathode voltage V_G held constant.

r_a has the dimensions of resistance and is therefore measured in ohms. It is the small-signal output resistance of the triode. This fact will be used later in developing small-signal equivalent circuits to help analyse the triode when used as an amplifier. r_a may be determined from the anode characteristics. It is the reciprocal of the slope at a given operating point of the I_A–V_A characteristic. For example, the anode slope resistance for the triode whose anode characteristics are sketched in fig. 10.6(a) at the

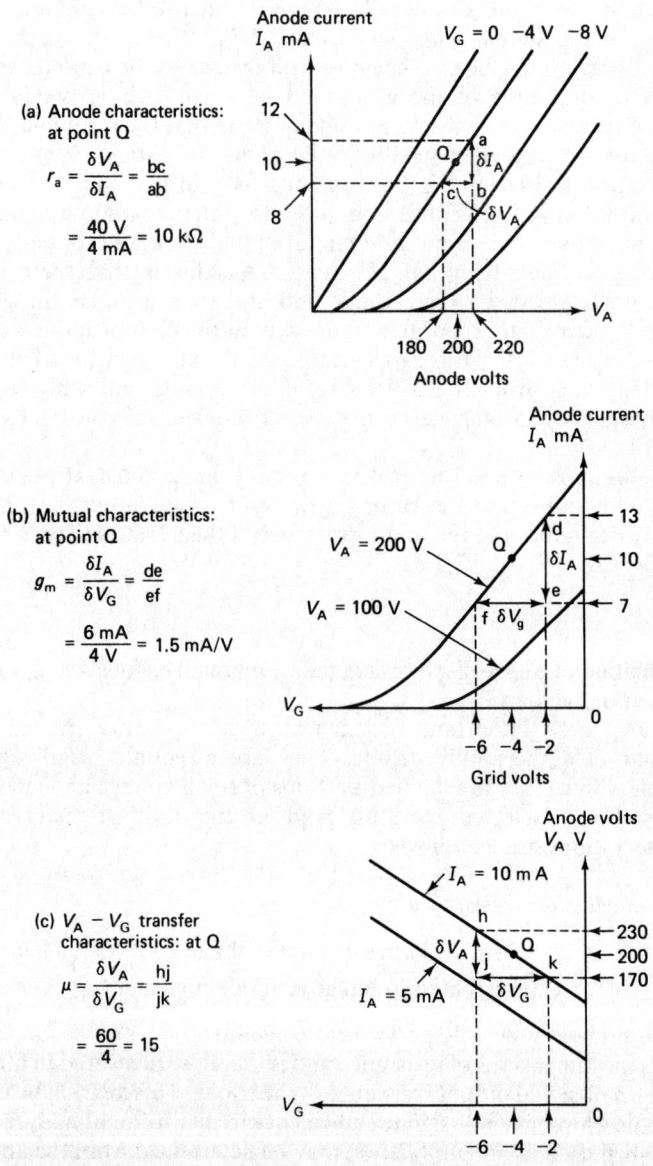

(a) Anode characteristics: at point Q

$$r_a = \frac{\delta V_A}{\delta I_A} = \frac{bc}{ab}$$

$$= \frac{40 \text{ V}}{4 \text{ mA}} = 10 \text{ k}\Omega$$

(b) Mutual characteristics: at point Q

$$g_m = \frac{\delta I_A}{\delta V_G} = \frac{de}{ef}$$

$$= \frac{6 \text{ mA}}{4 \text{ V}} = 1.5 \text{ mA/V}$$

(c) $V_A - V_G$ transfer characteristics: at Q

$$\mu = \frac{\delta V_A}{\delta V_G} = \frac{hj}{jk}$$

$$= \frac{60}{4} = 15$$

Fig. 10.6. Determination of anode slope resistance r_a, mutual conductance g_m and amplification factor μ from the slopes of the static characteristics.

point Q on the $V_A = -4$ V characteristic where $V_A = 200$ V, $I_A = 10$ mA is

$$r_a = \left(\frac{\delta V_A}{\delta I_A}\right)_{V_G = -4\,V} = \frac{bc}{ab}$$

$$= \frac{(220-180)\ V}{(12-8)\ mA} = \frac{40}{4 \times 10^{-3}} = 10\ k\Omega.$$

2. The mutual conductance g_m

$$g_m = \left(\frac{\text{small change in anode current, } \delta I_A}{\text{corresponding change in grid voltage, } \delta V_A}\right)$$

with the anode voltage V_A held constant.

g_m has the unit of conductance, that is siemens (S), although it is commonly quoted in mA/V; 1 mA/V = 1 mS. g_m is the slope at the given operating point of the I_A–V_G transfer characteristic. For example, the mutual conductance for the triode whose mutual characteristic is shown in fig. 10.6(b) at the point Q on the $V_A = 200$ V characteristic where $V_G = -4$ V, $I_A = 10$ mA is

$$g_m = \left(\frac{\delta I_A}{\delta V_G}\right)_{V_A = 200} = \frac{de}{ef} = \frac{(13-7)\ mA}{[-2-(-6)]} = \frac{6}{4} = 1 \cdot 5\ mA/V.$$

3. The amplification factor μ

$$\mu = \left(\frac{\text{small change in anode voltage, } \delta V_A}{\text{corresponding change in grid voltage, } \delta V_G}\right)$$

with the anode current I_A held constant.

μ has important practical significance in that it gives the maximum voltage amplification or gain that may be achieved when the triode is used as an amplifying element.

μ may also be obtained from the static characteristics. It is the magnitude of the slope of the V_G–V_A transfer characteristic at the given operating point. It will be noted that the slope of this characteristic in the mathematical sense is negative in value since, for example, a positive change in anode voltage must be compensated by making the grid voltage more negative to hold the anode current constant. In the present text, we will drop this negative sign. The magnitude of the slope at point Q, whose V_G–V_A transfer characteristic is sketched in fig. 10.6(c), is

$$\mu = \left(\frac{\delta V_A}{\delta V_G}\right)_{I_A = 10\,mA} = \frac{hj}{kj} = \frac{230-170}{6-2} = \frac{60}{4} = 15.$$

r_a, g_m, and μ are the basic parameters of the equivalent circuits used to analyse the small-signal behaviour of triode amplifiers. Two basic equivalent circuits, one incorporating r_a and μ and the other using r_a and g_m, are introduced in the two examples given at the end of the small-signal valve amplifier Chapter in Section 15.

10.8
The relationship between r_a, g_m, and μ
At a given operating point the values of anode slope resistance r_a, mutual conductance g_m, and the amplification factor μ, are related by the simple formula

$$r_a g_m = \mu$$

For example, we obtained from the anode and mutual static characteristics of fig. 10.6(a) and (b) respectively:

$$r_a = 10 \text{ k}\Omega, \, g_m = 1{\cdot}5 \text{ mA}$$

at the operating point $V_G = -4 \text{ V}, V_A = 200 \text{ V}, I_A = 10 \text{ mA}$; so on using the formula $\mu = r_a g_m$, we have for the amplification factor at the same operating point:

$$\mu = 10 \times 10^3 \times 1{\cdot}5 \times 10^{-3} = 15$$

which checks with μ obtained from the slope of the V_G–V_A characteristic of fig. 10.6(c).

Let us consider how the relation $r_a g_m = \mu$ might be verified in a general way.

(a) Suppose a small change in V_A, of δV_A, at constant V_G, results in a corresponding small change in I_A of δI_{A_1}. Then as

$$r_a = \frac{\delta V_A}{\delta I_{A_1}}, \text{ we have } \delta I_{A_1} = \frac{\delta V_A}{r_a}.$$

(b) Suppose also a small change in V_G, of δV_G, at constant V_A, results in a corresponding small change in I_A of δI_{A_2}. Then as

$$g_m = \frac{\delta I_{A_2}}{\delta V_G}, \text{ we have } \delta I_{A_2} = g_m \delta V_G.$$

(c) Thus if I_A is changed as a result of simultaneous small changes in both V_A and V_G, then we have for the total change in anode current,

$$\delta I_{A_1} + \delta I_{A_2} = \frac{\delta V_A}{r_a} + g_m \delta V_G.$$

(d) Finally consider the case when the change in I_A caused by the change in V_A is exactly offset by the change in V_G, that is

$$\delta I_{A_1} + \delta I_{A_2} = 0.$$

So

$$\frac{\delta V_A}{r_a} + g_m \delta V_G = 0$$

and therefore

$$\frac{\delta V_A}{r_a} = -g_m \delta V_G.$$

Hence

$$\frac{\delta V_A}{\delta V_G} \equiv \mu = -r_a g_m.$$

The negative sign results from the fact that if the anode voltage is *increased* the grid bias voltage has to be made more *negative* to maintain the anode current constant; so as far as magnitudes are concerned

$$r_a g_m = \mu$$

10.9
The determination of r_a, g_m, and μ from given characteristics
Remember at a given operating point, Q say, we define

$$r_a = \left(\frac{\delta V_A}{\delta I_A}\right)_{I_G} = \text{reciprocal of slope of anode characteristic at Q}$$

$$g_m = \left(\frac{\delta I_A}{\delta V_G}\right)_{I_A} = \text{slope of mutual transfer characteristic at Q}$$

$$\mu = \left(\frac{\delta V_A}{\delta V_G}\right)_{I_A} = \text{magnitude of slope of } V_G{-}V_A$$
$$\text{transfer characteristic at Q}$$

where δ is used to denote 'a small change in' and the subscript terms outside the brackets are used to denote that this third quantity must be held constant.

We showed previously in fig. 10.6 how r_a, g_m, and μ could be determined from the slopes of the static characteristics. The following example demonstrates how we may determine these parameters from a given set of output characteristics.

Example 10.2
Calculate the small-signal parameters r_a, g_m, and μ at the operating point $V_G = -2$ V, $V_A = 100$ V, $I_A = 65$ mA for the triode whose output characteristics are given in fig. 10.7.

Solution
First we locate the operating point on the output characteristics. This is

123

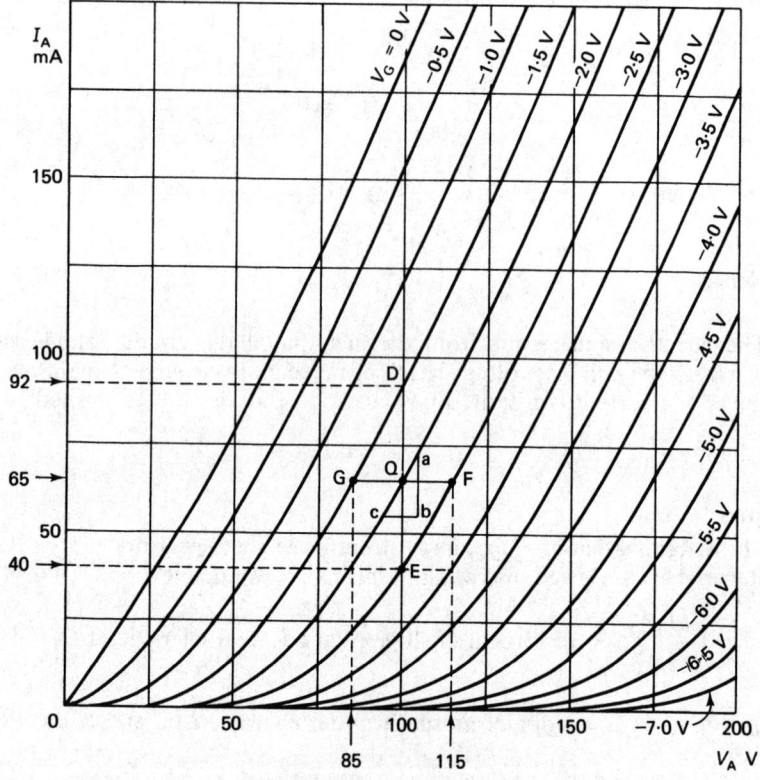

Fig. 10.7. Output characteristics for the triode of example 10.2.
(Courtesy of Mullard Ltd.).

marked in and denoted by Q in fig. 10.7. The anode slope resistance at Q
is

is $\quad r_a = \dfrac{1}{\text{(slope of characteristic at Q)}}$

$$= \frac{bc}{ad} = \frac{(105-94)V}{(74-55)mA} = \frac{11V}{19 \times 10^{-3}A} = \frac{11000}{19} \approx 580\Omega.$$

Now if we hold the anode voltage constant at the operating point value
of $V_A = 100$ V and make a small change in grid voltage from -2 V to
-1.5 V, i.e. $\delta V_G = 0.5$ V, then the corresponding change in anode
current, that is from point Q to D in fig. 10.7, is

$$\delta I_A = DQ = 92 - 65 = 27 \text{ mA}.$$

124

So the mutual inductance,

$$g_m = \left(\frac{\delta I_A}{\delta V_G}\right)_{V_A = 100 \text{ V}} = \frac{27 \text{ mA}}{0.5 \text{ V}} = 54 \text{ mA/V}.$$

Alternatively if we change the grid voltage from −2 V to −2·5 V, then the corresponding change in anode current from point Q to E is

$$\delta I_A = QE = 65 - 40 = 25 \text{ mA}$$

and so for these results

$$g_m = \frac{25 \text{ mA}}{0.5 \text{ V}} = 50 \text{ mA/V}.$$

We thus see that our calculated values of g_m differ slightly, depending on which direction we change the grid voltage. This shows that our 'small-signal' change δV_G is not really small enough. If we had made δV_a say about 0·2 V (we cannot here because we have no intermediate characteristics between −2 V and −1·5 V or −2·5 V) our calculated values should have been equal or at least approximately equal and within the experimental error in reading the small anode current changes from the characteristics. It is for this reason that we can only estimate g_m from the output characteristics. In our case we can state that at the operating point Q

$$g_m = 52 \text{ mA to within } 5\%$$

where we have taken the average of our two calculated values. We could, of course, use the output characteristics to determine pairs of I_A–V_G values at $V_A = 100$ V and then plot the transfer characteristic. g_m could then be obtained by finding the slope of this characteristic at point Q where $V_G = -2$ V, $I_A = 65$ mA.

Since we have calculated both r_a and g_m we can use the formula $\mu = r_a g_m$ to determine the amplification factor at Q, i.e.

$$\mu = 580 \times (52 \times 10^{-3}) = 30$$

Alternatively if we hold the anode current constant at the Q point value of $I_A = 65$ mA, we find from the output characteristics that if the grid voltage is changed from −2 V to −2·5 V, i.e. from Q to F, the corresponding change in anode voltage is

$$\delta V_A = FQ = 115 - 100 = 15 \text{ V}$$

So $\mu = \left(\frac{\delta V_A}{\delta V_G}\right)_{I_A = 65 \text{ mA}} = \frac{15}{0.5} = 30.$

125

Likewise changing the grid voltage from -2 V to $-1\cdot5$ V, i.e. from Q to G, gives

$$\delta V_{\mathrm{A}} = QG = 100 - 85 = 15$$

and hence
$$\mu = \frac{15}{0\cdot5} = 30.$$

Problems: B   Thermionic Valves

7.1 Describe the effect known as thermionic emission and give three examples of a thermionic emitter used in practice.

Describe the difference between a directly and an indirectly heated cathode in a thermionic valve.

8.1 Draw labelled diagrams showing the structure of an indirectly heated thermionic diode and its circuit symbol. Describe how current flow is produced in a thermionic diode.

8.2 Draw a test circuit for measuring the anode characteristics of a thermionic diode. Describe how you would use the circuit to obtain data to plot the anode characteristics for a series of different cathode temperatures.

8.3 Sketch a typical set of anode characteristics for a directly heated thermionic diode. Give a physical explanation of the two main forward-bias regions and the reverse-bias region of these characteristics.

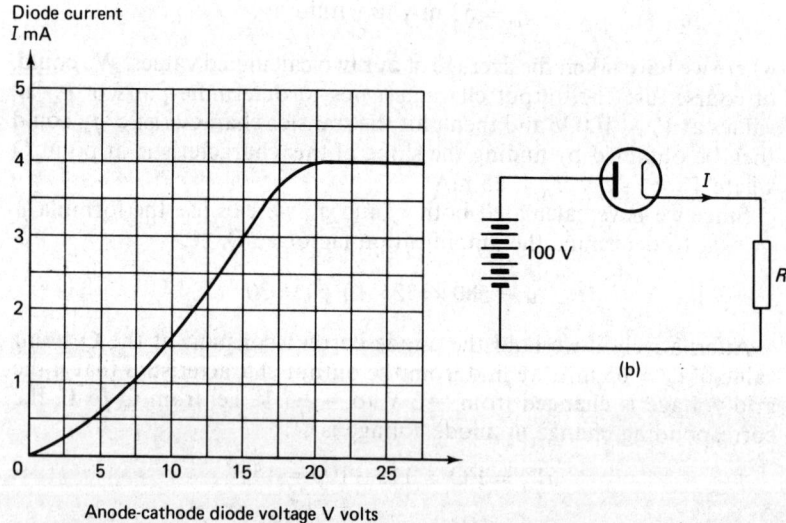

Fig. P8.4 for Problem 8.4

8.4 (a) Fig. P8.4 (a) shows the forward-bias characteristic of a thermionic diode. Use the characteristic to calculate the following:
   (i) the voltage dropped across the diode when the diode current is 3 mA;
   (ii) the minimum forward-bias voltage required to produce saturation.
(b) If the diode is used in the circuit of fig. P8.4(b) with a 100-V d.c. supply and the voltage dropped across the resistor R is 90 V, calculate the current I in the circuit and the value of R.

9.1 Sketch a typical characteristic for a silicon semi-conductor diode and a thermionic diode. Describe the rectifying action of each diode and compare their relative merits as rectifiers.

10.1 Draw labelled diagrams showing the structure of a thermionic triode and its circuit symbol.

Describe the effect of the control grid voltage on the anode current and include in your answer a circuit diagram which could be used to investigate its effect quantatively.

10.2 Describe, with the aid of a circuit diagram, procedures for obtaining data to plot the anode $I_A - V_A$ and mutual $I_A - V_G$ characteristics of a thermionic triode.

Sketch a typical set of anode and mutual characteristics for a triode.

10.3 The following results were obtained in an experiment to measure the mutual characteristics of a triode when the anode voltage was held constant at $V_A = 100$ V.

Grid voltage V_G volts	0	-1	-2	-3	-4	-5	-6	-7	-8	V
Anode current I_A mA	12	8·7	6·0	3·9	2·4	1·3	0·8	0·4	0·1	mA

Plot the mutual characteristic and use it to calculate the following:
(a) The anode current which would flow when $V_A = 100$ V and $V_G = -3.5$ V.
(b) The minimum value of grid bias voltage required to cut off the valve when $V_A = 100$ V.
(c) The mutual conductance g_m when $V_G = -2$ V and $V_A = 100$ V.

10.4 The following results were obtained in an experiment to measure the output characteristic of a triode when the grid bias voltage was held constant at $V_G = -2.5$ V:

Anode voltage V_A volts	0	20	40	60	80	100	120	140	160	V
Anode current I_A mA	0	0·2	0·5	1·2	2·8	4·6	7·0	9·6	12·7	mA

Plot the output characteristic and use it to determine the following:
(a) The anode current which would flow when $V_G = -2.5$ V and $V_A = 110$ V.
(b) The voltage dropped across the triode when $I_A = 11$ mA and $V_G = -2.5$ V.
(c) The anode slope resistance at the point $V_G = -2.5$ V and $I_A = 8$ mA.
10.5 Define the anode slope resistance r_a, the mutual conductance g_m, and the amplification factor μ, of a triode valve and state the relation between them.

The following results were obtained in an experiment to determine the small-signal parameters of a triode:
(a) On keeping the grid bias constant at $V_G = -5$ V and making a change in anode voltage from $V_A = 195$ V to 205 V, it was observed that the anode current changed from $I_A = 11.2$ mA to 13.4 mA.
(b) On keeping V_A constant at 200 V and making a change in V_G from -4.5 V to -5.5 V, it was observed that the anode current changed from $I_A = 14.3$ mA to 10.3 mA.

Use these results to calculate r_a, g_m, and μ at the point $V_A = 200$ V, $V_G = -5$ V. State any assumptions made in computing these parameters.
10.6 Describe an experiment to determine the mutual conductance of a triode valve at a given quiescent operating point.

Fig. P10.6 shows the I_A–V_G transfer characteristics of a triode. Use the characteristics to determine the following:

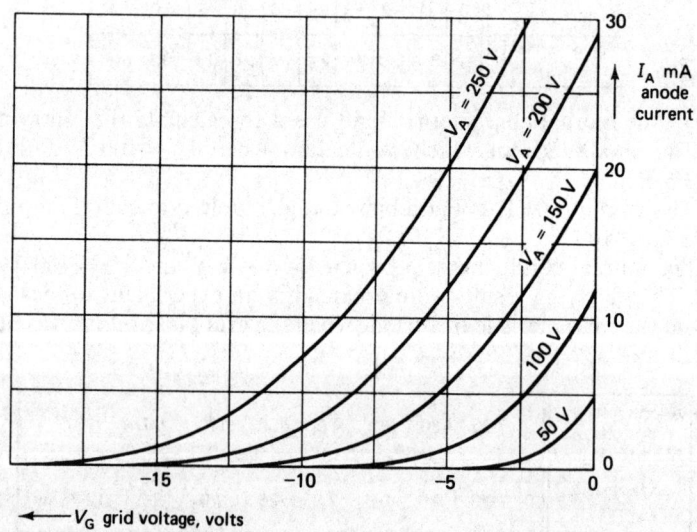

Fig. P10.6. Characteristics for Problem 10.6

(a) The anode current when $V_G = -5$ V and $V_A = 200$ V.
(b) The cut-off grid voltage when $V_A = 150$ V.
(c) The mutual conductance when $V_A = 250$ V and $I_A = 15$ mA.

10.7 Describe an experiment to determine the output characteristics of a triode valve. Explain how these may be used to calculate the anode slope resistance at a given operating point.

Fig. P10.7 shows the output characteristics of a triode. Use these to calculate r_a at the point $V_A = 250$ V, $I_A = 1·2$ mA and to estimate the mutual conductance g_m and amplification factor μ.

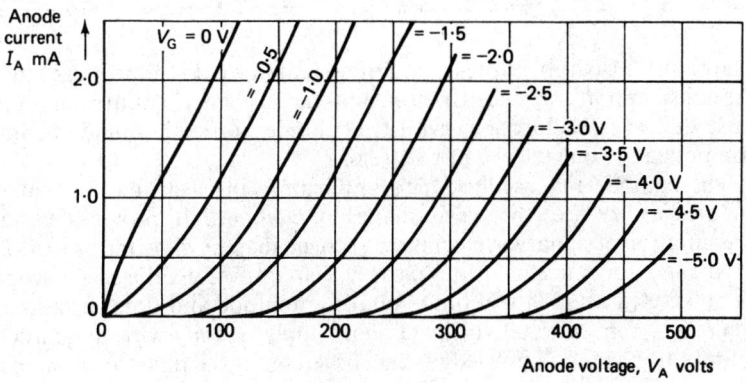

Fig. P10.7. Characteristics for Problem 10.7

Answers
8.4 (a) (i) 15 V   (ii) 20 V (b) $I = 1·6$ mA, $R = 56·25$ kΩ
10.3 (a) 3·1 mA   (b) $-8·5$ V   (c) 2·5 mA/V
10.4   (a) 5·6 mA   (b) 150 V   (c) 7·7 kΩ
10.5 $r_a = 4·5$ kΩ, $g_m = 4$ mA/V, $\mu = 18$
10.6 (a) 12 mA   (b) -12 V   (c) 2·7 mA/V
10.7 $r_a \approx 60$ kΩ, $g_m \approx 1·6$ mA/V, $\mu \approx 100$

C Cathode Ray Tube

Section 11: *The expected learning outcome of this section is that the student should know the principles of operation of a cathode ray tube.*

The cathode ray tube is the basic component in what is almost certainly the most important piece of electronic measurement equipment—the cathode ray oscilloscope—whilst at home it is the fundamental component in our television receivers.

The cathode ray oscilloscope's importance in electronics is that it enables us to 'see' the wave shapes of voltages. It provides visual presentation of signal waveforms on a screen, thereby allowing us to look at what is happening at various parts of a circuit. We use the oscilloscope to make measurements of both voltage amplitude and time variations, whereas a conventional voltmeter can supply us only with magnitude information and no knowledge whatsoever about the time variation of a signal.

There is a wide range of oscilloscopes available which may be used for measurement and design work in virtually all areas of electronics. We can 'observe' signals which vary in time from a few cycles per second to many megahertz. Cathode ray tubes find extensive application in many other fields. For example, in radar systems to provide a visual display of air and marine craft movements; in the display of computer data; in security systems; in measurement, test, and instrument systems.

11.1
A labelled diagram of a cathode ray tube
Fig. 11.1 shows the basic structure of a cathode ray tube which employs electrostatic deflection. The cathode ray tube, usually abbreviated to CRT, consists of the following basic parts housed in an evacuated glass or metallic envelope:

(a) *An electron gun*
The electron gun is designed to produce a narrow beam of electrons which move at a very high velocity and which may be focused to produce a small spot on the CRT screen. The electron gun section includes the cathode (which is the source of electrons), a control grid (which controls the electron beam intensity), and anode electrodes (which accelerate and

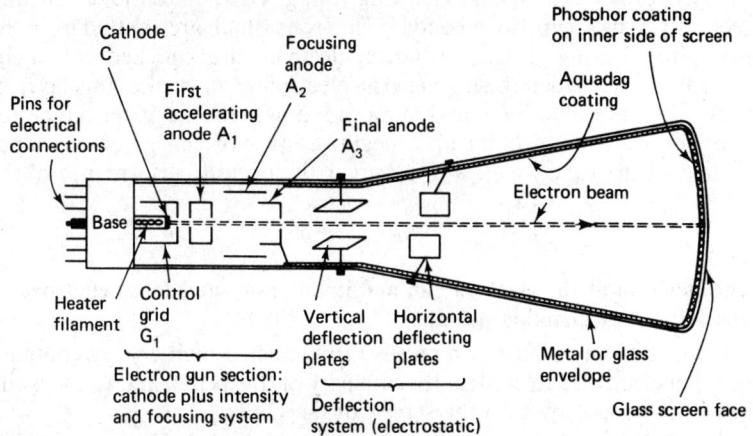

Fig. 11.1. Diagram of a cathode ray tube.

focus the beam). External electrical connections to the various electrodes in the gun are made at the pins in the CRT base.

(b) *Vertical and horizontal deflecting plates*

A voltage generator or signal connected across the vertical deflection plates deflects the beam vertically up, when the upper plate is positive with respect to the lower plate, or down when the polarity of the applied voltage reverses. A voltage connected across the horizontal deflection plates deflects the beam either left or right in a horizontal direction.

(c) *A phosphorescent screen*

The active layer of the screen consists of a thin coating of phosphor material deposited on the inner glass face of the CRT with a very thin backing layer of evaporated aluminium. When the electron beam strikes the screen, light is emitted from the phosphor coating at the point of impact; the aluminium backing reflects light that would otherwise be lost from the back of the screen. The phosphor continues to glow for a time after the beam has been removed. This time, known as the persistence time, depends on the type of phosphor material used and may vary from a few microseconds to several seconds. In the case of storage type CRT, even longer times can be obtained. The type of phosphor used also determines the colour of the light emitted and hence the colour of the beam trace on the screen.

The inside of the CRT from the end of the electron gun section to the screen is coated with a layer of colloidal graphite commonly known as aquadag. This layer is conducting and an electrical connection is made to the final anode of the electron gun section. It serves two purposes:

131

firstly, it provides a high accelerating voltage right up to the screen, and secondly, it picks up any secondary electrons which are emitted from the phosphor coating. These secondary electrons are knocked-out of the phosphor layer when the high energy electron beam strikes this layer. If they were not collected and removed down to the external power supplies, they could build up a negative layer of charge on the inner surface of the tube which would impair the overall performance of the CRT.

11.2
The function of the electron gun and its intensity and focus controls and the use of blanking pulses

The function of the electron gun is to produce a high energy compact beam of electrons, to control the intensity of this beam, and to focus the beam to a small spot on the screen of the CRT.

The cathode of the electron gun is shown diagrammatically in fig. 11.2(a). It consists of a nickel cylinder indirectly heated by a filament. The filament is normally supplied from a 6·3 V a.c. source obtained from appropriate secondary tapping points of a mains transformer. The end of the cylinder only is coated with a metal oxide layer, this layer acting as the electron source. This form of cathode produces electron emission almost entirely in the forward direction.

Surrounding the cathode is a metal cylinder—the control grid—which has a very small aperture in the centre of the end facing the screen. The function of the control grid is primarily to control the flow of electrons from the cathode through this small aperture. It also acts, in an analogous way to an optical lens, to provide a degree of focusing, concentrating the electrons into a beam. The grid, normally held at a negative voltage with respect to the cathode, controls the intensity of the electron beam and hence the brightness of the spot or trace formed by the beam on striking the screen. A basic diagram showing the essential features of the intensity control is drawn in fig. 11.2(b). The grid-cathode voltage is varied by the potentiometer, typically from a few volts negative to several tens of volts negative. The electron beam may be cut off by making the grid-cathode voltage sufficiently negative. Typical values of cut-off grid voltage vary from tens of volts negative up to about − 100 V, the actual value for a given gun obviously depending on the geometry and spacing of the cathode-grid structures.

Let us now consider the complete gun section shown in fig. 11.2(c) and its focusing action. The electrons which pass through the control grid aperture are accelerated by the initial accelerating anode A_1 which is at a high positive voltage with respect to the cathode, typically of the order of 2000 V or so. The focusing anode A_2 and the final anode A_3 act to focus electrons into a compact beam such that the beam strikes the screen in a well defined spot of small diameter. A_2 is at a lower voltage than A_3. Typically A_2 may be about 1500 V and this voltage may be varied by the potentiometer shown in (c). A_3 is at the same potential as A_1, i.e. about 2000 V. A_1, A_2, and A_3 are metal cylinders. A_1 has small centrally located apertures at both ends, A_2 is open at both ends, and A_3 open at one end only but with a small central aperture at its output side. During assembly at the factory, the anode cylinders must be carefully

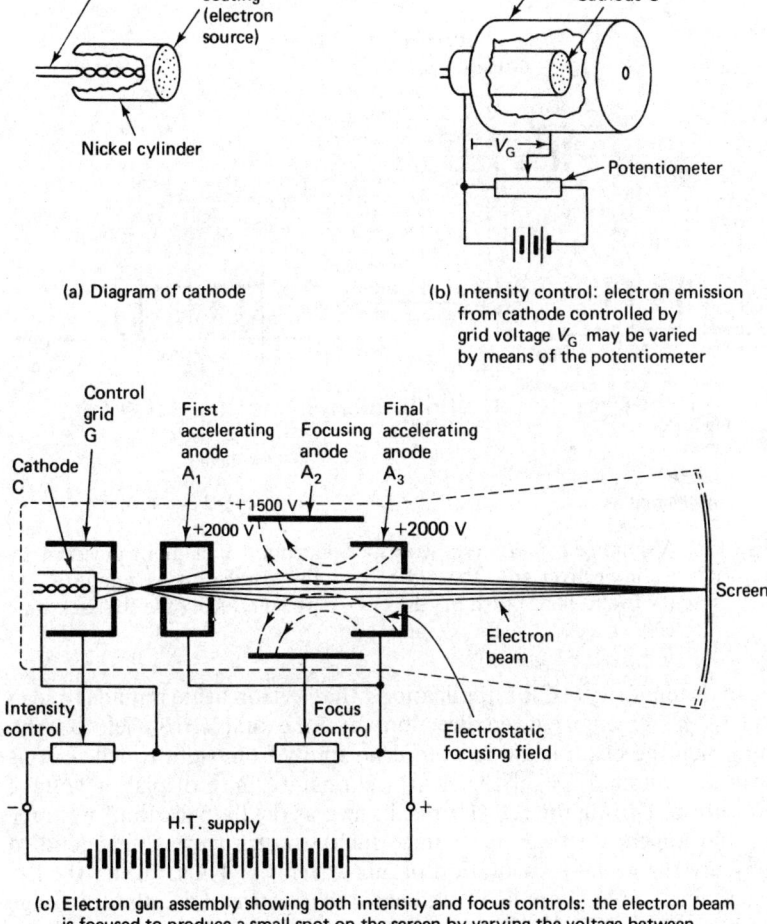

(a) Diagram of cathode

(b) Intensity control: electron emission from cathode controlled by grid voltage V_G may be varied by means of the potentiometer

(c) Electron gun assembly showing both intensity and focus controls: the electron beam is focused to produce a small spot on the screen by varying the voltage between anodes A_2 and A_3. Typical values of accelerating anode voltages for A_1 and A_3 may be of the order of 2000 V and for the focusing anode A_2, 1200 to 1500 V.

Fig. 11.2.

aligned with the cathode and control grid so that their axes, and hence apertures, all lie in a straight line coinciding with the axis of the CRT.

The voltage difference between A_2 and A_3 sets up an electrostatic field between the two anodes. The lines of force of this field are shown as dash lines in (c). The field tends to force any diverging electrons back into the centre of the beam. By adjustment of the voltage difference between A_2 and A_3, the beam can be focused to its minimal diameter at the point where it strikes the screen. The anodes A_2 and A_3 thus act in an analogous way to an optical lens but with the added advantage of a variable focus control.

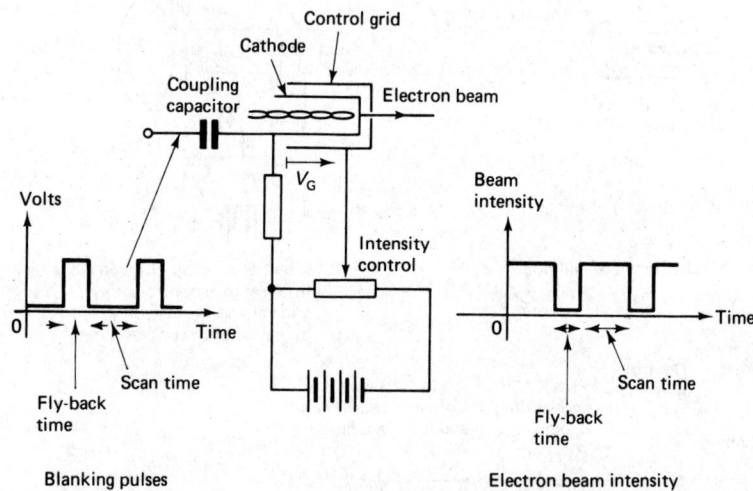

Fig. 11.3. A simplified diagram showing the application of blanking pulses to the cathode of an electron gun: the pulses bias the grid-cathode voltage V_G to beyond cut-off and hence turn off the electron beam during the fly-back time intervals.

In the majority of CRT applications, the electron beam is made to scan across the screen in a given direction, say, for example, from left to right, and then the beam is made to return rapidly from right to left. In this way, as we shall see in the next sections, we may display a voltage waveform. During the return time, known as the fly-back time, no trace should appear on the screen, since such a return trace would tend to obscure the visual presentation of our waveform. Thus, during the fly-back time, a pulse is applied to increase the negative grid-cathode voltage to beyond cut-off. This pulse, known as a blanking pulse, cuts off the electron beam and therefore no return trace is displayed on the screen. In practice, a series of blanking pulses is generated by internal circuits in the CRT, so that over every fly-back interval, the beam is cut off. A simple

134

circuit showing the application of blanking pulses to the cathode of an electron gun is given in fig. 11.3. The pulses shown are positive and hence raise the cathode voltage in the positive direction with respect to the control grid voltage. They are synchronized to occur and therefore cut off the electron beam during the fly-back time intervals.

11.3
The deflection of an electron beam by electric and magnetic fields

An electron beam may be deflected by both electric and magnetic fields. If the deflection is produced by an electric field, it is known as electrostatic deflection; when produced by a magnetic field, as magnetic deflection. Electrostatic deflection is invariably used in cathode ray oscilloscopes, that is for measurement applications, whereas magnetic deflection is used in television and radar display systems.

The CRT of an oscilloscope has both vertical and horizontal deflection plates so as to provide independent deflection in both vertical and horizontal directions. This arrangement allows us to display a two-dimensional trace of a signal waveform on the CRT screen. Electrostatic deflection is achieved by applying a voltage across the respective pairs of plates. These voltages generate electric fields between the plates. The

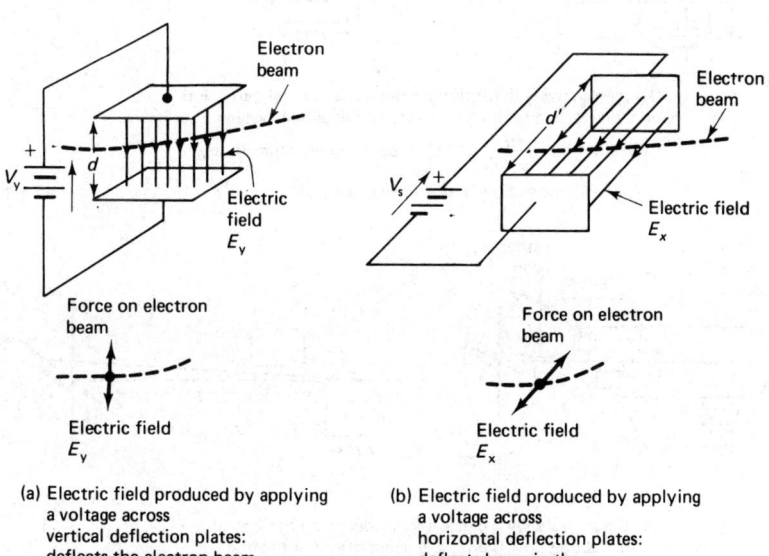

(a) Electric field produced by applying a voltage across vertical deflection plates: deflects the electron beam in the vertical plane

(b) Electric field produced by applying a voltage across horizontal deflection plates: deflects beam in the horizontal plane

Fig. 11.4.

135

direction of the electric field is from the positive polarity to the negative polarity plate. The magnitude of the electric field strengths are:

for the vertical deflection plates of fig 11.4(a), $E_y = \dfrac{V_y}{d}$ in the vertical plane.

for the horizontal deflection plates of fig. 11.4(b), $E_x = \dfrac{V_x}{d'}$ in the horizontal plane.

When an electron beam is passed between a given pair of plates, it will be attracted continually in a direction along the lines of electric force towards the respective positive plate. Thus, in passing axially between the vertical deflection plates, the beam will be deflected vertically, and in passing between the horizontal deflection plates, the beam will be

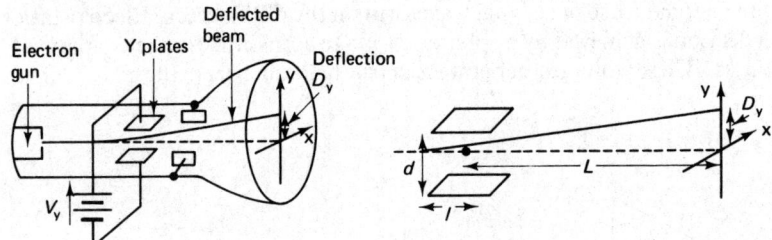

(a) Diagram showing deflection produced by vertical deflection or Y plates, and geometry of vertical deflection system.

Deflection $D_y = \dfrac{lL}{2dV_A} V_y$; V_A = beam accelerating voltage

(voltage of final anode in electron gun)

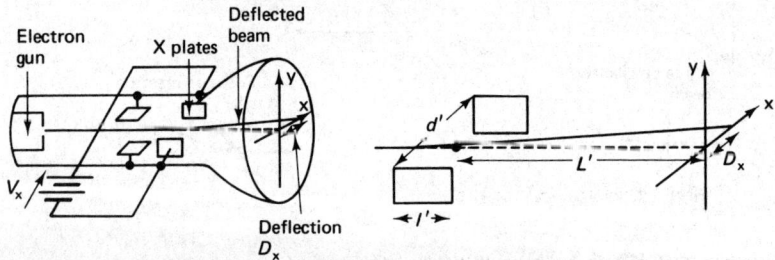

(b) Diagram showing deflection produced by horizontal deflection or X plates, and geometry of horizontal deflection system. Deflection $D_x = \dfrac{l'L'}{2d'V_A} V_x$

Fig. 11.5.

136

deflected horizontally. The passage of the beam between the plates follows a curved path. On emerging from the plates, the beam continues in a straight line, but this line is at an angle to the incident beam direction.

The amount of deflection of the electron beam increases as electric field is increased and, therefore, as the voltage difference between the plates increases and as the space between the plates is decreased. For the latter reason, the plates of an electrostatic system are normally located inside the CRT. The respective deflections produced by applying a voltage V_y to the vertical deflection plates (also called Y plates) and a voltage V_x to the horizontal deflection plates (also called X plates) are given by

$$D_y = \frac{lL}{2d\,V_A}V_y \quad \text{(see Fig. 11.5(a))}$$

$$D_x = \frac{l'L'}{2d'V_A}V_x \quad \text{(see Fig. 11.5(b))}$$

where l, l' = length of deflection plates,
L, L' = distance from centre of plates to screen,
d, d' = distance between plates,
V_A = accelerating voltage of electron beam,
V_y = voltage applied across Y plates,
V_x = voltage applied across X plates.

The ratios D_y/V_y and D_x/V_x are known as the deflection sensitivities of the respective pairs of plates. They express the deflection produced in metres per volt. The Y plates are placed further away from the screen than the X plates to give a greater value of L and hence a greater sensitivity, and therefore greater deflection for a given applied voltage. The reason for this is that the voltage whose waveform we wish to display is normally applied across the Y plates and hence it is advantageous that these plates should have the best possible sensitivity. A sawtooth time-base voltage which sweeps the beam across the CRT screen is normally applied to the X plates.

Fig. 11.6(a) and (b) show the deflection of an electron beam produced by magnetic fields. Note that the direction of the force on the beam, and hence the deflection produced, is at right angles to the magnetic field direction and the direction in which the beam is travelling. No force acts on a beam which travels parallel to the lines of magnetic force. In (a), a beam travelling horizontally and at right angles to a horizontal magnetic field is deflected vertically; in (b) a beam travelling horizontally and at right angles to a vertical magnetic field is deflected horizontally. If we reverse the direction of the magnetic field, the directions of the respective deflections are also reversed. In (c) the beam travels parallel to the magnetic field and experiences no force and, therefore, no deflection. Use

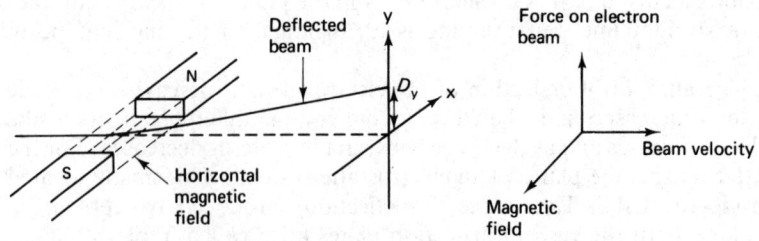

(a) Vertical deflection of an electron beam moving at right angles to a horizontal magnetic field

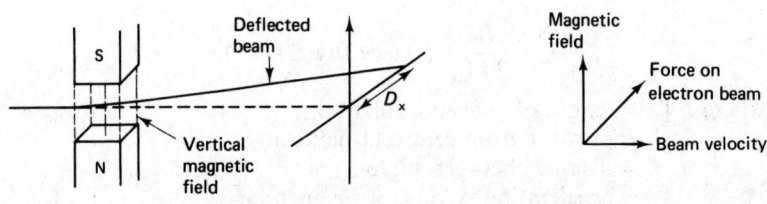

(b) Horizontal deflection of an electron beam moving at right angles to a vertical magnetic field

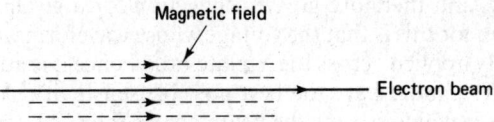

(c) An electron beam travelling parallel to a magnetic field is not deflected

Fig. 11.6.

is made of these properties in both magnetic deflection and magnetic focusing systems.

When magnetic deflection is used, the electron beam is normally passed through the magnetic field produced by pairs of coils. The amount of deflection produced is proportional to the magnetic field strength and hence to the current through the coils since the latter determines the strength of the magnetic field.

Some examples of the arrangements and types of coils used in

138

magnetic deflection systems are shown in fig. 11.7. (a) shows two pairs of coils which produce horizontal and vertical magnetic fields. (b) shows two pairs of coils located on the neck of the CRT. There is no need to separate the axial positions of the coils (as we do in the case of electrostatic deflection); they may be placed around a given section of the tube, as indicated in the diagram. This is an interesting point since it allows a CRT using magnetic deflection coils to be made much shorter in

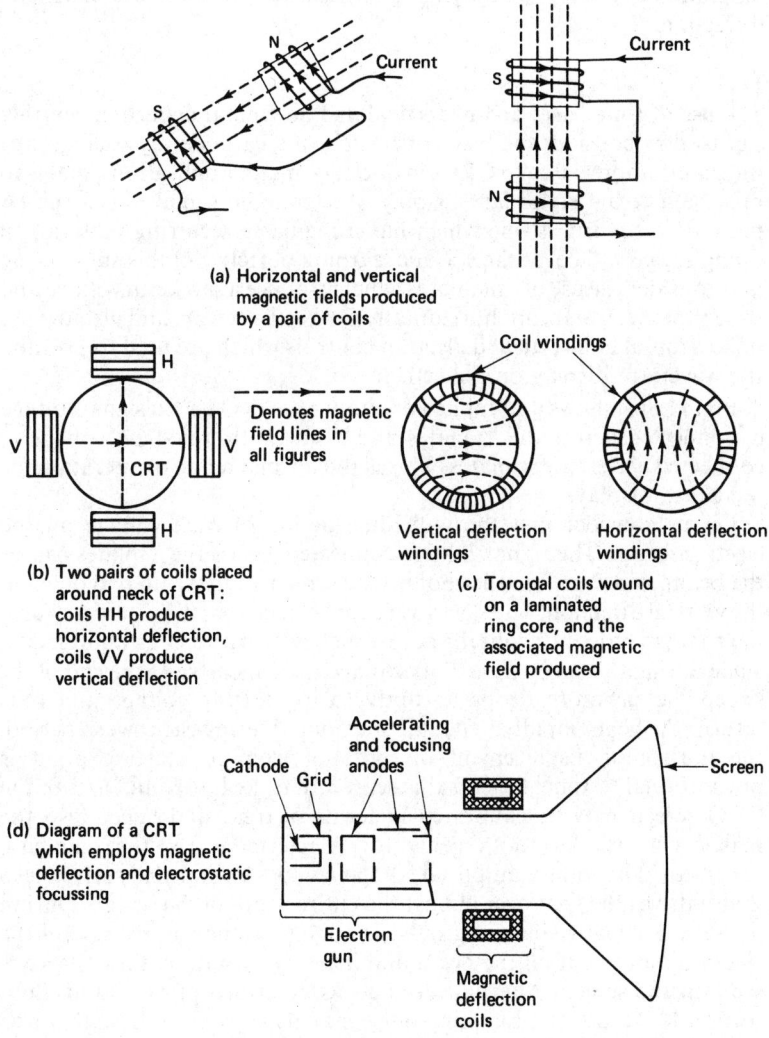

(a) Horizontal and vertical magnetic fields produced by a pair of coils

(b) Two pairs of coils placed around neck of CRT: coils HH produce horizontal deflection, coils VV produce vertical deflection

- - - → - - - Denotes magnetic field lines in all figures

(c) Toroidal coils wound on a laminated ring core, and the associated magnetic field produced

Vertical deflection windings

Horizontal deflection windings

Coil windings

(d) Diagram of a CRT which employs magnetic deflection and electrostatic focussing

Cathode Grid

Accelerating and focusing anodes

Screen

Electron gun

Magnetic deflection coils

Fig. 11.7.

length—a very important consideration in television receiver design. (c) shows an example of two pairs of coils wound as toroids; the first pair produces a horizontal magnetic field, and the second a vertical magnetic field. Non-uniform windings are used in both cases since these produce a uniform magnetic field in the plane of the cross-section of the toroid. In practice, both pairs are wound on the same toroidal ring, thus producing a very compact horizontal and vertical deflection system. (d) shows a diagram of a CRT employing electrostatic focusing and magnetic deflection.

11.4
The use of time-bases and of vertical and horizontal deflection controls
Let us now consider the basic operation of a cathode ray oscilloscope (normally abbreviated to CRO) in one of its major applications, that is to provide a visual stationary display of a periodic signal waveform. (A periodic waveform is one which has a regularly recurring pattern.) In doing so, we will cover the specific learning objective of this sub-section and consider the use of time-bases which produce a sawtooth voltage and sweep the electron beam horizontally across the screen, and also the use of horizontal and vertical deflection controls which are used to position the waveform display on the screen.

Fig. 11.8(a) shows a simplified diagram of a CRO with a signal voltage v_y connected across the Y plates and a sawtooth time-base voltage v_x connected across the X plates. This is the normal mode of operation for waveform display.

Let us consider first the individual action of each voltage on the electron beam. The signal voltage connected across the Y plates causes the beam, and therefore the spot on the screen, to move up and down in the vertical direction. The sawtooth connected across the X plates causes the beam to move horizontally across the screen with a constant velocity, since during the sweep time T_x its voltage rises linearly. At the end of the sweep the sawtooth drops abruptly to its starting voltage and thus returns the beam rapidly to its starting point. During each sweep period, the horizontal displacement of the spot from its starting point is proportional to time. It is for this reason that the horizontal axis of the CRO screen may be calibrated in terms of time, and hence also the reason why the sawtooth generator is referred to as the time-base generator. The voltage amplitude of the sawtooth is normally selected so that it drives the spot over almost the entire width of the screen. During fly-back, a negative blanking pulse is applied to the control grid of the electron gun to cut off the beam and hence to ensure that no retrace is seen on the screen. Alternatively and as described previously in Sub-section 11.2 a positive blanking pulse may be applied to the cathode to achieve beam cut-off during fly-back.

Now consider the combined effect of the sinusoidal signal voltage

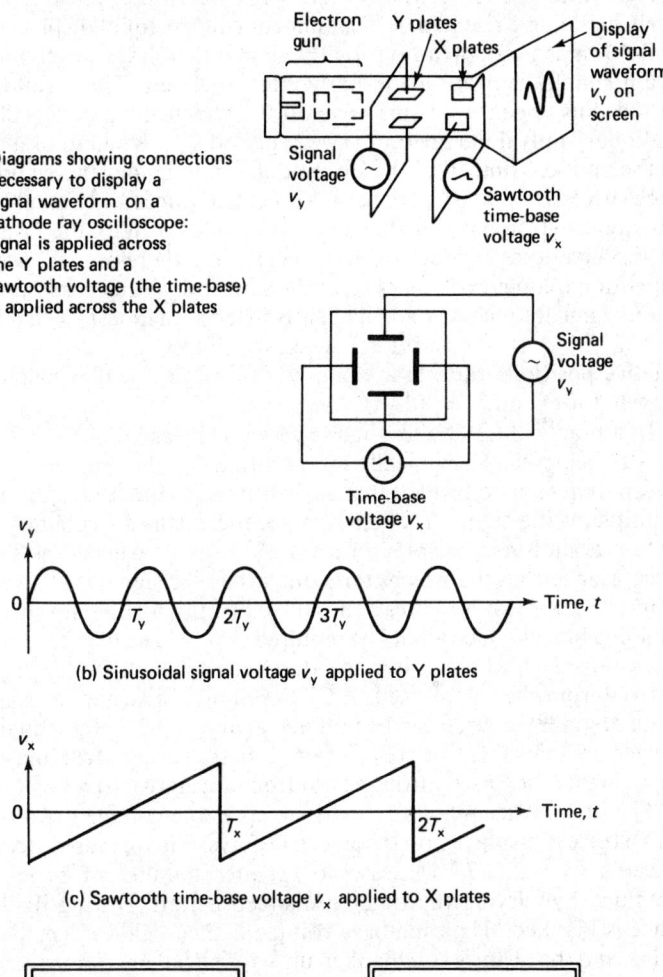

a) Diagrams showing connections necessary to display a signal waveform on a cathode ray oscilloscope: signal is applied across the Y plates and a sawtooth voltage (the time-base) is applied across the X plates

(b) Sinusoidal signal voltage v_y applied to Y plates

(c) Sawtooth time-base voltage v_x applied to X plates

(d) Trace of signal waveform v_y on screen when (i) time-base period is twice signal period, i.e. $T_x = 2T_y$ (ii) when time-base is not an exact integral (whole number) multiple of the signal period

Fig. 11.8.

141

shown in Fig. 11.8(b) applied to the Y plates, and the sawtooth voltage shown in (c) applied to the X plates. The sinusoidal voltage, which by itself would give rise to a vertical line moving an equal displacement up and down, is now spread out by the sawtooth voltage which moves the beam simultaneously in the horizontal direction. The resultant trace would thus appear as a sine wave on the screen. This trace will appear stationary only if the sawtooth sweep period T_x is equal to or a multiple of the periodic time T_y of the sinusoidal signal. Under these conditions, each successive sweep will produce coincident traces and thus the display will appear stationary, as shown in Fig. 11.8(d)(i). With the simple form of display described above, it is necessary, therefore, to adjust the repetition frequency, $f_x = 1/T_x$ of the sawtooth to an exact submultiple of the signal frequency; or equivalently in terms of time $T_x = nT_y$ where n = 1, 2, 3. . . . If this is not so, successive traces will occur in different relative positions causing a confused multi-trace display, such as that shown, for example, in (d)(ii).

In a practical CRO, life is made very much easier and a steady trace may be achieved electronically by synchronizing the time-base sawtooth sweep frequency to bring it into step with the signal frequency or a sub-multiple of the signal frequency. A second method frequently used to obtain a stationary display of a waveform is to trigger the time-base generator so that the sweep starts only when the signal is at a given part of its cycle. Fig. 11.9(a) gives a simplified functional diagram showing the building blocks which would be required to provide a trigger sweep mode operation of a CRO. In this mode, the sweep is initiated by a trigger pulse derived from the signal itself. A CRO will normally have a trigger level control so that a sweep can be initiated at any point on the signal voltage waveform. In fig. 11.9, we see an example of triggering occurring when the input signal increases through zero from a negative to a positive value.

The first trigger pulse starts the sweep as the voltage rises from zero and a trace is displayed on the screen (2·5 cycles in our case since we have selected $T_x = 2·5\ T_y$). A lesser or greater number of cycles can be obtained by varying the time-base sweep period. During fly-back the trace is blanked. The time-base voltage is then held off, i.e. at its start value, and the beam is still blanked until the signal once more reaches the same predetermined level, i.e. in our case rises through zero. At this point, another trigger pulse is generated which fires the time-base to produce the sweep and a second coincident trace. The whole sequence of events is then repeated with trigger pulses starting the sweep at the correct point on the signal waveform. In this way we produce a stationary waveform display. We do, however, rely on the persistence of vision time on the screen to be sufficient to hold a given trace until a subsequent trace reactivates the screen. For very slow sweep rates we can actually see the spot moving across the screen.

In addition to applying the signal to the Y plates and the sawtooth

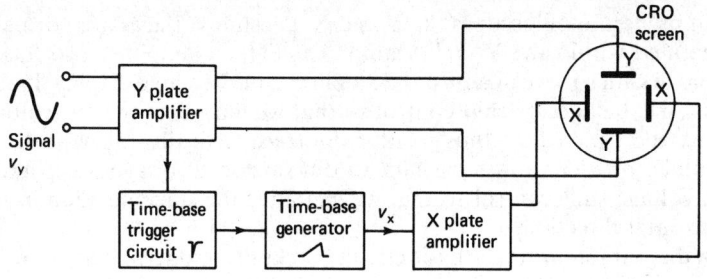

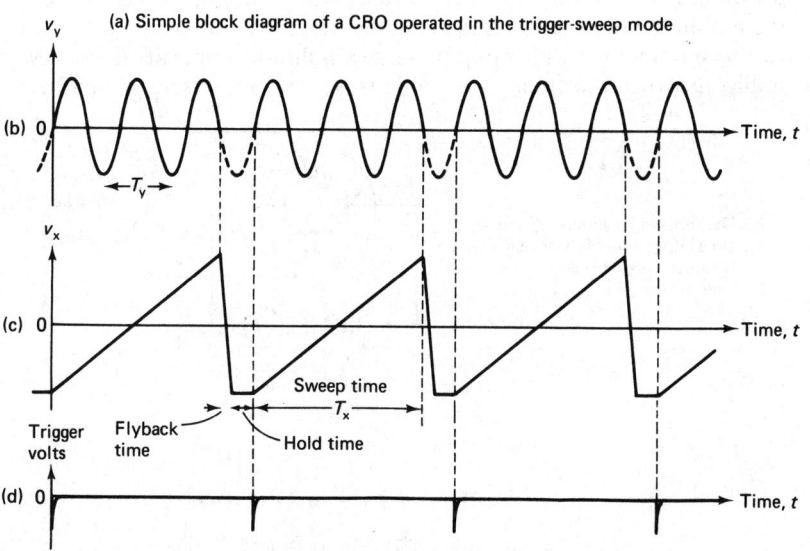

Fig. 11.9. (a) Simplified block diagram of a CRO operated in the trigger-sweep mode. (b) Signal waveform v_y (connected via an amplifier to Y-plates). (c) Time-base sawtooth voltage v_x (connected via an amplifier to X-plates). (d) Trigger pulses which are generated from signal v_y and used to start time-base sweep at a given signal level (or signal slope).

time-base voltage to the X plates it is necessary to be able to control the position of the trace produced to ensure that this is formed on the required part of the screen. This may be accomplished, as shown in principle in Fig. 11.10(a), by including an adjustable d.c. voltage in series with both the signal and time-base voltages. By varying the d.c. voltage V_y we can move the trace up and down, by varying the d.c. voltage V_x we can move the trace left or right. Fig. 11.10(b) gives a circuit showing how these variable d.c. voltages could be obtained by connecting a fraction of the constant d.c. voltage supply to the respective plates using potentiometers: the vertical shift potentiometer controls the value of V_y, the horizontal shift potentiometer the value of V_x.

In modern oscilloscopes the Y and X d.c. shift voltages are normally generated within the Y and X amplifiers. The Y-amplifier provides a variable gain so we can adjust the amplitude of the signal display. It also has a d.c. balance or shift control so that we may alter the d.c. voltage across the Y plates and thus position our trace vertically. Likewise the X-amplifier provides a variable gain for our sawtooth voltage sweep and a d.c. voltage shift control so that we can alter the trace position in the horizontal direction.

A diagram showing the basic circuit blocks of a modern oscilloscope is drawn in Fig. 11.11(a). A switched attenuator (a network which reduces the amplitude of a signal) is included before the Y-amplifier so that we can accommodate a wide range of signal amplitudes, typically from a few millivolts (when the attenuator would be set at zero) to several hundred

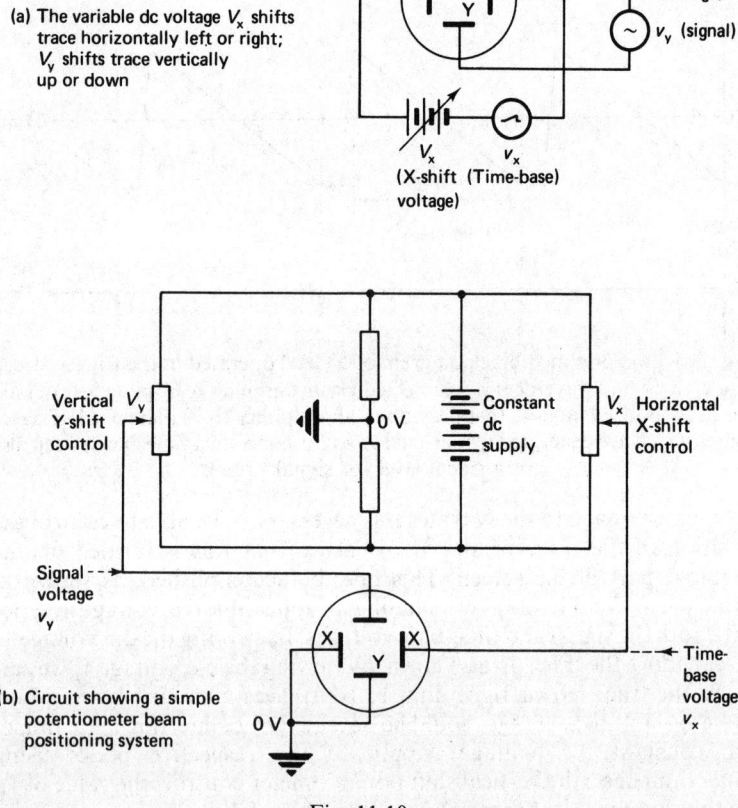

(a) The variable dc voltage V_x shifts trace horizontally left or right; V_y shifts trace vertically up or down

(b) Circuit showing a simple potentiometer beam positioning system

Fig. 11.10.

144

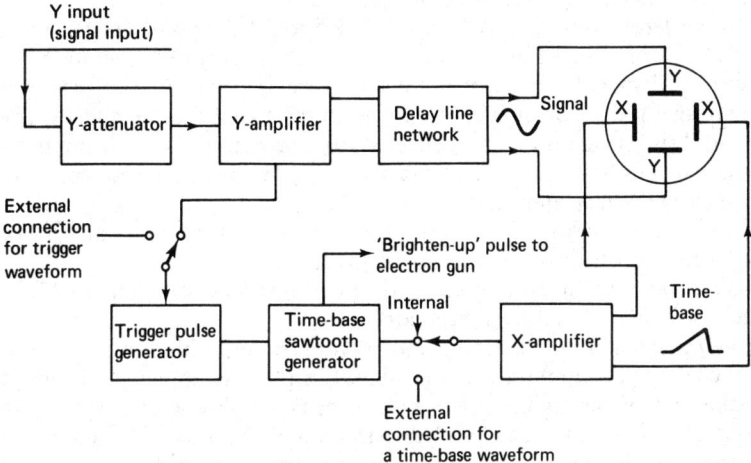

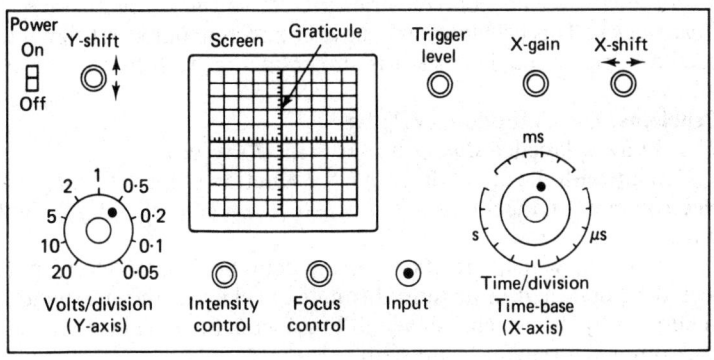

(a) Block diagram showing basic circuits in a modern cathode ray oscilloscope

(b) A typical front panel layout of controls in a modern cathode ray oscilloscope

Fig. 11.11.

volts. Fig. 11.11(b) shows a typical front panel layout of controls. Let us complete this section by reviewing their function.

Intensity control. This control is used to vary the intensity of the electron beam and hence the brightness of the trace.

Focus control. This control is used to focus the beam into a spot of small diameter.

Time/division (time-base control). This control may be switched to various ranges to select a given sweep time. It is provided with a second tuning control for fine variation of the sweep time, this being operated in practice once a suitable range is selected.

Trigger level control. This control is used to set the point on the signal waveform at which the time-base generator is triggered and starts its sweep to display the waveform, e.g. if the level is set at zero this results in the trigger pulse starting the display when the signal rises through its 0 V value. Triggering may also be effected from an external source and most oscilloscopes have a socket on the front panel to facilitate connection of an external triggering waveform.

X-gain control. This control varies the gain of the X-amplifier and hence controls the width of the time-base sweep.

X-shift control. This control varies the position of the trace left or right in the horizontal (time) axis direction.

Volts/division (Y-attenuator and amplifier control). This control may be switched to accommodate various signal amplitudes applied at the input socket, e.g. if the signal has a peak-to-peak amplitude of 8 V and the control is switched to the 1 volt/division range, the trace will extend over 8 divisions of the screen (each division being measured by the screen graticule) in the vertical direction. A second control for fine variation of the trace amplitude is also provided.

Y-shift control. This control varies the vertical position of trace.

Input socket. This provides a means of connecting our input signal to the oscilloscope. A coaxial socket is shown in Fig. 11.11(b).

Problems: C Cathode Ray Tube

11.1 Draw a labelled diagram of a cathod ray tube.

Explain with the aid of diagrams the function of the following: (a) the electron gun, (b) the intensity control, (c) the focus control, (d) blanking pulses.

11.2 State in which direction an incident horizontal electron beam would be deflected in passing through the electric and magnetic fields generated by the systems shown in Fig. P11.2. Sketch also for each case the complete electron beam path.

Describe with the aid of diagrams (a) a magnetic deflection system

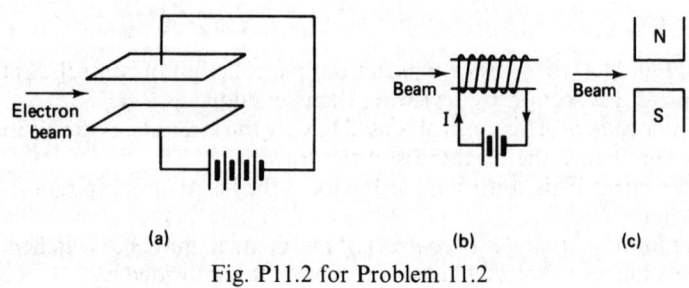

(a) (b) (c)

Fig. P11.2 for Problem 11.2

employing coils, (b) an electrostatic deflection system suitable for use in a cathode ray oscilloscope.

State one merit of an electrostatic and one merit of a magnetic deflection system.

11.3 (a) Describe how you would use a cathode ray oscilloscope to display the output voltage developed across the load R in the amplifier circuit of Fig. P11.3(a).

(b) A sinusoidal signal of 2·5 V peak-to-peak amplitude and 500 Hz frequency is applied to the Y plates of an oscilloscope. The Y amplifier control switch is set to 500 mV per division and the time-base sweep control to 1 ms per division. Sketch the waveform observed on the screen of the oscilloscope and label both horizontal and vertical axes. The screen graticule is shown in Fig. P11.3(b). Assume that the time-base is triggered when the signal rises through zero.

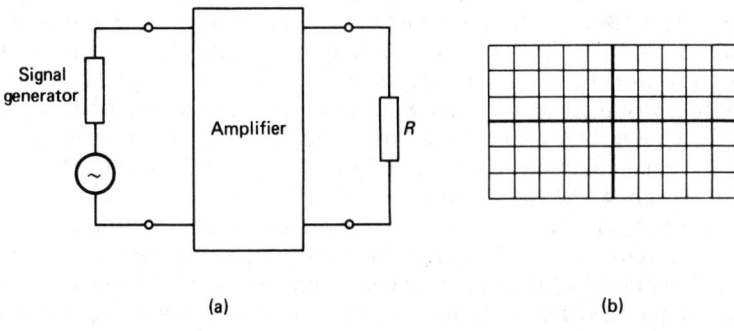

(a)   (b)

Fig. P11.3 for Problem 11.3

11.4 (a) Describe with the aid of suitable diagrams the function and operation of the vertical and horizontal deflection controls in a cathode ray oscilloscope.

(b) Sketch a practical time-base waveform and describe with the aid of a block diagram one method of synchronization used to obtain a stationary waveform display.

Answers
11. 2(a) Vertically down   (b) No deflection   (c) Horizontally (out of page).

D   Small-signal Amplifiers

Section 12: *The expected learning outcome of this section is that the student should know the circuit and operation of a small-signal common emitter amplifier.*

In Section 5 we first introduced the amplifier to help understand and explain the three basic modes of connection of a transistor and to give meaning to the important parameters of input and output resistance and current and voltage gain. In Section 6 we investigated the static characteristics of a transistor and showed how the small-signal input and output resistances and current gain could be calculated from the slopes of the characteristic curves. In this section we will concentrate on the small-signal common emitter amplifiers, our general learning objective being to know the circuit and its operation.

However, first let us explain why we use the qualification of small signal. Electronic devices, such as the transistor and the thermionic valve, only operate in a linear manner over a limited range, that is, they will only amplify a signal faithfully without distorting its waveform, provided the amplitude of the input waveform is restricted so that it does not drive the transistor or valve into the non-linear area of its characteristics. In fact, there are many other factors which determine the linear working of an active device, such as a transistor or valve. To be pedantic, even if we work in the linear area of the characteristics, the transistor or valve parameters do vary with bias currents and voltages. Therefore, to simplify our first discussion on amplifiers, we use the term small signal. 'Small signal' means that the transistor or valve amplifies linearly and its parameters, such as current gain, input and output resistances, do not change as the amplitude of the signal changes. They remain constant. Typically, we will be dealing with input signal base currents of the order of tens of microamps up to hundreds of microamps and corresponding base-emitter input voltages of the order of a few to tens of millivolts. These will produce a.c. changes in collector current up to a few milliamps and collector-emitter voltage of up to a few volts. An example of an amplifier operating in a linear and non-linear manner is shown in fig. 12.1. Amplifiers can, of course, operate with low distortion for much larger signal changes. At a later stage in our study of electronics, we shall, of course, be considering the large-signal operation of an amplifier.

148

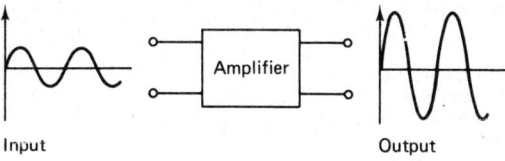

Input   Amplifier   Output

(a) Amplifier operating in a linear manner:
   output is a faithful reproduction of input signal

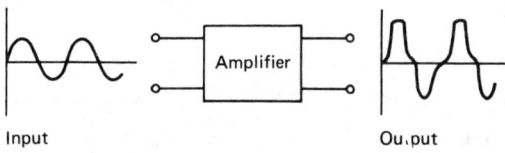

Input   Amplifier   Ou.put

(b) Amplifier not operating linearly:
   output waveform is distorted

Fig. 12.1.

12.1

The circuit diagram of a single-stage amplifier having a load resistor R_L

Fig. 12.2(a) shows the basic diagram of a single-stage common emitter amplifier. V_{CC} is the collector supply voltage. V_{BB} is the base-emitter bias voltage supply which, together with the resistor R_B' is used to set the value of base bias current required. The signal, represented by a source of voltage V_S and source resistance R_S, is connected between the input terminals 1-1'. R_L is the load across which the amplified signal is developed.

The source causes a varying signal current i_b to be injected into the transistor base. i_b is superimposed on the d.c. base bias current I_B. i_b is amplified by transistor action to produce an a.c. component of collector current $i_c \approx \beta i_b$, where β = small-signal common emitter current gain of the transistor. i_c is superimposed on the d.c. component of collector current I_C flowing in R_L. The voltage drop across R_L therefore consists of a d.c. component $R_L I_C$ and the amplified a.c. signal voltage $R_L i_c$. In practice, the a.c. component $R_L i_c$ is separated from the d.c. component $R_L I_C$ by use of a coupling or d.c. blocking capacitor, the capacitor C shown in the diagram. C acts so as to pass the signal component but to block the d.c. component. The output signal voltage of the amplifier stage may therefore be taken across the terminals 2-2'. In practice, two

149

(a) Basic circuit of a single-stage
common emitter amplifier:
ac signal output voltage $v_o = v_{ce}$
is taken between collector and
common emitter terminal; action
of C is to block dc components, C
passes the ac signal component

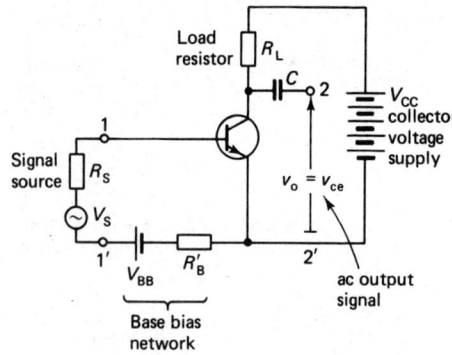

(b) Example of a single-stage
common emitter amplifier
where no bias battery is
required in the base circuit.
C' blocks dc from passing
through the signal source
but passes the ac
signal input current

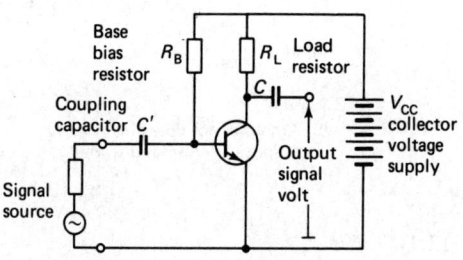

Fig. 12.2.

separate batteries for V_{BB} and V_{CC} are not used. Instead, we derive the
base-emitter bias voltage from one source, the collector supply V_{CC}. One
simple method of effecting this is shown in fig. 12.2(b). We will explain the
action of this circuit in Sub-section 12.4.

12.2
The collector supply voltage $V_{CC} = R_L i_C + v_{CE}$
Since the voltage applied in a series circuit must at all instants of time
equal the sum of the voltage drops across the individual components
making up the circuit, we have

collector supply voltage, V_{CC} = voltage drop across R_L
+ voltage drop across collector-
emitter terminals of the transistor

Thus if we denote the total collector current at any instant of time by i_C
and the total collector-emitted voltage by v_{CE}:

$$V_{CC} = R_L i_C + v_{CE}$$

Note: Although we have previously used upper and lower case subscripts

150

and symbols to denote d.c. and a.c. amplitudes, it is now perhaps a convenient point to summarize the notation used in electronics to indicate d.c. and time varying total values and signal components:

V_{CE}, I_C denote d.c. (quiescent) values of collector-emitter voltage and collector current,

V_{ce}, I_c denote a.c. amplitudes of collector-emitter voltage and collector current,

i.e. the peak positive or negative swing about the d.c. value or, and more usually, the root mean square (r.m.s.) values of the voltage and current. The peak and r.m.s. values are simply related for sinusoidal waveforms by a $\sqrt{2}$ factor:

$$\text{peak amplitude of an a.c. quantity} = \sqrt{2} \times \text{r.m.s. value}$$

$$\text{or r.m.s. value} = \frac{\text{peak amplitude}}{\sqrt{2}}$$

v_{ce}, i_c denote the instantaneous values of the varying components (e.g. the a.c. signal components) of collector-emitter voltage and collector current,

v_{CE}, i_C denote the total instantaneous values, that is, the d.c. + instantaneous components, so

$$v_{CE} = V_{CE} + v_{ce}, \; i_C = I_C + i_c$$

V_{CC}, V_{BB} denote collector, base supply voltages.

Likewise I_B, I_b, i_b, i_B denote respectively the d.c., r.m.s. (or peak signal), instantaneous signal, total amplitudes of base current; and V_{BE}, V_{be}, v_{be}, v_{BE} the d.c., r.m.s. (or peak signal), instantaneous signal, total amplitudes of base-emitter voltage.

Example 12.1

The signal generator connected across the input terminals of the single-stage common emitter amplifier circuit shown in Fig. 12.3 varies the collector current by $\pm 500 \; \mu A$ about the mean (i.e. d.c.) value of $I_C = 5$ mA. If the load resistor value $R_L = 2 \; k\Omega$ and the collector supply voltage $V_{CC} = 20$ V, calculate the collector-emitter voltage v_{CE} when the signal generator voltage is zero; at its maximum and at its minimum value. Calculate also the peak-to-peak voltage swing across R_L. If the signal generator produces sinusoidal variations of input current, calculate the r.m.s. value of the a.c. component of output voltage across R_L.

Solution

Now at all instants of time,

$$V_{CC} = R_L i_C + v_{CE}, \text{ i.e. } v_{CE} = V_{CC} - R_L i_C$$

where $V_{CC} = 20$ V, the collector supply voltage; and i_C and v_{CE} are the

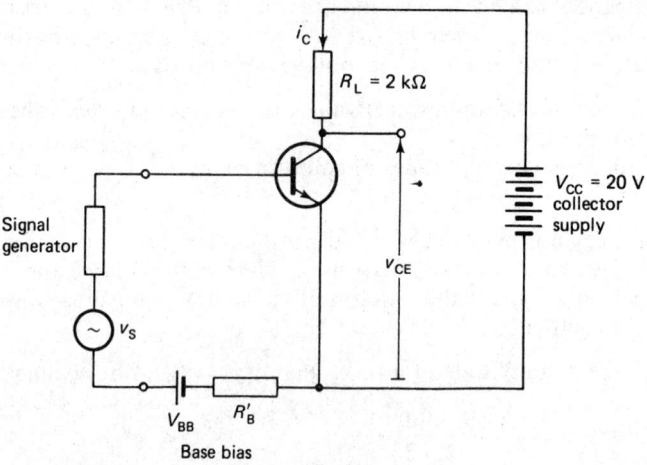

Fig. 12.3.

total instantaneous values of collector current and collector-emitter voltage; and $R_L = 2 \text{ k}\Omega$ is the load.

When the signal voltage is zero, $i_C = I_C = 5 \text{ mA}$.

So
$$v_{CE} = 20 - (2 \times 10^3) \times 5 \times 10^{-3}$$
$$= 20 - 10 = 10 \text{ V}.$$

When the signal voltage v_S is a maximum v_S aids V_{BB} to a maximum extent increasing the forward bias across the base-emitter junction. Thus i_B and consequently i_C increase to their maximum values. The maximum value of i_C is $5 \text{ mA} + 500 \, \mu\text{A} = 5 \cdot 5 \text{ mA}$, so the corresponding value of v_{CE} is

$$V_{CE} = 20 - (2 \times 10^3) \times (5 \cdot 5 \times 10^{-3})$$
$$= 20 - 11 = 9\text{V}.$$

When the signal voltage is at a minimum, i.e. maximum negative, v_S acts in opposition to V_{BB} to a maximum extent decreasing the forward-bias and therefore i_B and i_C. The minimum value of i_C is

$5 \text{ mA} - 500 \, \mu\text{A} = 4 \cdot 5 \text{ mA}$ and hence for this case
$$= 20 - (2 \times 10^3) \times (4 \cdot 5 \times 10^{-3}) = 20 - 9 = 11 \text{ V}.$$

The peak-to-peak voltage swing across R_L is 2 V, since v_{CE} varies between 9 and 11 V. The corresponding peak amplitude of the a.c. component of output voltage is $\frac{1}{2} \times 2 = 1$ V, and since the output voltage is sinusoidal (this is so because the input signal from the generator is given as a sine wave):

r.m.s. value of a.c. output voltage $= \dfrac{\text{peak amplitude}}{\sqrt{2}} = \dfrac{1}{\sqrt{2}} = 0 \cdot 707$ V.

12.3
Bias is required to give a selected quiescent operating point on the output characteristic

To use a transistor correctly so that it may operate as a linear amplifying element, we must bias the transistor so that it sits at an appropriate operating point. The quiescent operating point of a transistor connected in the common emitter mode is defined by the d.c. values of I_C, V_{CE}, and I_B, that is, the values of collector current, collector-emitter voltage, and base current flowing in the transistor electrodes with zero signal input. We select the quiescent operating point, usually denoted by Q, so as to satisfy the following:

(a) Q should be located in the 'straight line' region of the transistor characteristics.

(b) The position of Q should be such that the 'small-signal' changes in i_C and v_{CE} can be accommodated without running the transistor into the non-linear part of its characteristic, and also such that the maximum rated power dissipation of the transistor is not exceeded.

An example of a good choice of quiescent operating point Q for the transistor used in the common emitter amplifier circuit of fig. 12.4(a) is marked on its output characteristic in (b). The Q point selected is V_{CE} = 4 V, I_C = 21 mA, I_B = 200 μA, and lies well within the 'straight line' area of the characteristics.

The dashed curve shown in (b) is the locus of the maximum power dissipation P_C for the case P_C = 200 mW.

$$P_C = \text{collector current} \times \text{collector-base voltage}$$
$$= I_C V_{CB} \approx I_C V_{CE}$$

as $V_{CB} \approx V_{CE}$ (they differ by about 0·6 V for silicon, 0·2 V for germanium). Q must always lie under this curve so as to ensure that the maximum power dissipation allowed in the transistor is not exceeded. The approximate range of small-signal changes about Q is within the shaded box shown sketched in on the output characteristics of (b). Although the small-signal parameters of input and output resistance and current gain may vary, the transistor will operate approximately as a linear amplifying element within this 'box' region and the maximum power of dissipation in the transistor will not be exceeded.

Example 12.2

The output characteristics of the transistor used in the common emitter amplifier circuit of fig. 12.5(a) are given in (b). The collector supply voltage V_{CC} = 10 V. If a quiescent operating point of V_{CE} = 4 V, I_C = 5 mA is to be used, calculate (i) the value of base bias current required, (ii) the value of load resistance R_L required.

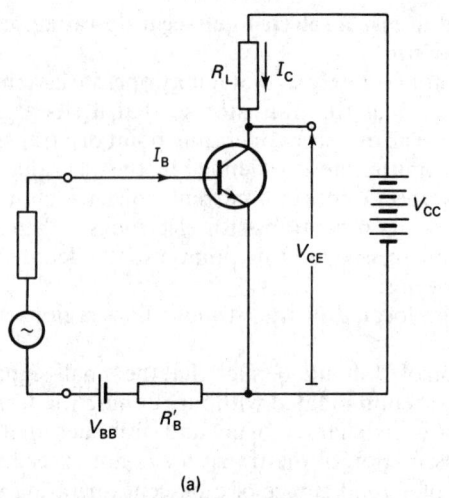

(a)

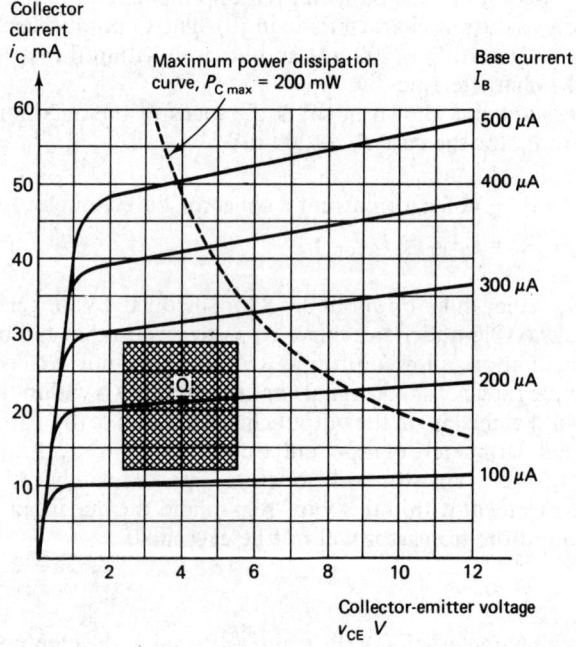

(b)

Fig. 12.4.

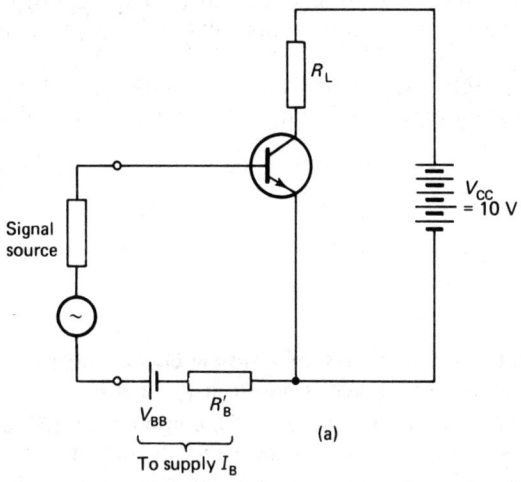

Signal source

R_L

V_{CC} = 10 V

V_{BB} R'_B

To supply I_B

(a)

Collector current
i_C mA

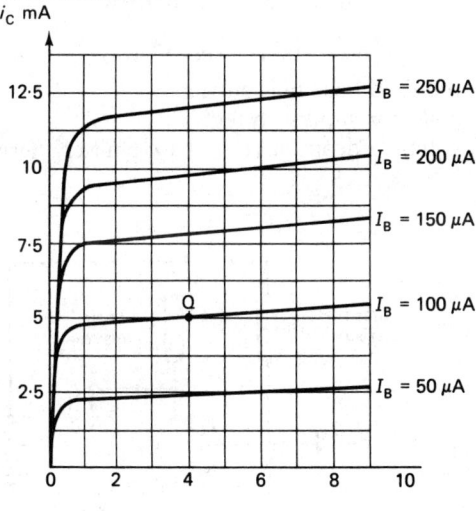

I_B = 250 μA

I_B = 200 μA

I_B = 150 μA

Q

I_B = 100 μA

I_B = 50 μA

Collector-emitter voltage v_{CE} V

(b) Output characteristic of transistor
used in circuit (a)

Fig. 12.5.

155

Solution

(i) The quiescent operating point Q corresponding to $V_{CE} = 4$ V, $I_C = 5$ mA is marked in on the output characteristics of fig. 12.5(b). Q lies on the $I_B = 100$ μA curve so we require this value of base bias current.

(ii) Now $V_{CC} = V_{CE} + R_L I_C$,

where in our problem $V_{CC} = 10$ V and we require $V_{CE} = 4$ V, $I_C = 5$ mA, so on substituting these values into the above equation we obtain:

$$10 = 4 + R_L \times 5 \times 10^{-3}$$

Hence $R_L \times 5 \times 10^{-3} = 10 - 4$

So $R_L = \dfrac{6}{5 \times 10^{-3}} = \dfrac{6000}{5} = 1200$ Ω

12.4
The circuit diagram and action of a simple bias arrangement consisting of a resistor connected between V_{CC} and base

Fig. 12.6 shows the circuit diagram of a common emitter amplifier in which the base bias current I_B is generated from the collector supply voltage V_{CC} by connecting the resistor R_B between V_{CC} and the base of the transistor. Let us calculate a design value for R_B to produce a given base bias current I_B. The voltage drop across R_B is

$$R_B I_B = V_{CC} - V_{BE}$$

where I_B = required base bias current,
    V_{CC} = collector supply voltage,
    V_{BE} = voltage drop across base-emitter terminals of the transistor.

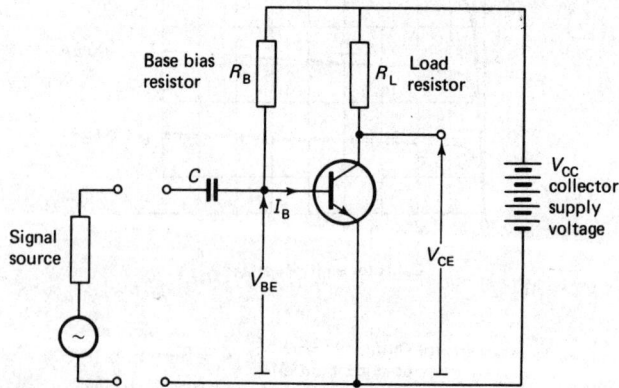

Fig. 12.6. A fixed-bias common-emitter amplifier circuit. R_B produces a fixed value of base bias current.

We may take $V_{BE} \approx 0.6$ V for a silicon transistor and $V_{BE} \approx 0.2$ V for a germanium transistor, since these are typical values for the forward-biased emitter junction of a transistor operating in the 'straight line' region (also known as active region) of its characteristics.

Thus $R_B = \dfrac{V_{CC} - V_{BE}}{I_B}$

$$= \frac{V_{EE} - 0.6}{I_B} \text{ for silicon, or } \frac{V_{CC} - 0.2}{I_B} \text{ for germanium.}$$

Since V_{CC} is usually much larger than V_{BE} we have a good 'engineering' approximation for R_B as

$$R_B \approx \frac{V_{CC}}{I_B}$$

R_B produces a given base bias current and does away with the necessity of having a separate bias supply in the base circuit. However, there is a serious practical drawback in using fixed bias current to establish the quiescent point. Even transistors of the same type have a large spread in their parameters and characteristics. These spreads inevitably occur during their production. For example, transistors allegedly the same may have values of h_{FE} (d.c. common emitter current gain) varying so widely that the maximum value may, in exceptional cases, be as much as 5 or more times the minimum. Thus if we were to design a fixed bias circuit with $I_B = 200 \ \mu A$ using a transistor of $h_{FE} = 100$ we would obtain a quiescent collector current of

$$I_C = h_{FE} I_B = 100 \times 200 \ \mu A = 20 \text{ mA}$$

If the transistor were to go faulty and we were to replace it by another of the same type but of $h_{FE} = 150$ then the quiescent current,

$$I_C = 150 \ I_B = 150 \times 200 \ \mu A = 30 \text{ mA.}$$

Thus a considerable change would be effected in the quiescent operating point and this might even be shifted into the non-linear region of the transistor. In the latter event the amplifier would no longer function.

In fig. 12.6 the signal is applied to the base circuit via a capacitor C. C blocks the flow of d.c. current and thus I_B flows via R_B into the base. If C were not present some of the current flowing through R_B would be shunted through the signal source itself and the base bias current would no longer be given by $(V_{CC} - V_{BE})/R_B$. C does, however, pass a.c. and thus the varying signal current is not impeded by C and enters the base.

Example 12.3
Suppose the transistor in the common emitter amplifier circuit of fig.

157

12.6 is silicon and it may be assumed that its base-emitter voltage V_{BE} = 0·6 V. If the collector supply voltage V_{CC} = 20 V and the quiescent operating point Q of the transistor is to be V_{CE} = 12 V, I_C = 2 mA when the base bias current I_B = 100 μA, calculate the values of R_L and R_B to set the transistor at this operating point Q. Calculate also the 'inferred' h_{FE} value of the transistor and the new quiescent point if we were to use a transistor of the same type but with an h_{FE} value of 40.

Solution
Since V_{CE} is required to be 12 V when I_C = 2 mA we have on applying

$$V_{CC} = V_{CE} + R_L I_C \text{ with } V_{CC} = 20V,$$
$$20 = 12 + R_L \times 2 \times 10^{-3}$$

So
$$R_L = \frac{20-12}{2 \times 10^{-3}} = 4000\Omega \text{ or } 4 \text{ k}\Omega.$$

Now we also require I_B to be set at 100 μA, so using

$$V_{CC} = R_B I_C + V_{BE} \text{ with } V_{BE} = 0·6V,$$

we have
$$20 = R_B \times 100 \times 10^{-6} + 0·6$$

So
$$R_B = \frac{20-0·6}{100 \times 10^{-6}} = \frac{19·4 \times 10^6}{100} = 19·4 \times 10^4 \Omega \text{ or } 194 \text{ k}\Omega.$$

Note that we normally take resistance values in design problems to the nearest preferred value since preferred values of resistance are the resistance values of practical resistors readily available. The nearest preferred value to 194 kΩ is 180 kΩ for $\pm 10\%$ tolerance resistors. Thus our engineering approximation,

$$R_B \approx \frac{V_{CC}}{I_B} = \frac{20}{100 \times 10^{-6}} = 20 \times 10^4 \ \Omega \text{ or } 200 \text{ k}\Omega$$

would lead to the selection of the same preferred value of 180 kΩ, the next nearest preferred value is 220 kΩ. The nearest $\pm 10\%$ preferred value for our load is 3·9 kΩ.

Note also that, although we did not state it in our problem, the h_{FE} value of the transistor is taken as

$$h_{FE} = \frac{I_C}{I_B} = \frac{2 \text{ mA}}{100 \text{ } \mu\text{A}} = \frac{2 \text{ mA}}{0·1 \text{ mA}} = 20.$$

If we were to replace the transistor with one having h_{FE} = 40 but kept R_B at the same value, then I_B would be unchanged but the new value of collector current would now be

$$I_C = h_{FE} I_B = 40 \times 100 \text{ } \mu\text{A} = 4000 \text{ } \mu\text{A or 4 mA}$$

and the corresponding quiescent value of collector-emitter voltage

$$V_{CE} = V_{CC} - R_L I_C = 20 - 4 \text{ k}\Omega \times 4 \text{ mA}$$
$$= 20 - 16 = 4 \text{ V}.$$

So the new quiescent point is $V_{CE} = 4$ V, $I_C = 4$ mA.

Example 12.4

Design a single resistor bias circuit for a common emitter amplifier such that the quiescent operating point is $V_{CE} = 8$ V, $I_C = 2$ mA. You are supplied with a fixed 15 V d.c. supply, and a silicon transistor with a d.c. current gain β_{DC} (or h_{FE}) value of 100 is used in the circuit. Take the emitter-base voltage $V_{BE} = 0.6$ V. Calculate also the value of load resistance that would be employed.

Solution
The circuit is exactly similar to that drawn in fig. 12.6, except that we make our fixed 15 V d.c. source the collector supply voltage, i.e. $V_{CC} = 15$ V. First let us calculate the value of load resistance R_L that would be used to give $V_{CE} = 8$ V, $I_C = 2$ mA the required quiescent operating point.

Since $\qquad V_{CC} = V_{CE} + R_L I_C$

we have $\qquad 15 = 8 + R_L \times 2 \times 10^{-3}$

So $\qquad R_L = \dfrac{15 - 8}{2 \times 10^{-3}} = 3.5 \times 10^3 \Omega$ or 3.5 kΩ

(nearest $\pm 10\%$ preferred value is 3.3 kΩ).

Now we are given

$$\beta_{DC} \equiv h_{FE} = \frac{I_C}{I_B} = 100$$

So the base current required to make $I_C = 2$ mA is

$$I_B = \frac{I_C}{\beta_{DC}} = \frac{2}{100} \, mA = 0.02 \text{ mA or } 20 \text{ } \mu A$$

We have also

$$R_B I_B = V_{CC} - V_{BE}$$

So on substituting $I_B = 20$ μA, $V_{CC} = 15$ V, $V_{BE} = 0.6$, we obtain the required value of resistance,

$$R_B = \frac{V_{CC} - V_{BE}}{I_B} = \frac{15 - 0.6}{20 \times 10^{-6}} = 0.72 \times 10^6 = 720 \text{ k}\Omega$$

(nearest $\pm 10\%$ preferred value is 680 kΩ).

12.5
The effect of a small sinusoidal current input on the quiescent condition

Fig. 12.7 shows a fixed bias common emitter amplifier, a typical quiescent operating point Q, $V_{CE} = 8$ V, $I_C = 20$ mA, $I_B = 200$ μA and the corresponding quiescent waveforms.

Let us now consider the effect on the quiescent values of collector current and collector-emitter voltage when a small sinusoidal current is fed into the amplifier input. Suppose the signal source in fig. 12.8(a) produces a small sinusoidal input base current i_b. The total current flowing into the base of the transistor is then

$$i_B = \text{signal current} + \text{base bias current} = i_b + I_B,$$

The corresponding waveforms are sketched in fig. 12.8(b).

The effect of the signal base current i_b is to cause the total collector current i_C to vary about its quiescent point value I_C:

$$i_C = \text{amplified signal current} + \text{quiescent current} = i_c + I_C$$

where $i_c \approx \beta i_b$ and β is the small-signal short-circuit common emitter current gain. The approximation sign is used here because the true current gain depends to some extent on the relative values of the load resistor R_L and the output resistance R_0 of the transistor. However, if $R_0 \gg R_L$, $i_c = \beta i_b$ to a good degree of approximation. Sketches of the total, a.c., and quiescent components of collector current are drawn in fig. 12.8(c). The a.c. variation in collector current is sinusoidal and a faithful replica of the input signal current to the base.

The total collector-emitter voltage v_{CE} also varies sinusoidally about its quiescent point value V_{CE}:

$$v_{CE} = \text{a.c. signal voltage across } R_L$$
$$+ \text{quiescent collector-emitter voltage}$$

$$= v_{ce} + V_{CE}$$

Now $v_{CE} = V_{CC} - R_L i_C$
$$= V_{CC} - R_L(I_C + i_c) = V_{CC} - R_L I_C - R_L i_c$$

But $V_{CC} - R_L I_C = V_{CE}$ (the quiescent collector-emitter voltage).

So $$v_{CE} = V_{CE} - R_L i_c.$$

showing that the small-signal variation across the collector-emitter terminals of the transistor is

$$v_{ce} = -R_L i_c$$

The $-$ sign indicates that v_{ce} decreases as i_c increases and vice versa. This is exactly as we would expect since the greater the collector current,

160

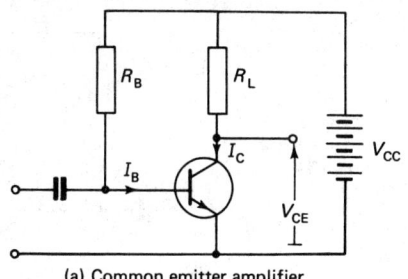

(a) Common emitter amplifier
with zero input signal

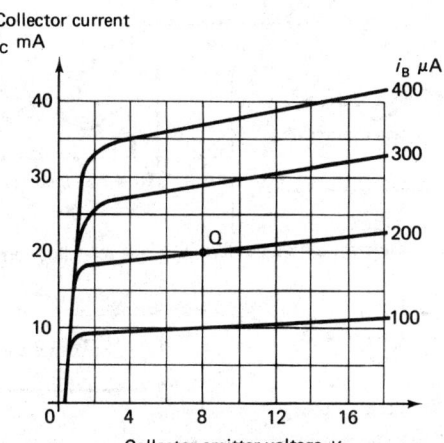

Collector current
i_C mA

(b) Output characteristics showing
quiescent operating point Q:
$V_{CE} = 8$ V, $I_C = 20$ mA, $I_B = 200$ μA

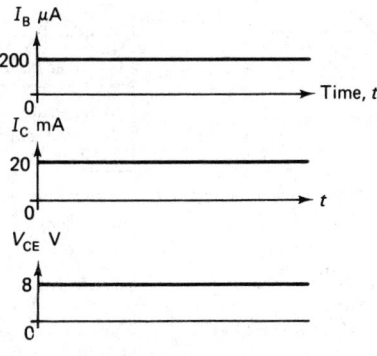

(c) Quiescent waveforms

Fig. 12.7.

161

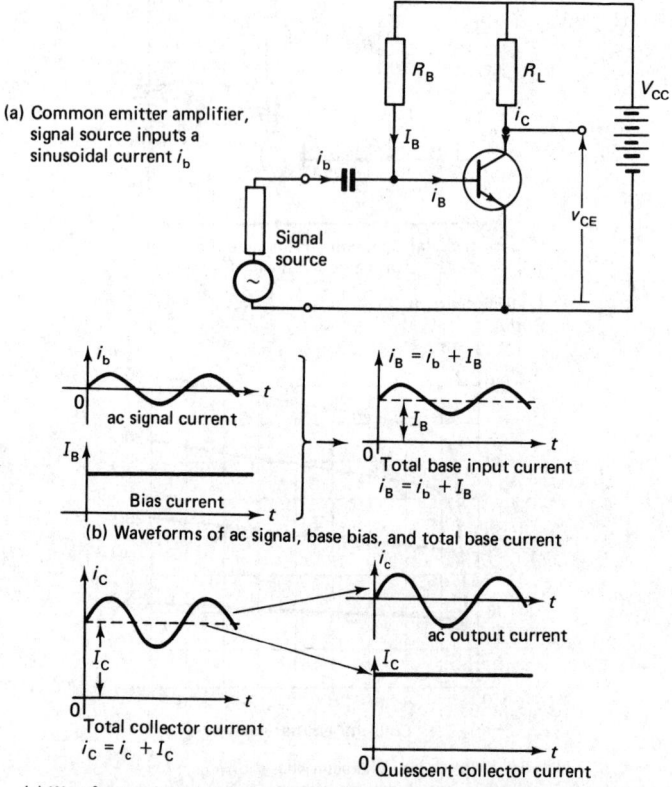

(a) Common emitter amplifier, signal source inputs a sinusoidal current i_b

Signal source

(b) Waveforms of ac signal, base bias, and total base current

ac signal current

I_B Bias current

Total base input current $i_B = i_b + I_B$

(c) Waveforms of total and ac and dc (quiescent) components of collector current

Total collector current $i_C = i_c + I_C$

ac output current

Quiescent collector current

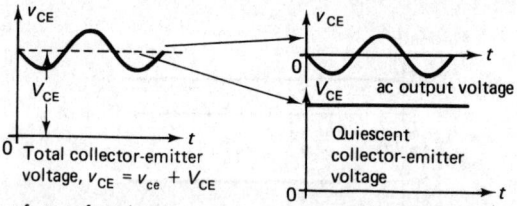

(d) Waveforms of total and ac and dc components of collector-emitter voltage

Total collector-emitter voltage, $v_{CE} = v_{ce} + V_{CE}$

ac output voltage

Quiescent collector-emitter voltage

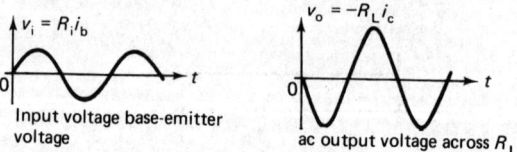

$v_i = R_i i_b$

Input voltage base-emitter voltage

$v_o = -R_L i_c$

ac output voltage across R_L

(e) Input voltage $v_i = v_{be}$ and output voltage $v_o = v_{ce}$ waveforms

Fig. 12.8.

162

the greater the voltage drop across R_L and hence the less the value of v_{CE}. The waveforms of total, quiescent, and signal collector-emitter voltage are drawn in fig. 12.8(d).

The a.c. input signal voltage between base-emitter of the transistor is

$$v_i = v_{be} = R_i i_b \text{ where } R_i = \text{input resistance looking into base-emitter}$$
$$\text{terminals of the transistor}$$

The a.c. output voltage is

$$v_o = v_{ce} = -R_L i_c$$

The corresponding a.c. input and output voltage waveforms are shown in fig. 12.8(e). Note that the a.c. output voltage taken between collector-emitter terminals is inverted with respect to the signal input voltage generated between base-emitter terminals.

The above discussion and waveform sketches show that sinusoidal input current causes sinusoidal variations of collector current and collector-emitter voltage about their quiescent value. A small change in base current causes a much greater change in collector current. Thus, current amplification occurs. Voltage amplification will also occur: an approximate value for the voltage gain is

$$A_v = \frac{v_o}{v_i} = \frac{-R_L i_c}{R_i i_b} \approx \frac{-R_L}{R_i} \beta$$

as the current gain $\dfrac{i_c}{i_b} \approx \beta$ (provided $R_o \gg R_L$).

12.6
Voltage phase inversion occurs between the input and output signals in a common emitter amplifier

The common emitter amplifier is an example of an amplifier in which voltage phase inversion occurs. Voltage phase inversion is shown diagrammatically in fig. 12.9(a). The sinusoidal output voltage is inverted with respect to the input sine wave signal voltage. Thus if

$$\text{the input voltage } v_i = V_i \sin \omega t$$

where $\omega = 2\pi f$ is the radian or angular frequency, units radian per second,

f = frequency in hertz,

then the output voltage,

$$v_o = -A v_i$$
$$= -A V_i \sin \omega t$$

163

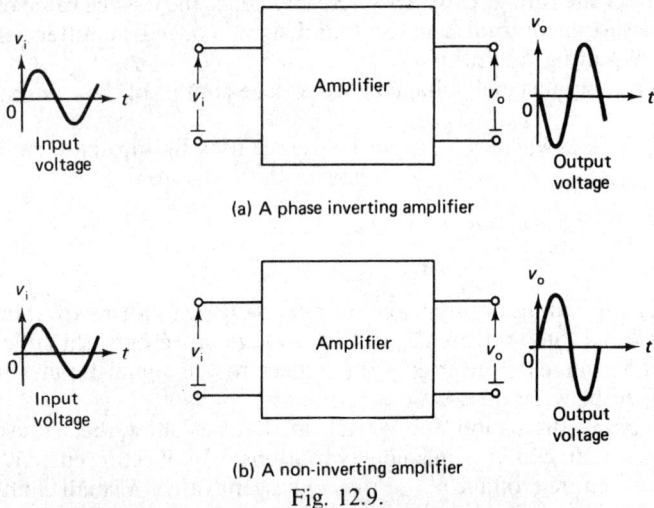

(a) A phase inverting amplifier

(b) A non-inverting amplifier

Fig. 12.9.

where A = positive number = magnitude of voltage gain of the amplifier. Now since $-\sin \omega t = \sin (\omega t + \pi)$

$$v_o = A V_i \sin (\omega t + \pi)$$

that is, the output voltage has a phase shift of π radians (180°) with respect to the input signal. Phase inversion or π phase shift (which is exactly the same thing) occurs between output and input voltages in the common emitter amplifier. No phase inversion of voltage occurs in the common base and common collector amplifier. The a.c. voltage output is in phase with the a.c. input, as shown in fig. 12.9(b).

Section 13: *The expected learning outcome of this section is that the student should be able to construct and use a d.c. load line on transistor characteristics.*

13.1
The construction of the load line on a set of output characteristics of a common emitter amplifier for a stated value of load resistance

The static output characteristics of a transistor connected in the common emitter modes are families of curves of I_C versus V_{CE} for a series of fixed values of I_B. They define the output behaviour of the transistor when different values of current are fed in at the base. We may use the output characteristics together with a load condition which relates the

164

collector-emitter voltage to collector current flowing in the load resistor to investigate the behaviour of an amplifier, or, as we shall see later in Section 21, transistor switching circuits.

However, it should be stressed that although manufacturers publish typical sets of characteristics for given types of transistors, these characteristics are average ones. The typical characteristics represent the average expected behaviour of a given type of transistor. The actual set of output characteristics for a given transistor can only be obtained by practical measurements (see, for example, Section 6). Although the general form of characteristics will be similar for similar type transistors, the actual quantitative values at any given point on a curve will vary. It should also be noted that the transistor characteristics are very dependent on temperature. Thus, in designing a transistor amplifier and investigating its performance using the static characteristics, we should always take care in making absolute quantitative predictions. However, the use of graphical techniques employing the transistor output characteristics gives an excellent insight into the voltage and current changes taking place in the transistor in the amplification process.

Let us now consider the basic common emitter amplifier circuit in fig. 13.1(a). The output characteristics of the transistor used are drawn in (b). The total collector current i_C and the total collector-emitter voltage v_{CE} are related by

$$v_{CE} = V_{CC} - R_L i_C$$

where V_{CC} = collector supply voltage,
$\quad\quad R_L$ = load resistor in collector circuit.

This equation is known as the load line equation since it relates the collector-emitter voltage to the current (collector current) flowing in the load. The load line equation plotted on a $v_{CE} - i_C$ graph is a straight line. Thus, if we plot this line on the output characteristics of our transistor, we may investigate the behaviour of the amplifier, as we have the transistor output current and voltage specified in the form of the load line equation, and the transistor behaviour itself specified implicitly by the output characteristics.

We shall now construct the load line for the amplifier of fig. 13.1(a) given the following data:
collector supply voltage V_{CC} = 9 V, load resistance R_L = 1500 Ω (1·5 kΩ). To draw in the load line it is sufficient to plot two points on the output characteristics and to draw a straight line through these two points. The two simplest points to calculate are:
(1) When v_{CE} = 0, we have

$$0 = V_{CC} - R_L i_C$$

So $\quad\quad i_C = \dfrac{V_{CC}}{R_L} = \dfrac{9}{1500} = 0{\cdot}006$ A = 6 mA in our case.

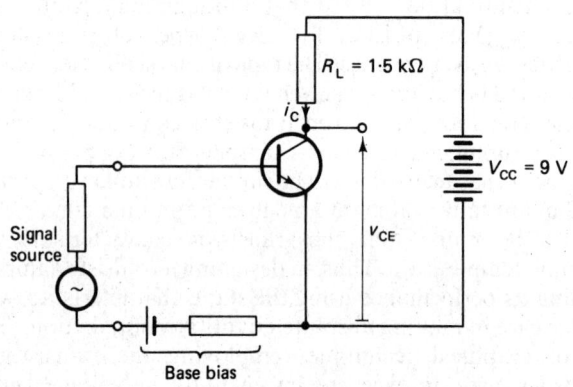

(a) Basic common emitter amplifier circuit

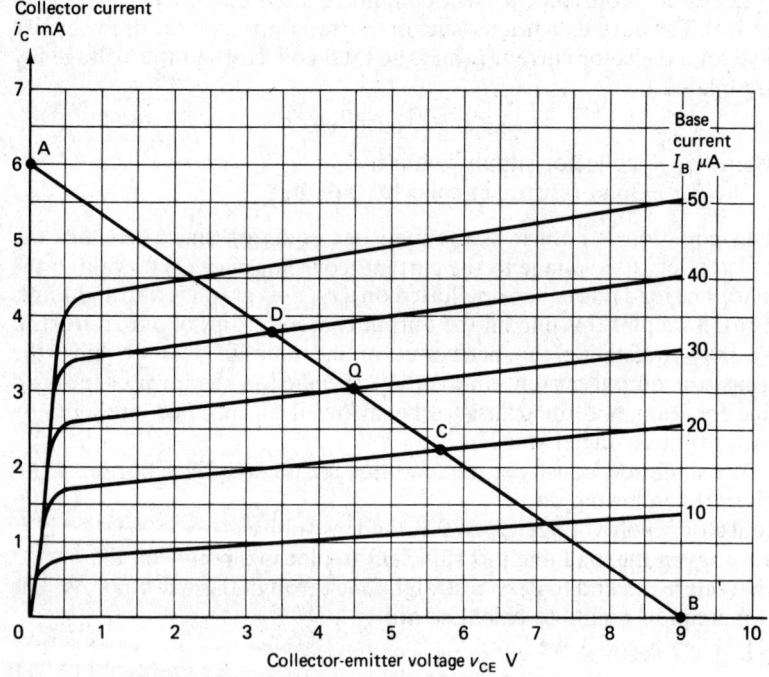

(b) Output characteristics onto which the load line $v_{CE} = V_{CC} - R_L i_C$ is drawn for the case: $R_L = 1500 \, \Omega$, $V_{CC} = 9$ V

Fig. 13.1.

We therefore mark in the point $v_{CE} = 0$, $i = 6$ mA. This point is shown as A in fig. 13.1(b).

(2) When $i_C = 0$, we have

$$v_{CE} = V_{CC} - R_L \times 0 = V_{CC}$$

We therefore mark in the point $v_{CE} = 9$ V, $i_C = 0$ and this is shown marked as B in fig. 13.1(b).

Finally, we join the points A and B by a straight line, thus constructing the load line for the case $R_L = 1500$ Ω, $V_{CC} = 9$ V.

We may now use the load line and output characteristics to calculate the values of v_{CE} and i_C for a given value of base bias current. For example:

(i) If $I_B = 20$ μA, the values of v_{CE} and i_C are given by the point of intersection of the load line with the output characteristic corresponding to $I_B = 20$ μA. For this case, the load line intersects the $I_B = 20$ μA characteristic at point C corresponding to $v_{CE} = 5.7$ V, $i_C = 2.2$ mA.

(ii) If $I_B = 40$ μA then $v_{CE} = 3.3$ V, $i_C = 3.8$ mA (point D).

(iii) If the base bias current $I_B = 30$ μA and an input signal causes the base current to swing by ± 10 μA, i.e. varies i_B between 40 μA and 20 μA, then the operating point of the transistor moves about its quiescent operating point Q: $I_B = 30$ μA, $V_{CE} = 4.5$ V, $I_C = 3.0$ mA, to $i_B = 40$ μA (point D) where $v_{CE} = 3.3$ V, $i_C = 3.8$ mA, and $i_B = 20$ μA (point C) where $v_{CE} = 5.7$ V, $i_C = 2.2$ mA. Note that the changes produced by the signal are:

Q to D: change in $v_{CE} = 4.5 - 3.3 = 1.2$ V
change in $i_C = 3.0 - 3.8 = -0.8$ mA
Q to C: change in $v_{CF} = 4.5 - 5.7 = -1.2$ V
change in $i_C = 3.0 - 2.2 = 0.8$ mA

showing, to the limit we can read off values accurately from our graph, that a signal changing the base current by ± 10 μA about $I_B = 20$ μA produces symmetrical changes in output voltage and current about the quiescent operating point Q.

13.2
Estimation of the r.m.s. voltage output from the load line for given quiescent conditions and a given input signal

The collector supply voltage and the load resistor for the common emitter amplifier circuit of fig. 13.2(a) are

$$v_{CC} = 10 \text{ V}, R_L = 2 \text{ k}\Omega$$

and thus the load line equation

$$v_{CE} = V_{CC} - R_L i_C$$

becomes $v_{CE} = 10 - 2000 i_C$ or $v_{CE} = 10 - 2i_C$ when i_C is quoted in mA. The load line is drawn in on the output characteristic of the transistor

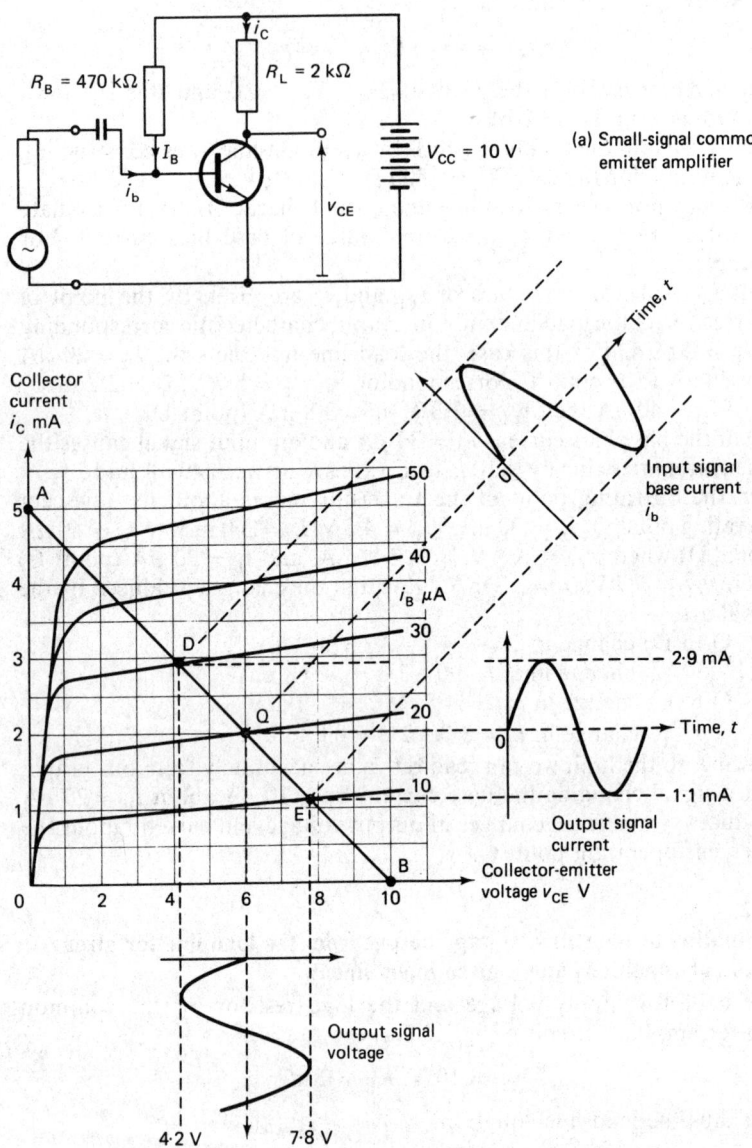

(a) Small-signal common emitter amplifier

(b) Load line for amplifier of (a) superimposed on the output characteristics: Q point is V_{CE} = 6 V, I_C = 2 mA when base bias current I_B = 20 μA; waveforms produced by a sinusoidal signal input current of peak amplitude 10 μA are also shown

Fig. 13.2.

used in the amplifier circuit in fig. 13.2(b), by joining points A and B, where

point A: $v_{CE} = 0$, $i_C = 10/2000$ A $= 5$ mA

point B: $i_C = 0$, $v_{CE} = V_{CC} = 10$ V.

The base bias resistor is 470 kΩ and is designed so that the base bias current $I_B = 20$ μA. We can check this result using the information given in Sub-section 12.4. Assuming that the transistor is silicon we have:

$$R_B = \frac{V_{CC} - 0.6}{I_B} = \frac{10 - 0.6}{20 \times 10^{-6}} = \frac{9.4 \times 10^6}{20} = 470 \text{ kΩ}.$$

The $I_B = 20$ μA characteristic intersects the load line at Q.

Thus, the quiescent operating point Q is $V_{CE} = 6$ V, $I_C = 2$ mA.

Suppose now our input signal injects a sinusoidal current of peak amplitude $I_b = 10$ μA into the base. This signal will swing the operating point of the transistor along the load line between:

$I_B + I_b = 20 + 10 = 30$ μA, the corresponding point being the point of intersection of the load line with the $i_B = 30$ μA characteristic, i.e. point D where $v_{CE} = 4.2$ V, $i_C = 2.9$ mA,

and $I_B - I_b = 20 - 10 = 10$ μA, the corresponding point being point E where $v_{CE} = 7.8$ V, $i_C = 1.1$ mA.

Thus the peak-to-peak swing in output collector-emitter voltage, that is between points E and D, is

$$7.8 - 4.2 = 3.6 \text{ V}$$

and the peak amplitude of the a.c. component of output voltage, that is from Q to D or Q to E, is

$$V_{CE} - 4.2 = 6 - 4.2 = 1.8 \text{ V (quiescent value, } V_{CE} = 6 \text{ V)}$$

or $7.8 - V_{CE} = 7.8 - 6 = 1.8$ V

which is half the peak-to-peak swing, exactly as we should expect for linear operation of the transistor.

The r.m.s. amplitude of the a.c. output voltage is

$$\text{r.m.s. output voltage} = \frac{\text{peak amplitude}}{\sqrt{2}} = \frac{1.8}{\sqrt{2}} = \frac{1.8}{1.414} = 1.27 \text{ V}.$$

Note that it is common practice to quote r.m.s. values rather than peak amplitudes when dealing with a.c. waveforms. One major reason for this is that the power dissipated in a resistor R by an a.c. voltage or current of r.m.s. values V_{rms}, I_{rms} is given by

$$P = \frac{V_{rms}^2}{R} \quad \text{or} \quad I_{rms}^2 R \text{ watts}$$

which compares directly with d.c. work. In the case of our amplifier the

169

power developed in the load R_L by the a.c. component of output voltage is

$$P = \frac{V_{rms}^2}{R_L}, \text{ and as } V_{rms} = 1.27 \text{ V}, R_L = 2000 \text{ } \Omega,$$

$$P = \frac{1.27^2}{2000} = 0.81 \text{ mW}.$$

If, out of interest, we wanted to calculate the total power dissipated in R_L we must also include the d.c. or quiescent component,

$$P_{dc} = \frac{(V_{CC} - V_{CE})^2}{R_L} = \frac{(10-6)^2}{2000} = 8 \text{ mW}$$

and so the total power dissipated in R_L is

$$P_T = P_{dc} + P = 8 + 0.81 = 8.81 \text{ mW}.$$

Also shown in fig. 13.2(b) are the waveforms of output collector-emitter voltage and collector current produced by the sinusoidal base signal current of peak amplitude 10 μA.

13.3
The determination of the voltage gain A_v from the static characteristics assuming a given input resistance
We continue our analysis of the common emitter amplifier given in fig. 13.2 by calculating its voltage gain. To be able to do this we must be given the small-signal input resistance R_i of the transistor appropriate to our quiescent operating point. Suppose this is specified as $R_i = 1.5$ kΩ, then the peak a.c. input voltage amplitude between base and emitter is

$$V_{be} = R_i \times \text{peak a.c. input current}$$
$$= R_i I_b = 1500 \times 10 \times 10^{-6} = 0.015 \text{ V or } 15 \text{ mV},$$

since the peak a.c. input current, $I_b = 10$ μA $= 10 \times 10^{-6}$ A.
Note that if we work with r.m.s. values the corresponding r.m.s. value of I_b and V_{be} are respectively $10/\sqrt{2} = 7.07$ μA and $15/\sqrt{2} = 10.6$ mV.

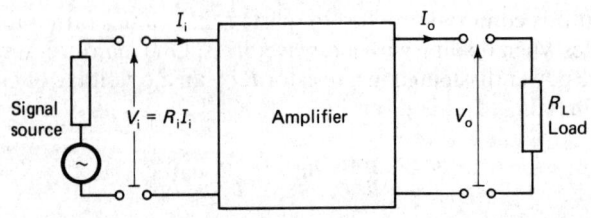

Fig. 13.3.

Now the small-signal voltage gain A_v of an amplifier (see fig. 13.3) is defined as

$$A_v = \frac{\text{a.c. output voltage amplitude}}{\text{a.c. input voltage amplitude}} = \frac{V_o}{V_i}$$

where the amplitudes may be quoted as peak or r.m.s. values. Either may be used but we must be consistent.

We found in the last sub-section that the r.m.s. a.c. output voltage amplitude generated across R_L (the same r.m.s. amplitude is generated across the collector-emitter terminals of the transistor) is

$$V_{ce} = 1 \cdot 27 \text{ V (r.m.s.) so } V_o = V_{ce} = 1 \cdot 27 \text{ V (r.m.s.)}$$

and we calculated above that the input voltage,

$$V_{be} = V_i = 10 \cdot 6 \text{ mV} = 0 \cdot 0106 \text{ V}.$$

Thus $A_v = \dfrac{1 \cdot 27}{0 \cdot 0106} = 120$ or working with peak values $A_v = \dfrac{1 \cdot 8}{0 \cdot 015}$ $= 120$.

To be absolutely correct the voltage gain of a common emitter amplifier is negative, the minus sign being included to take into account the fact that phase inversion occurs, so we should state:

$$A_v = -120$$

13.4
The determination of the current gain A_i from the static characteristics

The small-signal current gain of an amplifier (see again fig. 13.3) is defined as

$$A_i = \frac{\text{a.c. output current amplitude}}{\text{a.c. input current amplitude}} = \frac{I_o{}^*}{I_i}$$

and for the case of the common emitter amplifier we have:

$I_o = I_c$ = peak or r.m.s. amplitude of a.c. collector current,
$I_i = I_b$ = peak or r.m.s. amplitude of a.c. signal base current.

Thus on referring to fig. 13.2(b) we can read off the following and hence calculate A_i for the common emitter amplifier of (a):

*A small point as regards convention: in general 2-port theory we regard an amplifier as a 'black-box' with 2 pairs of terminals (the term port is used to describe a pair of terminals at which we drive (input) or take an output from a device). The convention adopted in 2-port theory is to take current directions as flowing into the ports. Thus in 2-port theory (you will meet this at a later stage where the transistor is defined by h-parameters) A_i would be defined with respect to current directions directed into the amplifier, so the definition given above would differ by a minus sign, but would have the same numerical value.

171

the peak-to-peak variation of collector current (point E to D) is from 1·1 mA to 2·9 mA, so the peak and r.m.s. amplitudes of the a.c. component of collector current are

$$I_c = \tfrac{1}{2}(2·9 - 1·1) = \tfrac{1}{2} \times 1·8 = 0·9 \text{ mA (peak)}$$
$$= 0·9/\sqrt{2} = 0·64 \text{ mA (r.m.s.)}$$

produced by the base signal current of amplitude,

$$I_b = 10 \ \mu A \text{ (peak) or } 10/\sqrt{2} = 7·07 \ \mu A \text{ (r.m.s.).}$$

So the current gain of the amplifier,

$$A_i = \frac{I_c}{I_b} = \frac{0·9 \text{ mA}}{10 \ \mu A} = \frac{0·9 \times 10^{-3}}{10 \times 10^{-6}} = 90$$

using peak values for I_c and I_b. Naturally if we used r.m.s. values we would obtain the same result.

13.5
The calculation of the power gain A_P of an amplifier
The power gain of an amplifier is defined as

$$A_p = \frac{\text{a.c. (signal) output power}}{\text{a.c. (signal) input power to amplifier}}$$

Now the a.c. output power of an amplifier,

$$P_o = V_o I_o \text{ watts}$$

where V_o = r.m.s. amplitude of a.c. output voltage,
I_o = r.m.s. amplitude of a.c. output current.

And the a.c. input power to the amplifier from the signal source,

$$P_i = V_i I_i \text{ watts}$$

where V_i = r.m.s. amplitude of a.c. input voltage,
I_i = r.m.s. amplitude of a.c. input current.

So we have,

$$A_p = \frac{V_o I_o}{V_i I_i}$$

$$= \frac{V_o}{V_i} \times \frac{I_o}{I_i} = A_v A_i$$

where A_v = voltage gain, A_i = current gain of the amplifier.
Thus using the values previously calculated in Sub-section 13.3 for A_v and 13.4 for A_i we have for the power gain of the common emitter amplifier of fig. 13.2(a):

$$A_P = A_V \times A_i = 120 \times 90 = 10\,800$$

Now electronic engineers normally quote power gain in units known as decibels, abbreviation dB. The power gain expressed in decibels is

$$\text{power gain in dB} = 10 \log_{10} \frac{P_o}{P_i} = 10 \log_{10} A_P \text{ dB}$$

So expressing $A_P = 10\,800$ (a number ratio) in decibels (a logarithmic ratio) we have,

$$\text{Power gain} = 10 \log_{10} 10\,800 = 10 \times 4.0334 = 40.3 \text{ dB}$$

Let us give a little more explanation to the decibel. Decibels are sub-multiple units of the bel, abbreviation b. The bel is the unit used to express the logarithnic ratio of two powers, e.g. $\log_{10} \frac{P_o}{P_i}$ b; deci- is the decimal prefix for one-tenth, thus 1 bel = 10 decibels. Hence the reason why the 10 appears in the power gain expressed in dB. We work in dB rather than number ratios for two main reasons: firstly because gains expressed as number ratios may be very large and thus by taking the log

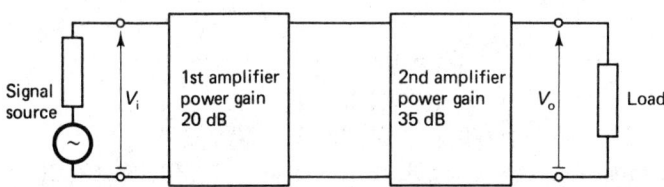

Fig. 13.4. Two amplifiers in cascade: total power gain = 20+30 = 55dB.

we will be dealing with lower values; secondly and more importantly we can add log gains to determine the total gain of amplifiers joined in cascade. For example the total power gain of the two amplifiers in cascade shown in fig. 13.4 is

$$A_P = \frac{V_o}{V_i} = A_{P_1} \times A_{P_2}$$

where A_{P_1} and A_{P_2} are the individual amplifier gains.
The total power gain expressed in dB is

$$10 \log A_P = 10 \log A_{P_1} + 10 \log A_{P_2} \text{ dB}$$

So if the gain of the first amplifier is 20 dB ($A_{P_1} = 100$) and the second 35 dB ($A_{P_2} = 3162$) the power gain,

$$10 \log A_P = 20+35 = 55 \text{ dB}.$$

Checking this result using number ratios we have

$$A_P = 100 \times 3162 = 316\,200$$

$$10 \log 316\,200 = 10 \times 5{\cdot}50 = 55 \text{ dB}.$$

We may also introduce into the power gain expression the input resistance R_i to the amplifier and the load resistance R_L in which the output power is developed:

The signal power input to the amplifier, $P_i = V_i I_i$
$$= V_i^2/R_i = I_i^2 R_i$$

since $V_i = R_i I_i$.
The output power, $P_o = V_o I_o = V_o^2/R_L = I_o^2 R_L$
since $V_o = R_L I_o$.
Thus the power gain,

$$A_p = \frac{P_o}{P_i} = \frac{V_o^2}{R_L} \bigg/ \frac{V_i^2}{R_i} = \frac{V_o^2}{V_i^2} \times \frac{R_i}{R_L} = A_v^2 \frac{R_i}{R_L}$$

$$\text{or} \quad A_p = I_o^2 R_L / I_i^2 R_i = \frac{I_o^2}{I_i^2} \times \frac{R_L}{R_i} = A_i^2 \frac{R_L}{R_i}$$

So a special case arises if $R_i = R_L$ when we have

$$A_P = A_V^2 = A_i^2$$

and the power gain in dB is then

$$10 \log A_p = 10 \log A_v^2 = 20 \log A_v \text{ dB}$$
$$\text{or } 10 \log A_p = 10 \log A_i^2 = 20 \log A_i \text{ dB}$$

(Remember $\log A^2 = 2 \log A$, so $10 \log A^2 = 2 \times 10 \log A = 20 \log A$). Sometimes the voltage gain of an amplifier is quoted in dB as

$$\text{voltage gain dB} = 20 \log A_V = 20 \, log \, \frac{V_o}{V_i} \text{ dB}$$

It is very important to understand that the voltage gain expressed in dB is $20 \times \log (V_o/V_i)$ and is not equal to the power gain unless $R_i = R_L$. For example, if the voltage gain of an amplifier is 30 dB then

$$30 = 20 \log A_V$$

So $\qquad \log A_V = \dfrac{30}{20} = 1{\cdot}5$ and on taking antilogs

$$A_V = 31{\cdot}6 = V_o/V_i$$

Further, if we are given that the input resistance $R_i = 1 \text{ k}\Omega$ and the load resistance $R_L = 10 \text{ k}\Omega$ we have

$$A_P = A_V^2 \times \frac{R_i}{R_L} = 31{\cdot}6^2 \times \frac{1}{10} \approx 100.$$

So power gain in dB, $10 \log A_P = 10 \, log \, 100 = 10 \times 2 = 20 \text{ dB}$.

Example 13.1

(a) The power gain of an amplifier is quoted as 36 dB. Calculate the ratio of signal output to input power. If the r.m.s. input current to the amplifier is 100 μA and the input resistance is 2 kΩ, calculate the signal input and output powers.

(b) The voltage gain of an amplifier $A_V = 60$. The input resistance R_i to the amplifier is 1·2 kΩ and the load resistance R_L is 5 kΩ. If the r.m.s. input current to the amplifier is 20 μA, calculate (i) the a.c. r.m.s. input and output (load) voltages, (ii) the current gain of the amplifier, (iii) the power gain of the amplifier in dB.

Solution

(a) The power gain quoted in dB means

$$10 \log A_P = 36 \text{ dB} \quad \text{or} \quad \log A_P = 3·6.$$

So
$$A_P = \frac{P_o}{P_i} = 3981 \text{ (on taking antilogs)}$$

where P_o = output power, P_i = input power.
Thus the ratio $P_o:P_i = 3981:1 \approx 4000:1$.
The signal input power,

$$P_i = I_i^2 R_i$$

where I_i = r.m.s. signal input current = 100 μA,
R_i = input resistance = 2 kΩ.
Thus $P_i = (100 \times 10^{-6})^2 \times 2000 = 10^{-8} \times 2000$
$\qquad = 20 \times 10^{-6}$ W or 20 μW.
And the signal output power,

$$P_o = A_P P_i \approx 4000 \times 20 \ \mu\text{W} = 80 \text{ mW}.$$

(b) (i) The r.m.s. input voltage,

$$V_i = R_i I_i$$

where I_i = r.m.s. input current = 20 μA,
R_i = input resistance = 1·2 kΩ.
Thus $V_i = 1200 \times 20 \times 10^{-6}$ V = 24 mV.
And the r.m.s. output voltage,

$$V_o = A_V V_i = 60 \times 24 = 1440 \text{ mV} = 1·44 \text{ V}$$

(ii) Since the load resistance $R_L = 5$ kΩ and the r.m.s. output voltage V_o = 1·44 V, the r.m.s. output current,

$$I_o = \frac{V_o}{R_L} = \frac{1·44}{5 \text{ k}\Omega} = 0·288 \text{ mA} = 288 \ \mu\text{A}.$$

175

So the current gain,

$$A_i = \frac{I_o}{I_i} = \frac{288 \ \mu A}{20 \ \mu A} = 14\cdot4.$$

(iii) The power gain,

$$A_P = A_V A_i = 60 \times 14\cdot4 = 864$$

and $10 \log A_P = 10 \log 864 = 10 \times 2\cdot9365 = 29\cdot4 \ dB$

Check: Input power, $P_i = I_i^2 R_i = (20 \times 10^{-6})^2 \times 1200 = 0\cdot48 \ \mu W.$
Output power, $P_o = I_o^2 R_L = (288 \times 10^{-6})^2 \times 5000 = 414\cdot7$ $\mu W.$

So $A_P = \dfrac{P_o}{P_i} = \dfrac{414\cdot7}{0\cdot48} = 864.$

Example 13.2

The output characteristics of the transistor used in the common emitter amplifier circuit of fig. 13.5(a) are drawn in (b). The collector supply voltage $V_{CC} = 8 \ V$, the load resistance $R_L = 150 \ \Omega$, and the bias resistor R_B is adjusted to give a quiescent base current of $I_B = 300 \ \mu A$.

Construct the load line and hence determine the values of collector current and voltage at the quiescent operating point.

If an a.c. signal of peak current amplitude of $100 \ \mu A$ is fed into the base and the small-signal base-emitter input resistance $R_i = 1 \ k\Omega$, determine (i) the peak and r.m.s. values of a.c. output voltage and current, (ii) the voltage, current, and power gains of the amplifier, (iii) the a.c. power output in the load resistor.

Solution

The load line equation,

$$v_{CE} = V_{CC} - R_L i_C$$
$$= 8 - 150 \ i_C$$

as $V_{CC} = 8 \ V$, $R_L = 150 \ \Omega$ in our problem.
The load line is constructed by joining:

point A: $v_{CE} = 0$, $i_C = \dfrac{8}{150} \ A = 53\cdot3 \ mA,$

and point B: $v_{CE} = 8 \ V$, $i_C = 0,$

by a straight line.

The quiescent operating point is given by the point of intersection of

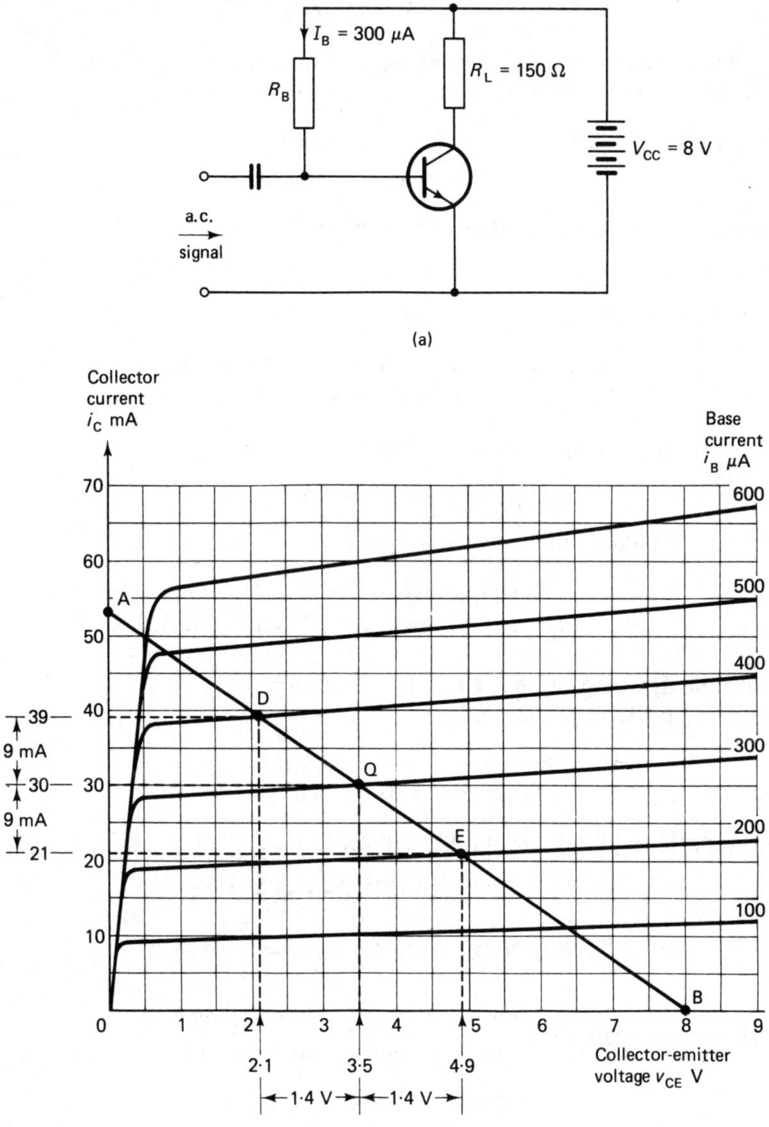

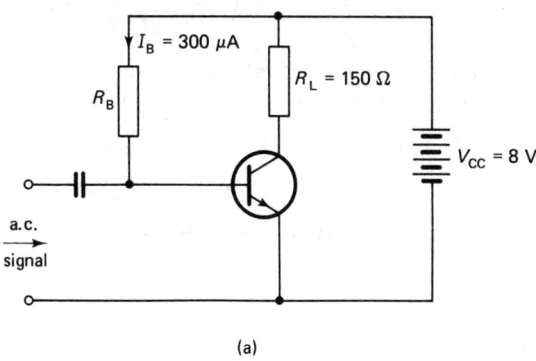

I_B = 300 μA
R_B
R_L = 150 Ω
V_{CC} = 8 V
a.c.
signal
(a)
Collector
current
i_C mA
Base
current
i_B μA
70
60
50
40
30
20
10
0
1
2
3
4
5
6
7
8
9
600
500
400
300
200
100
A
D
Q
E
B
39
30
21
9 mA
9 mA
2·1
3·5
4·9
1·4 V
1·4 V
Collector-emitter
voltage v_{CE} V
(b)

Fig. 13.5.

the load line with the $I_B = 300\ \mu A$ characteristic. Thus from fig. 13.5(b), where the quiescent point is marked Q, we obtain

$$i_C = I_C = 30\,\text{mA},\ v_{CE} = V_{CE} = 3\cdot5\ \text{V}.$$

The signal input current of peak amplitude $100\ \mu A$ moves the operating point about Q along the load line between:

point D: $i_B = I_B + 100 = 300 + 100 = 400\ \mu A,$

where $\qquad\qquad\qquad v_{CE} = 2\cdot1\ \text{V}, i_C = 39\ \text{mA},$

and point E: $i_B = I_B - 100 = 300 - 100 = 200\ \mu A,$

where $\qquad\qquad\qquad v_{CE} = 4\cdot9\ \text{V},\ i_C = 21\ \text{mA}.$

(i) Peak amplitude of a.c. output voltage

$$= V_{CE} - v_{CE}\ \text{(point D)} = 3\cdot5 - 2\cdot1 = 1\cdot4\ \text{V}$$
or $\qquad = v_{CE}\ \text{(point E)} - V_{CE} = 4\cdot9 - 3\cdot5 = 1\cdot4\ \text{V},$

and the r.m.s. amplitude of a.c. output voltage

$$= \text{peak}/\sqrt{2} = 1\cdot4/\sqrt{2} = 1\cdot4/1\cdot414 \approx 1\ \text{V}.$$

Peak amplitude of a.c. output current

$$= i_C\ \text{(at point D)} - I_C = 39 - 30 = 9\ \text{mA}$$
or $\qquad = I_C - i_C\ \text{(at point E)} = 30 - 21 = 9\ \text{mA},$

and the r.m.s. output current $= 9/\sqrt{2} = 6\cdot4\ \text{mA}.$
(ii) The peak amplitude of input voltage,

$$V_{im} = R_i \times (\text{peak input current})$$
$$= 1000 \times (100 \times 10^{-6}) = 10^{-1} = 0\cdot1\ \text{V}.$$

The peak amplitude of output voltage,

$$V_{om} = 1\cdot4\ \text{V (computed in(i))}.$$

So the voltage gain, $A_V = \dfrac{V_{om}}{V_{im}} = \dfrac{1\cdot4}{0\cdot1} = 14.$

The current gain, $A_i = \dfrac{\text{peak or rms ac output current}}{\text{peak or rms ac input current}}$

$$= \frac{9\ \text{mA}}{100\mu A} = \frac{9000\ \mu A}{100\ \mu A} = 90.$$

The power gain $A_P = A_V A_i = 14 \times 90 = 1260$
or, in dB, $10 \log A_P = 10 \log 1260 = 31\cdot0$ dB.

(iii) The a.c. input power to the amplifier,

$$P_i = \text{r.m.s. input current} \times \text{r.m.s. input voltage}$$
$$= [(100/\sqrt{2}) \times 10^{-6}] \times [0 \cdot 1/\sqrt{2}]$$
$$= \tfrac{1}{2} \times 100 \times 10^{-6} \times 10^{-1} = 5 \times 10^{-6} \text{ W}.$$

So the a.c. output power developed in the load,

$$P_o = A_p \times P_i = 1260 \times 5 \times 10^{-6}$$
$$= 6300 \times 10^{-6} \text{ W} = 6 \cdot 3 \text{ mW}.$$

Check: We can compute P_o directly since

$$P_o = (\text{r.m.s. output current}) \times (\text{r.m.s. output voltage})$$
$$= (6 \cdot 4 \text{ mA}) \times (1 \text{ V}) = 6 \cdot 4 \text{ mW}$$

taking the values directly from (i). The small discrepancy between 6·3 and 6·4 mW occurs because in (i) we calculated the r.m.s. values to only one decimal place.

13.6
The description of thermal runaway in a transistor

The maximum average power that a transistor may safetly dissipate without causing permanent damage depends on its mode of construction and also on how the transistor is mounted in use so that the heat dissipated within the device may be conducted away. The value of maximum power is principally determined by the maximum temperature that the collector-base junction region can withstand. These temperatures are typically within the range 150°C to 250°C for silicon and 50°C to 100°C for germanium. The range of maximum power varies from tens to a few hundreds of milliwatts for small-signal transistors up to a few hundred watts for high-power transistors.

There are two factors which determine the operating temperature of a transistor: obviously the first is the surrounding temperature, the second is the power dissipated within the transistor itself. The internal power dissipated in the transistor is principally due to that dissipated at the collector junction. If the collector-base voltage is V_{CB} and the collector current flowing is I_C, then the power dissipated in the collector junction region is $P_C = V_{CB}I_C$. P_C causes the junction temperature to rise, which in turn increases the collector current since more 'intrinsic' current carriers are generated. This produces an increased power dissipation in the transistor and consequently a further rise in junction temperature. Unless adequate cooling is provided or the transistor has built-in temperature compensation circuits to prevent excessive collector current rise, the junction temperature will continue to rise until the maximum allowable value of junction temperature is exceeded. If this situation occurs, the transistor will be permanently damaged. The unstable condition where, owing to rise in temperature, the collector current rises and continues to increase, is known as thermal runaway. Thermal runaway must always be avoided. If it occurs, permanent damage is caused and the transistor must be replaced.

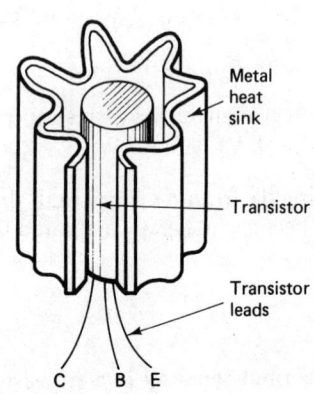

(a) Simple heat sink which can be 'sprung' over transistor can

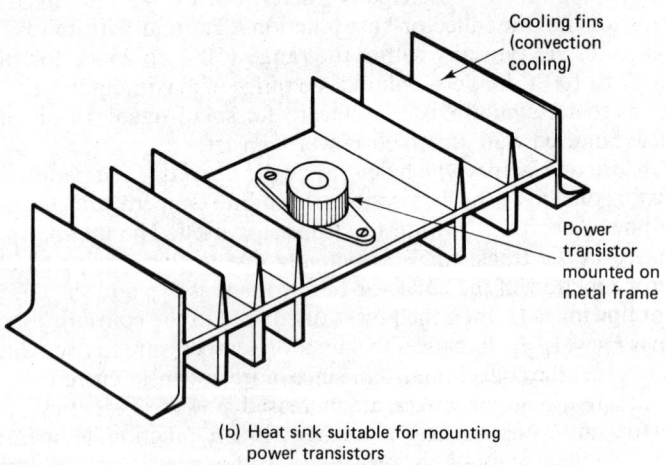

(b) Heat sink suitable for mounting power transistors

Fig. 13.6.

13.7
The reason for the use of heat sinks

Heat sinks are used to provide means of conducting, convecting, and radiating away the heat generated within the transistor. They are an important direct practical means of combating undesirable thermal effects which may ensue owing to rise in collector junction temperature.

Two practical examples of heat sinks are shown in fig. 13.6. In both cases the transistor is in contact with a relatively large mass of metal which acts as a good conductor of heat. The metal heat sink is often painted black to assist cooling by radiation. Improved cooling rates are effected by convection by corrugating the metal as in (a) and by the use of fins as in (b). It should be realized that the use of heat sinks alone may not be sufficient to prevent thermal runaway under all conditions. In designing a transistor circuit consideration should also be given to the choice of quiescent operating point, the value of ambient temperatures which are likely to be encountered, and the type of transistor used (for example, metal case transistors are more readily cooled by conduction than plastic ones). Circuits may also be designed to compensate automatically for temperature changes and thus stabilize the operation of the transistor components.

Section 14: *The expected learning outcome of this section is that the student should know the circuit and operation of a small-signal thermionic triode amplifier.*

Thermionic triodes are now mainly used for specialist applications, having been replaced by solid-state devices, which are normally more robust, cheaper, and more reliable. Thus the following note is included in the Technician Education Council's syllabus: 'college staff are advised that the treatment of thermionic devices and applications should depend on industrial requirements'. However, in high-power broadcasting transmitters and industrial radio frequency heating, thermionic valves capable of handling several kilowatts, and even up to megawatts, of power are invariably used.

14.1
The circuit diagram of a single-stage thermionic triode voltage amplifier having a load resistor

Fig. 14.1 shows the basic diagram of a single-stage triode amplifier. The input signal is applied across the terminals 11′ and varies the voltage between grid and cathode about the d.c. grid bias voltage V_G provided by the grid bias battery shown in the grid circuit. It is important to note that the latter is connected so that its negative terminal is connected to the

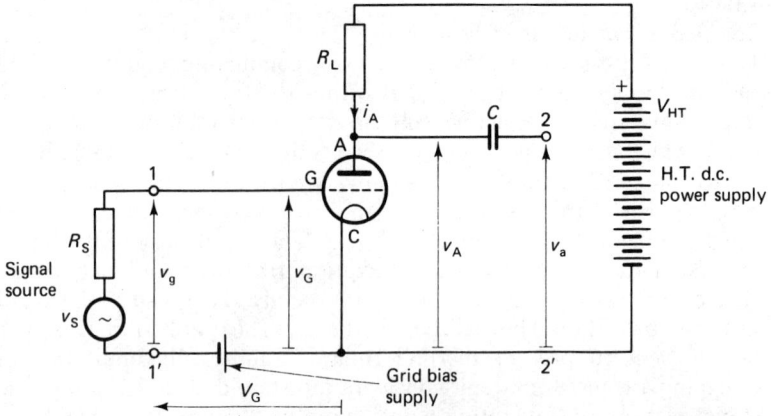

Fig. 14.1. Circuit diagram of a single-stage thermionic triode amplifier.

grid side of the input circuit and its positive terminal is joined to the cathode, so that V_G biases the grid negative with respect to the cathode. The amplitude of the signal should always be less than the magnitude of V_G to ensure that the grid remains negatively biased with respect to the cathode. The load resistor is R_L and the anode is maintained at a high positive voltage (typically in the range 50 V to 350 V, much higher than the collector voltage supply in transistor circuits) with respect to the cathode by the H.T. power supply also shown in the circuit diagram. The total anode voltage v_A is, of course, less than the H.T. voltage by the voltage drop across R_L, i.e.

$$v_A = V_{HT} - R_L i_A$$

where i_A = total anode current flowing through the triode,
V_{HT} = voltage of the H.T. power supply.

The total values of anode voltage v_A, current i_A, and control grid voltage v_G, are marked in on the circuit diagram. The total values consist of the steady d.c. values plus a.c. components produced by the input signal, i.e.

$$v_A = V_A + v_a, \quad i_A = I_A + i_a, \quad v_G = V_G + v_g$$

where V_A, I_A, V_G are the respective quiescent values (i.e. values in the absence of a signal or when the signal is instantaneously zero) and v_a, i_a, v_g are the small-signal changes effected by the signal varying the control grid voltage. Since the grid is always held negative with respect to the cathode, the grid current is normally negligible, and hence no current flows through the signal source resistance R_s. Thus the a.c. component and the control grid voltage $v_g = v_s$; where v_s is the e.m.f. of the source.

The amplified signal is developed across the terminals 22′. C acts as a

182

d.c. blocking capacitor which passes a.c. but blocks d.c. Thus the output voltage across 22' equals v_a, the alternating component of anode voltage. The magnitude of the voltage gain of the amplifier equals $V_a/V_g = V_a/V_s$ where V_a, $V_g = V_s$ are the amplitudes of v_a, v_g, v_s respectively.

Example 14.1
The load resistor and H.T. supply in the triode circuit of fig. 14.2(a) have the values $R_L = 2$ kΩ, $V_{HT} = 250$ V. The following measurements were recorded for the anode current I_A when the grid bias voltage V_G was varied:

when $V_G = -2$ V, $I_A = 50$ mA
$V_G = -5$ V, $I_A = 30$ mA
$V_G = -8$ V, $I_A = 10$ mA
$V_G = -12$ V, $I_A = 0$.

Calculate the corresponding anode-cathode voltage (V_A) values.

If the potentiometer is set at $V_G = -5$ V and an a.c. signal of peak amplitude 3 V is applied in series in the grid circuit, as shown in fig. 14.2(b), calculate the peak-to-peak a.c. voltage swing across the triode.

Solution
The anode-cathode voltage is given by

$$V_A = V_{HT} - R_L I_A$$
$$= 250 - 2000 I_A, \text{ since } V_{HT} = 250 \text{ V}, R_L = 2000 \ \Omega$$

or $V_A = 250 - 2I_A$, when I_A is in milliamperes.

Thus when $V_G = -2$ V, $I_A = 50$ mA, we have

$$V_A = 250 - 2 \times 50 = 150 \text{ V},$$

when $V_G = -5$ V, $I_A = 30$ mA, so $V_A = 260 - 2 \times 30 = 190$V,
when $V_G = -8$ V, $I_A = 10$ mA, so $V_A = 250 - 2 \times 10 = 230$ V,
when $V_G = -12$ V, $I_A = 0$, so $V_A = 250 - 2 \times 0 = 250$ V.

The a.c. signal will vary the total grid bias voltage between

$v = V_G + 3 = -5 + 3 = -2$ V on 'positive' half-cycles,

and $v_G = V_G - 3 = -5 - 3 = -8$ V on 'negative' half-cycles,

i.e. the signal will combine with the steady bias of -5 V and vary v_G by ± 3 V (3 V = peak amplitude of signal) between the values of -2 V and -8 V. The corresponding variations in anode voltage are:

when $v_G = -2$V, $v_A = 150$ V,

when $v_G = -8$ V, $v_A = 230$ V,

these results having already been calculated in the first part of our problem. Thus the peak-to-peak a.c. voltage swing across the triode is

$$230 - 150 = 80 \text{ V}.$$

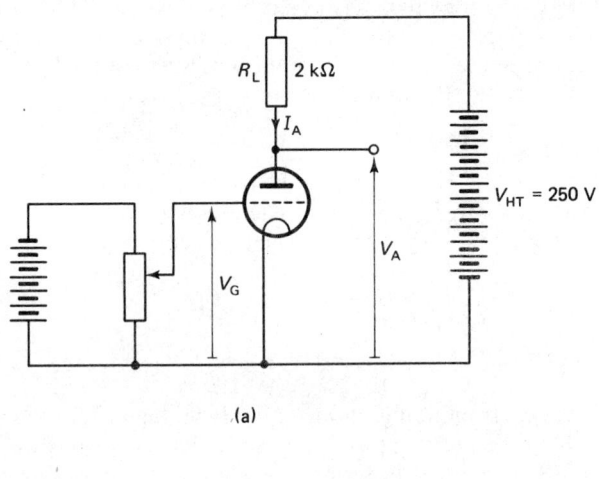

(a)

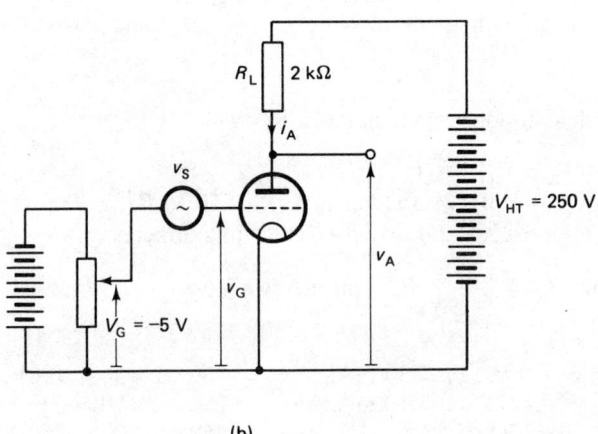

(b)

Fig. 14.2. Circuits for example 14.1.

14.2
Bias is required to obtain a selected quiescent operating point on the output characteristics

To use a thermionic triode correctly so that it may operate as a linear amplifying element, we must bias the triode so that it sits at an appropriate quiescent operating point. Remember that a suitable operating point was also required to operate a transistor amplifier in a linear manner. The quiescent operating point of a triode is defined by the d.c. values of grid bias voltage V_G, anode current I_A, and anode voltage V_A, that is, the values of grid-cathode voltage, anode current, and anode-cathode voltage with zero signal input.

184

We select the quiescent operating point, which we will denote by Q, with the following in mind:

(a) Q should be located in the 'straight line' region of the $v_A - i_A$ output characteristics of the triode.

(b) The position of Q should be such that the small-signal changes in i_A and v_A can be accommodated without running the triode into the curved region of the output characteristics—since this would lead to distortion—and also such that the maximum rated anode power dissipation P_A is not exceeded.

An example of an appropriate choice of Q point which could be used for the triode in the voltage amplifier circuit of fig. 14.3(a) is marked in on its anode characteristics in (b). The Q point selected is

$$V_G = -2 \text{ V}, V_A = 120 \text{ V}, I_A = 7 \text{ mA}.$$

The dashed curve shown in (b) is the locus of maximum anode power dissipation for the case $P_A = 1 \cdot 6$ W. This curve may easily be plotted by working out i_A values for set values of v_A as follows:

the power dissipated in the anode, $P_A = i_A v_A$,

so $i_A = \dfrac{P_A}{v_A} = \dfrac{1 \cdot 6}{v_A}$ when $P_A = 1 \cdot 6$ W,

hence when $v_A = 100$ V, $i_A = \dfrac{1 \cdot 6}{100}$ A $= 16$ mA;

$$v_A = 120 \text{ V}, i_A = \frac{1 \cdot 6}{120} \text{ A} = 13 \cdot 3 \text{ mA};$$

and so on.

We then plot the points ($v_A = 100$ V, $i_A = 16$ mA); ($v_A = 120$ V, $i_A = 13 \cdot 3$ mA) . . . and join these points up by a smooth curve.

Note that the Q point must always lie beneath the P_A curve to ensure that the maximum power dissipation allowed in the triode is not exceeded. The Q point should always be positioned well away from the curved part of the characteristic to avoid distortion in the amplifier. Thus, for example, points Q′ and Q″, also shown in fig. 14.3(b), should not be used:

Q′, where $V_G = -6$ V, $V_A = 150$ V, $I_A = 2 \cdot 1$ mA, lies in the curved part of the characteristic, so an operating point here would lead to distortion, i.e. the signal output v_a from the amplifier would not be a faithful replica of the input signal applied in the grid circuit;

Q″, where $V_G = -4$ V, $V_A = 200$ V, $I_A = 12$ mA, lies above the maximum anode dissipation curve, so an operating point here would lead to excessive heating in the triode and might cause permanent damage.

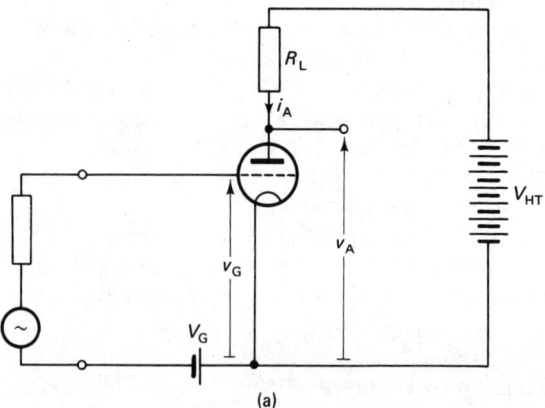

(a)

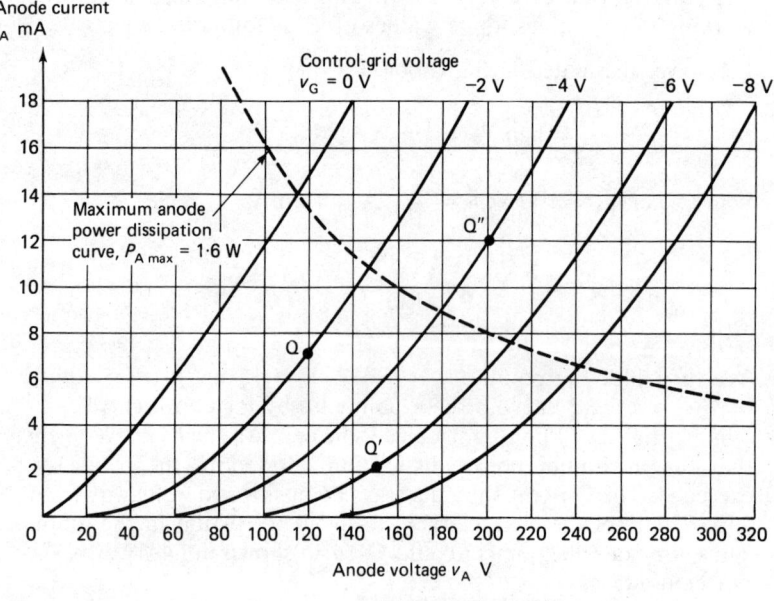

Anode current
i_A mA

Control-grid voltage
v_G = 0 V   −2 V   −4 V   −6 V   −8 V

18

16

14

Maximum anode
power dissipation
curve, $P_{A\,max}$ = 1·6 W

12

10

8

Q

6

4

Q′

2

0   20   40   60   80   100   120   140   160   180   200   220   240   260   280   300   320

Q″

Anode voltage v_A V

(b)

Fig. 14.3. A thermionic triode amplifier and the output characteristics of the triode used. Point Q is a suitable quiescent point.

14.3
The effect of a small sinusoidal voltage input on the quiescent condition

Suppose the quiescent operating point Q of the triode used in the

186

amplifier of fig. 14.4(a) is that shown marked in on its output characteristic (b); Q corresponds to $V_G = -5$ V, $V_A = 250$ V, $I_A = 20$ mA (we will consider how the Q point may be obtained in practice in Section 15 where we draw the load line $v_A = V_{HT} - R_L i_A$ on the output characteristics and find where this line intersects the characteristic corresponding to our given grid bias value V_G). The quiescent waveforms are sketched in fig. 14.4(c).

Now if a small sinusoidal voltage v_S is applied to the grid input circuit, it will cause the total grid-cathode voltage v_G to vary sinusoidally about the quiescent value V_G: $v_G = V_G + v_g = V_G + v_S$; the fact that $v_g = v_S$ follows, since, if we assume that no current flows in the grid input circuit, then no voltage drop occurs across the signal source resistance R_S and hence $v_g = v_S$. This state of affairs is closely approximated to in practice provided the grid is always held negative with respect to the cathode. Under these conditions the input resistance to the grid-cathode terminals is very high and the grid input circuit current negligible. The waveforms of the input signal $v_S = v_g$, the grid bias voltage V_G and the total grid voltage v_G are sketched in fig. 14.4(d).

Since the grid voltage controls the anode current, the variation in grid voltage produced by the sinusoidal input signal will also cause the anode current to vary sinusoidally about its quiescent value. When the input signal acts so as to make the grid less negative than it is in the quiescent condition, then v_G becomes less negative and consequently i_A increases. When the input signal acts in the same direction as the grid bias voltage, v_G becomes more negative and hence i_A decreases. The input signal voltage is in-phase with the a.c. component of anode current, i.e. i_a follows v_S, reaching its maxima (and minima) at exactly the same instants of time. The waveforms of total anode current i_A and its quiescent and a.c. signal components are shown in fig. 14.4(e).

The total anode voltage, $v_A = V_{HT} - R_L i_A$, also varies sinusoidally about its quiescent value V_A, reaching a minimum value when the anode current is at a maximum and therefore the voltage drop across R_L a maximum, and a maximum when the anode current is at its lowest value. The waveforms of total, quiescent, and a.c. signal components of anode voltage are sketched in fig. 14.4(f). (g) compares the input signal voltage waveform with the a.c. component waveform of anode voltage. Note that they are in anti-phase, i.e. when v_S reaches a maximum, v_a falls to a minimum and vice versa.

Thus, to summarize, the effect of a small sinusoidal input voltage on the quiescent condition causes a sinusoidal variation in anode current and anode voltage about their quiescent values. Voltage gain is effected because relatively small grid voltage changes can very effectively control the electron current flowing in the triode and thus produce relatively large variations in this current and hence in the anode voltage, since with a resistive load the latter is given by $v_A = V_{HT} - R_L i_A$.

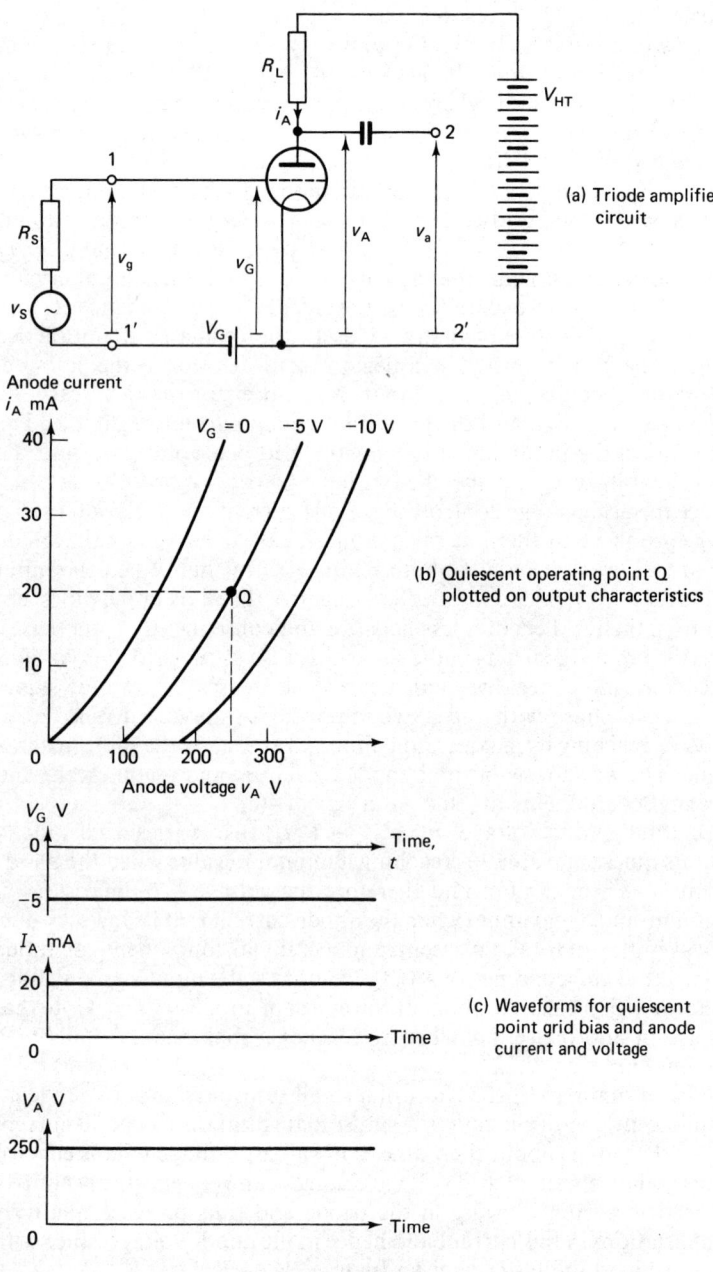

(a) Triode amplifier circuit

(b) Quiescent operating point Q plotted on output characteristics

(c) Waveforms for quiescent point grid bias and anode current and voltage

188

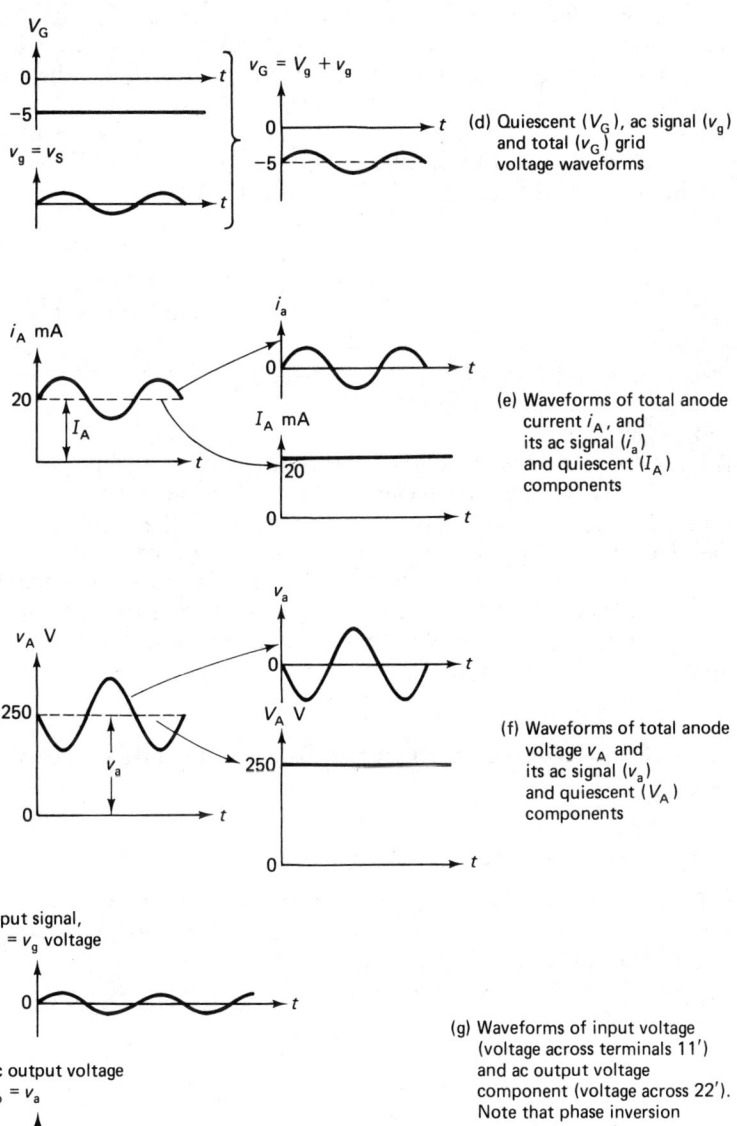

Fig. 14.4. The effect of a small sinusoidal voltage input on the quiescent condition.

189

14.4
Voltage phase inversion occurs between input and output signals
Just as in the case of the common emitter transistor amplifier, voltage phase inversion also occurs in the triode voltage amplifier where the triode is connected with its cathode common to both input and output circuits. The a.c. output signal v_a taken across terminals 22' in fig. 14.4(a) is inverted with respect to the input signal applied to the grid circuit as terminals 11'. This is illustrated graphically in fig. 14.4(g).

Section 15: *The expected learning outcome of this section is that the student should be able to construct and use a d.c. load line on the characteristics of a thermionic triode.*

15.1
The construction of the load line for a stated value of load resistance on a given set of output characteristics of a triode amplifier.
Let us consider the construction of the load line on the output (anode) characteristics of the triode amplifier shown in fig. 15.1(a). The manufacturers' output characteristics for the triode used are given in (b); the stated value of load resistance is $R_L = 10$ kΩ and the H.T. anode supply voltage $V_{HT} = 200$ V. The load line equation relating anode voltage v_A to anode current i_A is

$$v_A = V_{HT} - R_L i_A$$

and since we are considering the case where $R_L = 10$ kΩ, $V_{HT} = 200$ V, we have

$v_A = 200 - 10\,000\,i_A$
or $v_A = 200 - 10\,i_A$ when i_A is in milliamperes.

To construct the load line, it is sufficient to plot two points on the output characteristics and to draw a straight line through these two points. The two easiest points to calculate are
(1) When $v_A = 0$ we have

$$0 = V_{HT} - R_L i_A$$

So
$$i_A = \frac{V_{HT}}{R_L} = \frac{200}{10} = 20 \text{ mA}$$

We can now mark in the point $v_A = 0$, $i_A = 20$ mA. This point is shown as A in fig. 15.1(b).
(2) When $i_A = 0$ we have

$$v_A = V_{HT} - R_L \times 0 = V_A = 200 \text{ V}$$

190

So we mark in the point $v_A = 200$ V, $i_A = 0$ mA. This point is shown as **B** in fig. 15.1(b).

Finally, we join the points A and B by a straight line, thus constructing the load line on the output characteristics for the case $R_L = 10$ kΩ, $V_{HT} = 200$ V.

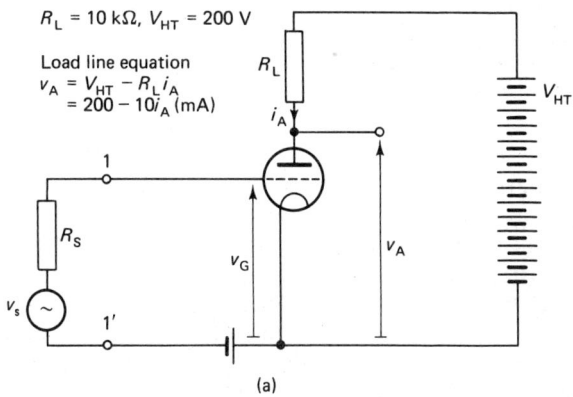

$R_L = 10$ kΩ, $V_{HT} = 200$ V

Load line equation
$v_A = V_{HT} - R_L i_A$
$= 200 - 10 i_A$ (mA)

(a)

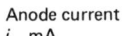
Anode current
i_A mA

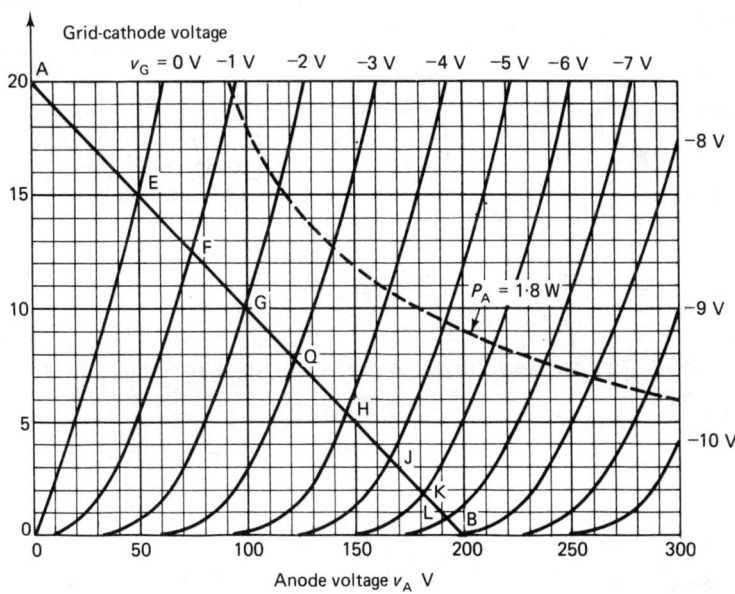

(b)

Fig. 15.1. A triode amplifier and the construction of the load line on the output characteristics of the triode.

15.2

The selection of a suitable bias voltage to obtain a desired quiescent point on the output characteristics

We can use the load line drawn in on the output characteristics of fig. 15.1(b) to determine the values of anode current and anode voltage for each of the values of grid voltage for which characteristics are plotted, i.e. for $v_G = 0, -1$ V, -2 V . . . etc. Each pair of v_A and i_A values is read off at the point of intersection of the load line with the respective output characteristic curve. These values are given in Table 15.1 for the amplifier of fig. 15.1(a) where we have $R_L = 10$ kΩ, $V_{HT} = 200$ V.

We may now use the table of values to plot curves of i_A versus v_G and v_A versus v_G for the amplifier. These curves are drawn in fig. 15.2. The curves are known as dynamic characteristics since they give i_A and v_A as v_G is varied for the actual amplifier and take into account the presence of the load resistor in the anode circuit; whereas the static transfer characteristics of $I_A - V_G$ for example, are plots of I_A against V_G for constant anode voltage (see Section 10.5). Note also that the dynamic characteristics refer to a given value of load R_L and H.T. voltage supply V_{HT}. If either one, or both, of these values are varied, a new load line must be drawn in on the output characteristics. This load line can then be used to construct a new table of $i_A - v_A$ values. These values in turn may be used to draw the new dynamic characteristics.

We can use the dynamic characteristics directly to select a suitable quiescent point for our amplifier. For example, suppose we require a quiescent point with a value of anode current $I_A = 6$ mA, we find the required value of grid bias voltage from the $i_A - v_G$ dynamic characteristic of fig. 15.2(a). When $i_A = 6$ mA, $v_G = -3.7$ V. This point is marked in on the characteristic as point N. Thus, the grid bias supply in the amplifier circuit of fig. 15.1 should be set at $V_G = -3.7$ V. The

Table 15.1 *Dynamic values for triode amplifier for* $R_L = 10$ kΩ, V_{HT} = 200 *V*

Grid voltage v_G (volts)	Anode voltage v_A	Anode current i_A (mA)	Corresponding point marked on fig. 15.1(b)
0	49	15·1	E
−1	74	12·5	F
−2	99	10·1	G
−3	122	7·75	Q
−4	145	5·4	H
−5	165	3·4	J
−6	191	1·8	K
−7	193	0·7	L

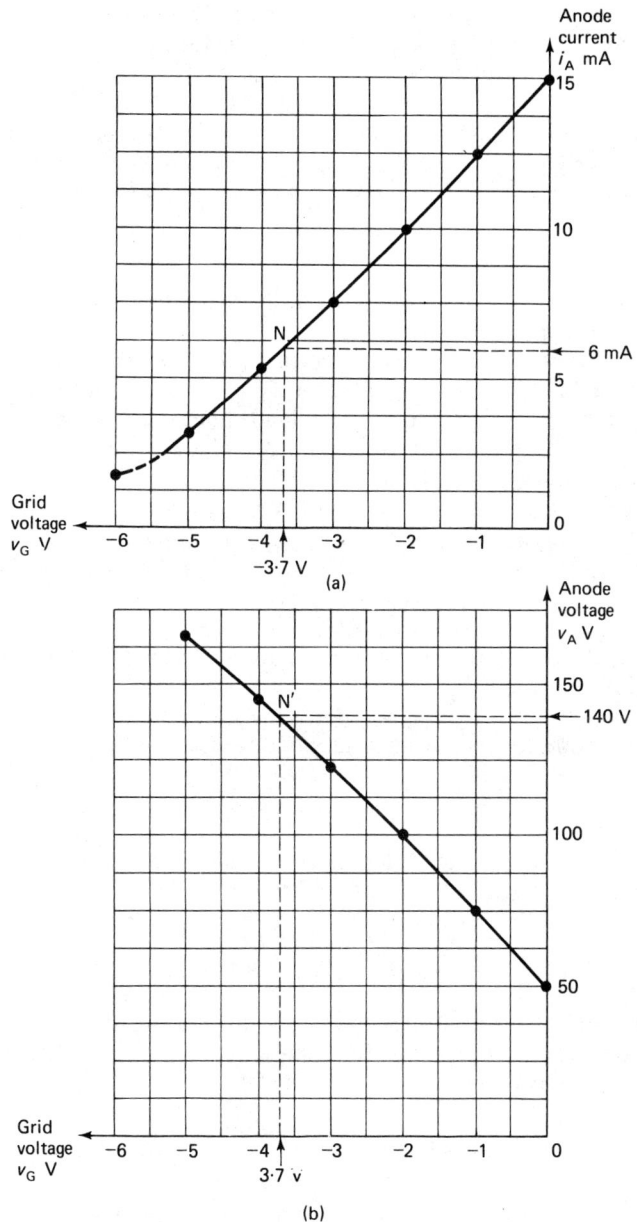

Fig. 15.2. Dynamic characteristics for triode amplifier of Fig. 15.1 for the case $R_L = 10$ kΩ, $V_{HT} = 200$ V.

corresponding quiescent value of anode voltage, $V_A = V_{HT} - 10 I_A = 200 - 10 \times 6 = 140$ V. V_A may also be obtained from the $v_A - v_G$ dynamic characteristic of fig. 15.2(b), by finding the value of v_A corresponding to $v_G = -3.7$ V, i.e. point N'.

Check for yourself that a quiescent point of $V_A = 100$ V, $I_A = 10$ mA may be obtained for the amplifier if the grid bias voltage is set at $V_G = -2.0$ V. Determine, also, the quiescent point obtained if the grid bias is set at $V_G = -2.5$ V. (The answer is $V_A = 112.5$ V, $I_A = 8.75$ mA.)

15.3
Estimation of the r.m.s. voltage from the load line for given quiescent conditions and input signal

Suppose we select a quiescent operating point for our amplifier of fig. 15.1 by setting the grid bias voltage to $V_G = -3$ V. On reading off from the point of intersection of the load line with the $v_G = -3$ V characteristic we have for the quiescent values of anode voltage and current (point Q in fig. 15.1(b): $V_A = 122$ V, $I_A = 7.75$ mA.

Now suppose we apply a sinusoidal voltage of peak amplitude $V_s = 1$ V (r.m.s. amplitude $1/\sqrt{2} = 0.707$ V) across the input terminals $11'$ of the amplifier. This signal will vary the grid voltage between

$$v_G = V_G + V_s = -3 + 1 = -2 \text{ V and } v_G = V_G - V_s = -3 - 1 = -4 \text{ V}$$

and therefore move the operating point along the load line about Q from
point G where $v_A = 99$ V, $i_A = 10.1$ mA, $(v_G = -2$ V);
to point H where $v_A = 145$ V, $i_A = 5.4$ mA, $(v_G = -4$ V).
Thus the peak-to-peak swing in anode voltage, that is, the voltage change between H and G is

$$145 - 99 = 46 \text{ V}$$

and the peak amplitude of the a.c. component of anode voltage, that is the voltage change from Q to G or from Q to H is

$$V_A - 99 = 122 - 99 = 23 \text{ V}$$
or $$145 - V_A = 145 - 122 = 23 \text{ V}$$

which is half the peak-to-peak swing—exactly as we should expect for linear operation of the triode. The corresponding r.m.s. amplitude of the a.c. component of anode voltage is

$$\text{r.m.s. amplitude} = \frac{\text{peak amplitude}}{\sqrt{2}} = \frac{23}{\sqrt{2}} = 16.3 \text{ V.}$$

The actual waveforms of output anode current and voltage may be obtained graphically using the dynamic characteristics as shown in fig. 15.3

194

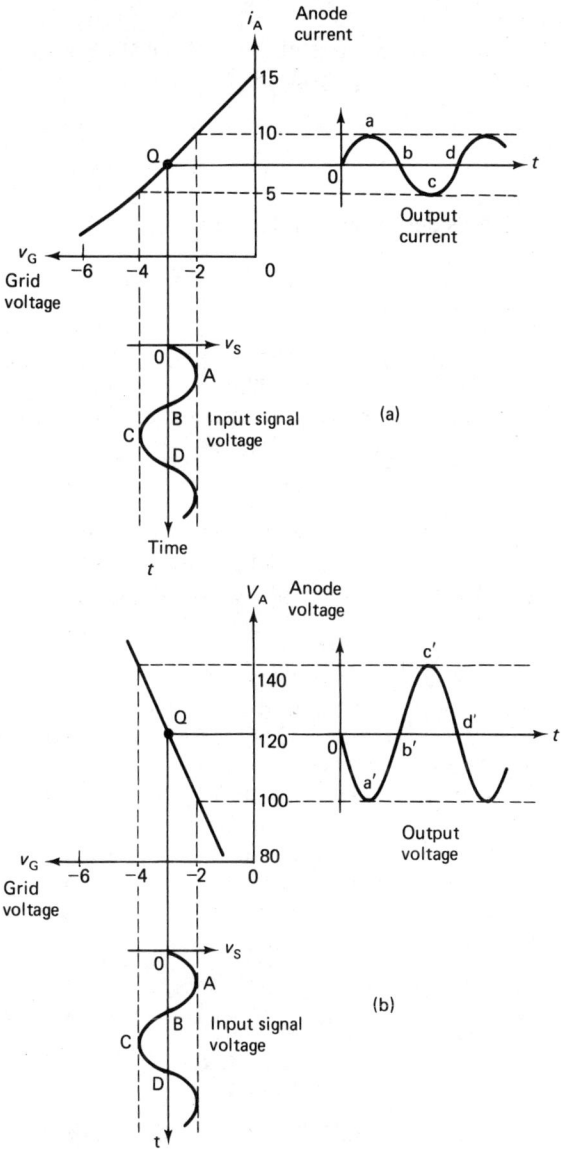

Fig. 15.3. Output current and voltage waveforms for triode amplifier of Fig. 15.1 obtained using dynamic characteristics: in (a) points O, a, b, c, d of the output current waveform correspond to points O, A, B, C, D of the input signal $v_s = v_g$ of peak amplitude $V_s = 1V$; in (b) points O, a′, b′, c′, d′ of the output voltage correspond to points O, A, B, C, D of the input signal. Q-point of amplifier: $V_G = -3V$, $V_A = 122$ V, $I_A = 7.75$ mA.

15.4
The determination of the voltage gain A_V of the stage
The magnitude of the voltage gain of the stage is

$$\frac{\text{peak or r.m.s. amplitude of a.c. component of output voltage}}{\text{peak or r.m.s. amplitude of a.c. signal input voltage}}$$

$$= \frac{23}{1} = 23 \text{ considering peak values, or}$$

$$= \frac{16 \cdot 3}{0 \cdot 707} = 23 \text{ considering r.m.s. values.}$$

Since voltage phase inversion occurs in the amplifier, the voltage gain A_V in these cases includes a minus sign. Thus we may state that the voltage gain of our triode amplifier stage at an operating point Q of $V_G = -3$ V, $V_A = 122$ V, $I_A = 7 \cdot 75$ mA is

$$A_V = -23.$$

We include the specification of the Q-point used to qualify A_V since the small-signal parameters of r_a, g_m, and μ which, together with the load resistance R_L, determine the value of A_V depend on the Q-point values.

Example 15.1
The manufacturers' output characteristics for the triode used in the amplifier of fig. 15.4(a) are given in (b). Construct the load line for the case when the load resistance $R_L = 2$ kΩ and the H.T. supply voltage $V_{HT} = 180$ V, and determine the quiescent values of anode voltage and current when the grid bias supply is set to $V_G = -2 \cdot 5$ V. Calculate also the power dissipation in the anode at the quiescent point.

If a sinusoidal signal of peak amplitude 1 V is applied at the grid circuit input terminals 11', determine the peak and r.m.s. values of the a.c. components of anode voltage and current about the above quiescent point and the voltage gain of the stage.

Solution
The anode voltage for the amplifier circuit,

$$v_A = V_{HT} - R_L i_A$$
$$= 180 - 2i_A$$

as $V_{HT} = 180$, $R_L = 2$ kΩ, and where i_A is in milliamperes. The load line is constructed by joining:

$$\text{point A: } v_A = 0, i_A = \frac{180}{2} = 90 \text{ mA,}$$

and point B: $v_A = 180$, $i_A = 0$,
by a straight line as shown in fig. 15.4(b).

196

The quiescent operating point when $V_G = -2.5$ V is given by the point of intersection of the $v_G = -2.5$ V output characteristic with the load line, i.e. point Q in fig. 15.4(b). On reading off the values of anode voltage

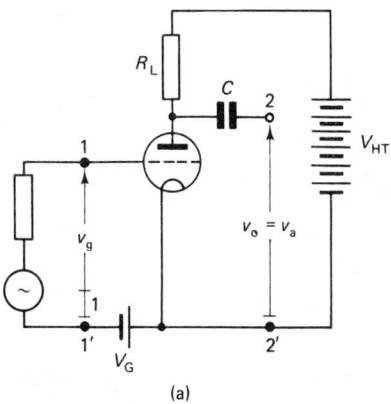

(a)

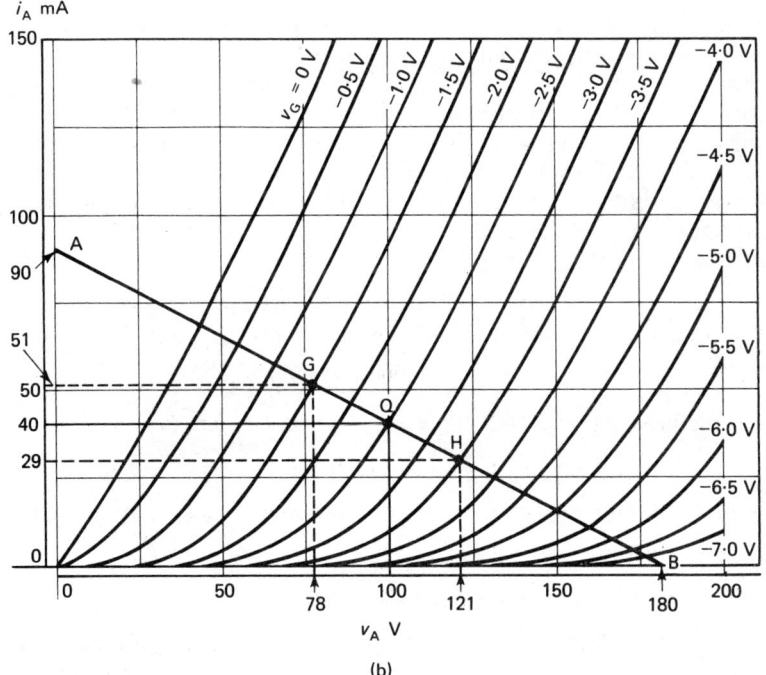

(b)

Fig. 15.4. Triode amplifier and the triode output characteristics used in example 15.1.

and current at this point we obtain the quiescent values of
$$V_A = 100 \text{ V}, I_A = 40 \text{ mA}.$$
The power dissipation in the anode at the Q-point is
$$P_A = I_A V_A = (40 \times 10^{-3}) \times (100) = 4 \text{ W}.$$
The sinusoidal signal of peak amplitude 1 V moves the operating point about Q along the load line between:

point G where $v_G = V_G + 1 = -2 \cdot 5 + 1 = -1 \cdot 5$ V and
$$v_A = 78 \text{ V}, i_A = 51 \text{ mA},$$

and point H where $v_G = V_G - 1 = -2 \cdot 5 - 1 = -3 \cdot 5$ V and
$$v_A = 121 \text{ V}, i_A = 29 \text{ mA}.$$

Thus the peak-to-peak swing in anode voltage between H and G is
$$121 - 78 = 43 \text{ V}$$
and hence the peak amplitude of the a.c. component of anode voltage is
$$\tfrac{1}{2} \times 43 = 21 \cdot 5 \text{ V}$$
Note: anode voltage change Q to G is $100 - 78 = 22$ V

anode voltage change Q to H is $121 - 100 = 21$ V

and ideally these should be equal for distortion-free amplification, However, there is a practical limit as to how accurately we can construct our load line and read off values of v_A and i_A at points of intersection. I think it is for this reason that we have the small discrepancy for peak voltage amplitude. As a compromise let us take the peak amplitude as the mean of 22 V and 21 V, i.e. 21·5 V. The corresponding r.m.s. amplitude of the a.c. component of anode voltage is
$$21 \cdot 5 / \sqrt{2} = 21 \cdot 5 / 1 \cdot 414 = 15 \cdot 2 \text{ V}.$$
The peak-to-peak swing in anode current between H and G is
$$51 - 29 = 22$$
and the peak amplitude of the a.c. component of anode current is
$$\tfrac{1}{2} \times 22 = 11 \text{ mA}$$
agreeing in this case with the peak swings about the quiescent point, i.e.

anode current change Q to G is $51 - 40 = 11$ mA

anode current change Q to H is $40 - 29 = 11$ mA.

The corresponding r.m.s. value of the a.c. component of anode current is
$$11 / \sqrt{2} = 0 \cdot 707 \times 11 = 7 \cdot 8 \text{ mA}.$$
Note: A peak swing of 11 mA would give a corresponding peak swing in

anode volts of $R_L \times 11$ mA $= 2 \times 11 = 22$ V. However there is no valid reason why we should modify our previous answer except that it would be better to say that the peak voltage amplitude is almost certainly within ± 1 V of 21·5 V.

The magnitude of the voltage gain of the stage is

$$\frac{\text{a.c. output voltage amplitude}}{\text{signal input voltage amplitude}} = \frac{21\cdot5}{1}$$

So including the minus sign to indicate voltage phase inversion we have, $A_V = -21\cdot5$.

Example 15.2

The small-signal parameters r_a and g_m of the triode used in the amplifier of the previous example at the Q-point $V_G = -2\cdot5$ V, $V_A = 100$ V, $I_A = 40$ mA are quoted by the manufacturer as

anode slope resistance $r_a = 600\ \Omega$
mutual conductance $g_m = 50$ mA/V.

An equivalent circuit which may be used to analyse the small-signal behaviour of a thermionic triode amplifier is shown in fig. 15.5. The input signal v_g is applied across terminals 11′ and the output signal is developed across the load resistance R_L connected across terminals 22′. The polarity of the a.c. voltage generator μv_g is taken in the direction shown so that the circuit takes into account the fact that voltage phase inversion occurs between input and output signals. μ is the amplification factor of the triode. Note that the equivalent circuit omits all details of bias suppliers. It is used only to analyse the small-signal behaviour of amplifiers.

Use the equivalent circuit of fig. 15.5 to derive a formula for the voltage gain of the amplifier and calculate the gain when $R_L = 2$ kΩ.

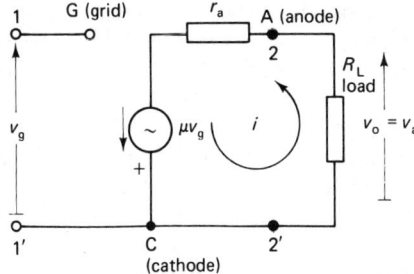

Fig. 15.5. A small-signal equivalent circuit for a triode amplifier, incorporating a voltage source μv_g.

Solution

The current i flowing in the output part of the circuit is

$$i = \frac{\text{generator voltage}}{\text{total resistance}} = \frac{\mu v_g}{r_a + R_L}$$

and the output voltage developed across R_L.

$$v_o = -iR_L = \frac{-\mu v_g}{(r_a + R_L)} \times R_L$$

(the minus sign signifies the following: i flows from 2' to 2, so voltage drop from 2' to 2 is iR_L, the voltage drop considered from 2 to 2' is equal but of opposite sign, hence $v_o = -iR_L$).

Thus the voltage gain of the amplifier,

$$A_V = \frac{\text{output voltage}}{\text{input voltage}} = \frac{v_o}{v_g} = \frac{-\mu R_L}{r_a + R_L}$$

Now the amplification factor of the triode μ is related to the other two small-signal parameters r_a and g_m by the formula:

$$\mu = r_a g_m$$

So on substituting $r_a = 600\,\Omega$, $g_m = 50\,\text{mA/V} = 50 \times 10^{-3}\,\text{A/V}$ we have,

$$\mu = 600 \times 50 \times 10^{-3} = 30.$$

So the voltage gain of the amplifier when $R_L = 2\,\text{k}\Omega$ is

$$A_V = -\frac{30 \times 2000}{600 + 2000} = -23.$$

Example 15.3

Fig. 15.6 shows a second equivalent circuit which is also frequently used to represent the small-signal behaviour of a triode amplifier. This circuit contains a current generator $g_m v_g$ which will drive current through both the internal resistance of the triode (i.e. the anode slope resistance r_a) and the load resistance R_L connected across the output terminals 22'. As in fig. 15.5, the input signal volage is applied across terminals 11', i.e. effectively between grid and cathode terminals. Note also that no resistance is connected across those terminals, showing that we assume the grid input current to be zero, or equivalently that the input resistance of the triode is infinite.

Use the equivalent circuit of fig. 15.6 to derive a formula for the voltage gain v_o/v_g and calculate the value of the latter when $r_a = 600\,\Omega$, $g_m = 50\,\text{ma/V}$, $R_L = 2\,\text{k}\Omega$.

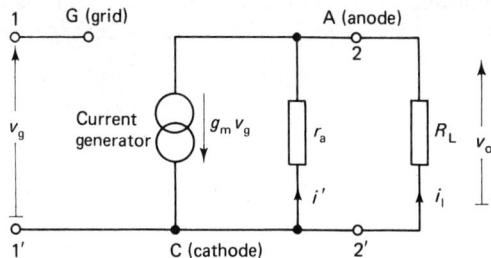

Fig. 15.6. A second form of small-signal equivalent circuit for a triode amplifier, incorporating a current source $g_m v_g$.

Solution

Now the generator current,

$$g_m v_g = i' + i_1$$

where i' = current flowing in $r_a = -\dfrac{v_o}{r_a}$

i_1 = current flowing in $R_L = -\dfrac{v_o}{R_L}$

(the minus sign occurring because we have taken i' and i_1 to flow from $2'$ to 2, in the opposite direction to the voltage drop v_o marked on the diagram from 2 to $2'$). Thus on substituting for i' and i_1 we obtain the following equation linking v_g to v_o:

$$g_m v_g = -\frac{v_o}{r_a} - \frac{v_o}{R_L}$$

$$= -v_o\left(\frac{1}{r_a} + \frac{1}{R_L}\right)$$

So on solving for the ratio v_o/v_g we have,

$$\text{the voltage gain, } A_V = \frac{v_o}{v_g} = \frac{-g_m}{\left(\dfrac{1}{r_a} + \dfrac{1}{R_L}\right)} \quad \dots \ (1)$$

On substituting $r_a = 600$, $1/r_a = (1\cdot67)10^{-3}$; $R_L = 2\,\text{k}\Omega$, $1/R_L = (0\cdot5)10^{-3}$, and $g_m = 50 \times 10^{-3}\,\text{A/V}$ we obtain

$$A_V = -\frac{50 \times 10^{-3}}{(1\cdot67+0\cdot5)10^{-3}} = \frac{-50}{2\cdot17} = -23$$

Note that if we multiply top and bottom of expression (1) by $r_a R_L$ (or simplify the denominator) we can show that our expression for A_v is identical to the one we derived in our previous example, i.e.

$$A_v = \frac{-g_m r_a R_L}{\left(\dfrac{1}{r_a} + \dfrac{1}{R_L}\right) r_a R_L} = \frac{-g_m r_a R_L}{(R_L + r_a)}$$

$$= \frac{-\mu R_L}{(R_L + r_a)} \quad \text{since } r_a g_m = \mu.$$

Section 16: *The expected learning outcome of this section is that the student should understand the automatic biasing of small-signal amplifiers.*

16.1
An explanation, with the aid of diagrams, of methods of biasing a triode and a transistor amplifier stage.

(a) *The automatic biasing of a triode amplifier stage*
In a practical triode amplifier, the grid bias voltage is generated from the H.T. anode voltage supply rather than using a separate grid bias battery.

Let us see how this may be achieved by reference to fig. 16.1.
(a) shows a circuit in which the grid bias is generated using a separate source. Suppose the circuit is designed to operate with a load resistor R_L = 10 kΩ from a H.T. supply of voltage V_{HT} = 200 V and to have a quiescent operating point of: grid-cathode voltage, V_G = -4 V; anode voltage, V_A = 150 V; anode current, I_A = 5 mA.
(b) shows a circuit which generates an automatic bias by connecting a resistor R_C in the cathode lead.

The circuit of fig. 16.1(a) provides a situation in which, under no signal conditions, the grid is held at a potential of 4 V negative with respect to the cathode. Raising the potential of the cathode 4 V relative to the grid clearly has the equivalent effect. The resistor R_C in the cathode lead in fig. 16.1(b) is included to achieve this. The anode current I_A will also pass through R_C and develop a voltage across it, raising the potential of the cathode above the H.T. negative line by $R_C I_A$. Thus to develop the required grid bias voltage of 4 V, we have

$$\text{voltage dropped across } R_C = R_C I_A = 4 \text{ V}$$

So $\qquad R_C = \dfrac{4 \text{ V}}{5 \text{ mA}} = \dfrac{4 \text{ V}}{0 \cdot 005 \text{ A}} = 800 \ \Omega, \text{ as } I_A = 5 \text{ mA}.$

In making this calculation we have assumed that the anode current does

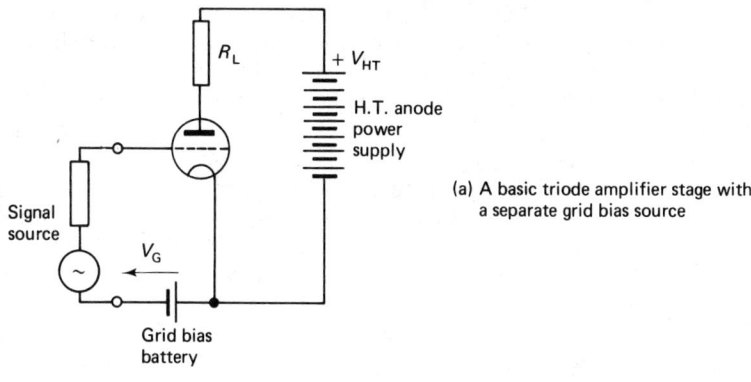

(a) A basic triode amplifier stage with a separate grid bias source

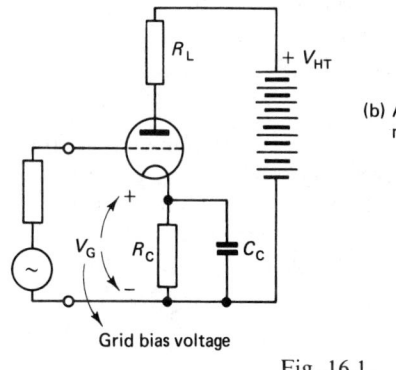

(b) A triode amplifier with automatic bias network, $R_C - C_C$, in cathode lead

Grid bias voltage

Fig. 16.1.

not change when R_C is included in the cathode lead. This is obviously not quite true since we have

$$V_A = V_{HT} - R_L I_A \text{ for circuit (a)},$$

but
$$V_A = V_{HT} - (R_L + R_C) I_A \text{ for circuit (b)}.$$

However, $R_C = 800\ \Omega$ is relatively small compared to the load resistor value $R_L = 10\ 000\ \Omega$, and our approximation should not be too much in error, especially as resistors with tolerances of $\pm 10\%$ are often used in amplifier circuits.

A capacitor, C_C in fig. 16.1(b), is normally included in parallel with R_C. The value of C_C is selected so that it has a very low reactance and thus provides an effective short-circuit or 'by-pass' to alternating current in the cathode lead. C_C is known as a by-pass capacitor. Let us consider some values of C_C and its corresponding shunting effect. The reactance of

C_C varies with frequency and is given by $1/(2\pi f C_C)$ ohms, where f is the frequency in hertz. Thus if the 800 Ω bias resistor were shunted by a 0·1 μF capacitor, the latter would provide a reactance of only 1·6 Ω when, for example, $f = 1$ MHz; so the presence of the 800 Ω resistor has a negligible effect as far as a 1 Mhz signal is concerned. However, at 50 Hz the same capacitor would have a reactance of about 32 kΩ, large compared not only with $R_C = 800$ Ω but also with $R_L = 10$ kΩ. Unfortunately it is difficult to state a general criterion for the value of C_C. It is not sufficient just to state that the reactance of C_C should be much less than R_C at the lowest frequency at which the amplifier is to be used. Typical values of C_C are of the order of 50 μF and higher for satisfactory operation without severe reduction in gain, down to frequencies of a few tens of hertz. Note that we can use our amplifier with R_C alone but the omission of C_C will reduce the gain of our amplifier. The absence of C_C does however improve the stability of the amplifier (see Section 18.1 and the comments made on negative feedback). In practice the design value of C_C will depend on the lowest frequency at which our amplifier is to be used, the valve parameters r_a and μ, and the values of R_L and R_C as demonstrated in the following example.

Example 16.1
A triode amplifier employing automatic bias has a load resistor, $R_L = 8$ kΩ and an anode power supply of 200 V. The amplifier is to be designed to have a quiescent grid bias voltage of -5 V. It is known from drawing a load line on the anode characteristics of the triode that at $V_G = -5$ V, $I_A = 10$ mA, $V_A = 120$ V. Calculate the value of R_C (the bias resistor in the cathode lead) required to achieve the grid bias of -5 V.

Solution
The voltage drop across R_C is to be 5 V, so

$$R_C I_A = 5$$

and neglecting the effect of the inclusion of R_C on the anode current we can assume $I_A = 10$ mA. Thus

$$R_C = \frac{5}{0\cdot010} = 500 \ \Omega.$$

Further discussion on the effects of R_C and C_C on the amplifier gain:
The voltage gain of an amplifier when a grid bias battery is used rather than automatic bias is given by

$$A_o = \frac{\mu R_L}{r_a + R_L}$$

where R_L = load resistance, and r_a and μ are the anode slope resistance
204

and amplification factor of the triode. Thus, if we are given in this case $r_a = 1 \text{ k}\Omega$ $\mu = 30$ and remembering $R_L = 8 \text{ k}\Omega$, we obtain

$$A_o = \frac{30 \times 8}{1+8} = \frac{240}{9} = 26 \cdot 7.$$

It can be shown that, with R_C providing bias, instead of a battery, the voltage gain is given by

$$A = \frac{A_o}{1 + \dfrac{(\mu+1)R_C}{r_a + R_L}}$$

Note that, as we would expect, putting $R_C = 0$ in this equation gives $A = A_o$. On substituting the values already given, including $R_C = 500 \, \Omega$, we obtain

$$A = \frac{26 \cdot 7}{1 + \dfrac{31 \times 500}{9000}} = \frac{26 \cdot 7}{1 + 1 \cdot 72} = 9 \cdot 8$$

thus showing what a large reduction in gain results from introducing R_C on its own without a by-pass capacitor.

It can also be shown that, if C_C is now connected across R_C, the voltage gain A' is given by

$$A' = \frac{A_o \sqrt{(1 + \omega^2 C_C^2 R_C^2)}}{\sqrt{\left[\left(1 + \dfrac{(\mu+1)R_C}{r_a + R_L} \right)^2 + \omega^2 C_C^2 R_C^2 \right]}}$$

where $\omega = 2\pi f$ and $f = $ frequency of signal to be amplified. It will be seen that, when $C_C = 0$, this equation becomes the equation given in (c).

Taking the values given for μ, r_a, R_L, R_C and selecting $C_C = 50 \, \mu\text{F}$ we have on considering the performance of the amplifier at the frequency $f = 10 \text{ Hz}$ ($\omega = 2\pi \times 10$):

$$A' = \frac{26 \cdot 7 \sqrt{[1 + (20\pi \times 50 \times 10^{-6} \times 500)^2]}}{\sqrt{\left[\left(1 + \dfrac{31 \times 500}{9000} \right)^2 + (20\pi \times 50 \times 10^{-6} \times 500)^2 \right]}}$$

$$= \frac{26 \cdot 7 \sqrt{(1 + 2 \cdot 47)}}{\sqrt{(2 \cdot 97 + 2 \cdot 47)}} = \frac{26 \cdot 7 \sqrt{(3 \cdot 47)}}{\sqrt{(5 \cdot 44)}} = 21 \cdot 3;$$

and if $f = 100 \text{ Hz}$, $\omega = 2\pi \times 100$ we have for the gain,

$$A' = \frac{26 \cdot 7 \sqrt{(1 + 247)}}{\sqrt{(2 \cdot 97 + 247)}} = 26 \cdot 5$$

Thus we see the effect of C_C in maintaining the gain close to A_o.

(b) *Automatic biasing of a transistor amplifier stage*

(i) *Fixed bias and collector to base bias circuits*

We discussed the fixed bias circuit shown in fig. 16.2(a) in Sub-section 12.4. We showed that the design value of R_B to produce a given quiescent base current of value I_B was

$$R_B = \frac{V_{CC} - V_{BE}}{I_B}$$

where V_{CC} = collector voltage supply and $V_{BE} \approx 0.6$ V for silicon, 0.2 V for germanium transistors. We also stated that this circuit does not provide good stability, the quiescent point changes with temperature and also is strongly dependent on the current gain (h_{FE}) of the transistor.

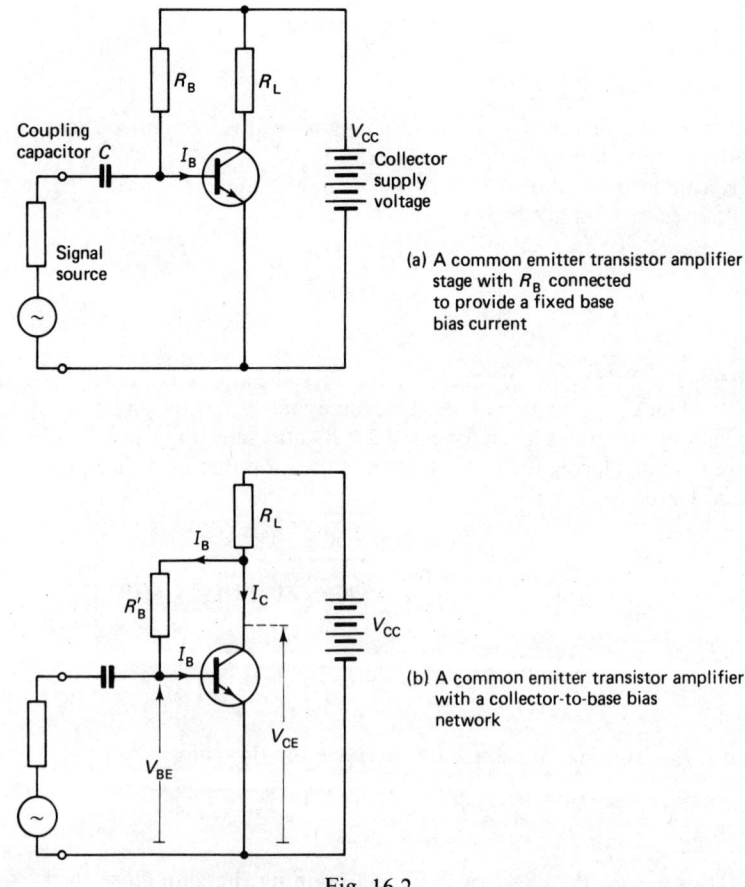

(a) A common emitter transistor amplifier stage with R_B connected to provide a fixed base bias current

(b) A common emitter transistor amplifier with a collector-to-base bias network

Fig. 16.2.

An improvement in stability may be obtained by using the bias arrangements shown in fig. 16.2(b). Here the bias resistor R_B' connected to the base is returned to the collector rather than the battery terminal. The value of R_B' to set a given d.c. base bias current I_B and d.c. collector-emitter voltage V_{CE} is determined as follows:

$$\text{voltage drop across } R_B', \quad R_B'I_B = V_{CE} - V_{BE}$$

So
$$R_B' = \frac{V_{CE} - V_{BE}}{I_B}$$

where V_{BE} is taken as 0.6 V for silicon, 0.2 V for germanium transistors.

We have two further equations from which we can determine the value of the load resistor R_L to give our required quiescent point, that is, the value of V_{CE}. The quiescent value of the collector current is determined by the d.c. current gain of the transistor, i.e.

$$I_C = h_{FE}I_B$$

and since the total current in R_L is $I_B + I_C$, we have for the collector circuit:

$$V_{CC} = R_L(I_B + I_C) + V_{CE}$$

So $R_L = \dfrac{V_{CC} - V_{CE}}{(I_B + I_C)}$

$$= \frac{V_{CC} - V_{CE}}{I_B(1 + h_{FE})} \text{ or } \frac{V_{CC} - V_{CE}}{I_C(1 + 1/h_{FE})}, \text{ since } I_C = h_{FE}I_B.$$

Invariably h_{FE} is large, typically several tens to hundreds, so there is normally little error in neglecting I_B in comparison with I_C and hence

$$R_L \approx \frac{V_{CC} - V_{CE}}{I_C} \quad \text{or} \quad \frac{V_{CC} - V_{CE}}{h_{FE}I_B}$$

The physical reason as to why the collector-to-base bias circuit of fig. 16.2(b) is an improvement on the fixed bias circuit of (a) may be argued as follows. If the collector current tends to increase owing to either a temperature rise or because the transistor has been replaced by one with a larger h_{FE} value, then the collector-emitter voltage

$$V_{CE} = V_{CC} - R_L(I_C + I_B) \approx V_{CC} - R_LI_C$$

tends to decrease and hence

$$I_B = \frac{V_{CE} - 0.6}{R_B'} \text{ (silicon). } I_B = \frac{V_{CE} - 0.2}{R_B'} \text{ (germanium),}$$

also tends to decrease. Consequently so must I_C, since I_C is controlled by I_B. The net result is that the collector current does not increase as much as it would in the fixed bias circuit. This effect is demonstrated quantitatively in the following example.

Example 16.2
Calculate (a) the value of R_B in the fixed bias circuit of fig. 16.2(a), (b) the value of R'_B in the collector-to-base bias circuit of fig. 16.2(b) required to produce a quiescent collector current of $I_C = 1$ mA. The transistor used in both circuits is silicon with a d.c. current gain at $I_C = 1$ mA of $h_{FE} = 50$. Assume the base-emitter voltage $V_{BE} = 0.6$ V. The load resistor and collector supply voltage for both circuits are $R_L = 5$ kΩ, $V_{CC} = 15$ V.

If the transistor were changed in both circuits to one with $h_{FE} = 100$, calculate the new quiescent operating points obtained.

Solution

(a)
$$R_B = \frac{V_{CC} - V_{BE}}{I_B}$$

and we are given $V_{CC} = 15$ V, $V_{BE} = 0.6$ V, $I_C = 1$ mA, $h_{FE} = 50$. So $I_B = I_C/h_{FE} = 1/50 = 0.02$ mA, and we have

$$R_B = \frac{(15 - 0.6)\ \text{V}}{(0.02)\ \text{mA}} = \frac{14.4}{0.02} = 720\ \text{kΩ}.$$

Note that the nearest $\pm 10\%$ tolerance preferred value resistor is 680 kΩ (the next is 820 kΩ), and thus in practice we would use 680 kΩ for R_B.

(b)
$$R'_B = \frac{V_{CE} - V_{BE}}{I_B}$$

We know $I_B = 0.02$ mA is the required base current and $V_{BE} = 0.6$ V. We can calculate V_{CE} using

$$\begin{aligned}
V_{CE} &= V_{CC} - R_L(I_C + I_B) \\
&= 15 - (5\ k\Omega)\,(1 + 0.02)\ mA \\
&= 15 - 5 \times 1.02 = 9.9\ V.
\end{aligned}$$

So
$$R'_B = \frac{(9.9 - 0.6)\ \text{V}}{(0.02)\ \text{mA}} = \frac{9.3}{0.02} = 465\ \text{kΩ}.$$

The nearest preferred resistance value is 470 kΩ.

Now let us work out the new operating points when the transistor is changed to one with $h_{FE} = 100$.
(1) In the fixed bias circuit the base current remains unchanged so $I_B = 0.02$ mA, but

$$I_C = h_{FE}I_B = 100 \times 0.02 = 2\ \text{mA},$$

and $V_{CE} = V_{CC} - R_L I_C = 15 - (5\ k\Omega) \times (2\ mA) = 5$ V.

Thus the new quiescent operating point is $V_{CE} = 5$ V, $I_C = 2$ mA, a change of 100% when compared to the original one where $V_{CE} = 10$ V, $I_C = 1$ mA.

(2) Let us work from first principles and write down the equations relating I_C, I_B and V_{CE} for the collector-to-base bias circuit of fig. 16.2(b). First we know $I_C = h_{FB}I_B$, so $I_C = 100\ I_B$.

On equating the sum of the voltage drops across R_L and the transistor we have

$$V_{CC} = R_L(I_C + I_B) + V_{CE}$$

i.e.
$$15 = 5 \times 101 I_B + V_{CE} \qquad \dots \ (1)$$

where I_B is in mA.

We also have, the voltage drop across R'_B, ($R'_B = 465$ kΩ),

$$R'_B I_B = V_{CE} - V_{BE}$$

so
$$465 I_B = V_{CE} - 0.6$$

or
$$V_{CE} = 465 I_B + 0.6 \qquad \dots \ (2)$$

On substituting in (1) for V_{CE} using (2) we obtain

$$15 = 505\ I_B + 465\ I_B + 0.6$$

So
$$(505 + 465) I_B = 15 - 0.6$$

$$I_B = \frac{14.4}{970} = 0.0148 \text{ mA}.$$

Thus $I_C = 100\ I_B = 1.48$ mA
and $V_{CE} = 465\ I_B + 0.6 = 7.5$ V
i.e. our new quiescent operating point is (7·5 V, 1·48 mA) showing a far less change from the original than the fixed bias circuit.

(ii) *Emitter bias*

A very popular automatic biasing circuit which provides a stable operating point is shown in the circuit diagram of fig. 16.3. This method of biasing is known as emitter biasing since it incorporates a resistor R_E in the emitter lead of the transistor. There are two other components making up the biasing circuit; these are the resistors R_1 and R_2.

The voltage necessary to forward-bias the base-emitter junction of the transistor is supplied from the collector voltage supply V_{CC} by the $R_1 - R_2$ network. V_{CC} supplies both the collector current I_C and the current I_1 which flows through R_1. Part of I_1 supplies the base bias current I_B and the balance $I_1 - I_B$ flows through R_2 back to the supply. The voltage developed across R_2 is $V_2 = R_2(I_1 - I_B)$. However, since emitter current flows in R_E, a voltage $R_E I_E$ is developed across R_E. This voltage acts to

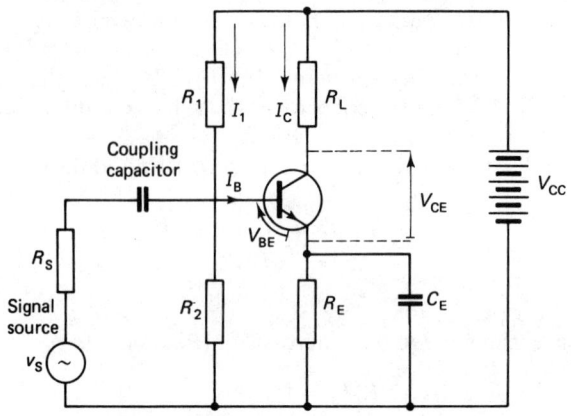

Emitter bias circuit: a very popular bias circuit which produces good stability of the quiescent point for temperature and h_{FE} changes

Fig. 16.3.

reduce the forward-bias. Thus, the resultant forward-bias voltage across the base-emitter junction is

$$V_{BE} = V_2 - R_E I_E$$
$$= R_2(I_1 - I_B) - R_E(I_B + I_C), \text{ as } I_E = I_B + I_C.$$

Now, with this result in mind, we can give a physical reason for the good stability of the circuit. If the collector current tends to increase owing to rise in temperature, or because the transistor is replaced by another of larger h_{FE}, the current in R_E increases, i.e. if I_C increases, so does I_E, and hence so does the voltage drop across R_E. Thus, the forward-bias voltage across the base-emitter junction tends to decrease, with the net result that any increase in I_C tends to be curtailed. Likewise, if I_C tends to fall, I_E will also tend to fall, and hence so does the voltage drop across R_E. Thus the forward-bias voltage across the base-emitter junction will tend to rise so that the original decrease in I_C is not nearly as great as it would have been in the absence of the emitter resistor R_E.

The following example demonstrates the calculations to be made to determine the values of R_E, R_1, and R_2 to obtain a given quiescent point. The example also considers what changes occur in the quiescent point values when a change is made in h_{FE}. The results show the superiority of the emitter bias circuit over the two previous methods.

One final point is that a by-pass capacitor C_E is normally connected in parallel with R_E so as to provide an effective short-circuit to alternating current in the emitter lead. To ensure that the voltage gain of the

210

amplifier is not significantly reduced at low frequencies—for example, down to a frequency f, the value C_E should be such that

$$C_E > \frac{1 + R'/R}{2\pi f_1 R_E} \text{ farads}$$

where $R' = (1 + h_{fe})R_F$,

and $R = h_{ie} + R_S$

h_{fe} and h_{ie} are respectively the common emitter small-signal current gain and input resistance; R_S is the resistance of the signal source feeding the amplifier.

If C_E is omitted, the voltage gain of the amplifier will be reduced since the a.c. component of current in R_E provides negative feedback (see Subsection 18.1).

Example 16.3

Calculate the values of R_1, R_2, and R_E in the emitter bias circuit of fig. 16.3 to provide a quiescent operating point of $I_C = 1$ mA, $V_{CE} = 10$ V. The transistor used in the circuit is silicon with a d.c. current gain at 1 mA of $h_{FE} = 50$. Assume the base-emitter voltage $V_{BE} = 0.6$ V. The load resistor $R_L = 5$ kΩ and the collector supply voltage $V_{CC} = 16$ V. The above specification does not lead to unique values for R_1 and R_2. We may therefore specify one further factor which in practice may take the form of a stability condition, or an input resistance condition, or a supply current condition. Suppose we select the last condition and impose the restriction that the current through resistor R_1, I_1 should be $0.1 I_C$, i.e. $I_1 = 0.1$ mA.

If the transistor were replaced by one with $h_{FE} = 100$, calculate the new quiescent point.

Solution

Let us work once again from first principles. First mark in currents I_C, I_B, and I_1 on the circuit diagram. Then write down the circuit equations:

(i) The condition imposed by the transistor itself is

$$I_C = h_{FE}I_B$$

so as the required $I_C = 1$ mA and $h_{FE} = 50$ we have

$$I_B = I_C/h_{FE} = 1/50 = 0.02 \text{ mA}$$

(ii) On equating the voltage drops across R_L, the transistor, and R_E to the collector supply voltage, we obtain

$$V_{CC} = R_L I_C + V_{CE} + R_E(I_C + I_B)$$

Note that the current flowing in R_E is $I_B + I_C$. On substituting values

$$16 = 5 \text{ k}\Omega \times 1 \text{ mA} + 10 + R_E \text{ k}\Omega \times (1 + 0.02) \text{ mA}$$

i.e. $$16 = 5 + 10 + 1.02 R_E,$$

we have $R_E = \dfrac{1}{1.02} \approx 1 \text{ k}\Omega.$

(iii) On equating the voltage drop across R_2 (and noting $I_1 - I_B$ flows in R_2) to the base-emitter voltage drop across the transistor plus the voltage drop across R_E, we obtain

$$R_2(I_1 - I_B) = V_{BE} + R_E(I_B + I_C) \quad \cdots \cdots (1)$$

i.e. $$R_2(0.1 - 0.02) = 0.6 + \frac{1}{1.02}(1 + 0.02)$$

$$0.08 R_2 = 0.6 + 1$$

So $$R_2 = \frac{1.6}{0.08} = 20 \text{ k}\Omega.$$

(iv) Finally on equating the sum of the voltage drops across R_1 and R_2 to the collector supply voltage we obtain

$$V_{CC} = R_1 I_1 + R_2(I_1 - I_B) \qquad \cdots \cdots \quad (2)$$

So $$16 = R_1 \times 0.1 + 20(0.1 - 0.02)$$

$$= 0.1 R_1 + 1.6$$

Hence $$R_1 = \frac{16 - 1.6}{0.1} = 144 \text{ k}\Omega.$$

Finally we are required to determine the new quiescent point when the transistor is changed to one with $h_{FE} = 100$. Equations (1) and (2) are quite general, so on substituting the values we determined for R_1, R_2, and R_E, i.e. $R_1 = 144 \text{ k}\Omega, R_2 = 20 \text{ k}\Omega, R_E = 1 \text{ k}\Omega$, and $V_{BE} = 0.6 \text{ V}$, equations (1) and (2) give, respectively:

$$20(I_1 - I_B) = 0.6 + 1(I_B + I_C) \quad \cdots \cdots (3)$$

$$16 = 144 I_1 + 20(I_1 - I_B) \quad \cdots \cdots (4)$$

But we also know for our new transistor

$$I_C = h_{FE} I_B = 100 I_B$$

So on substituting for I_C in (3) we obtain

$$20(I_1 - I_B) = 0.6 + (I_B + 100 I_B)$$

i.e. $$20 I_1 - 121 I_B = 0.6 \quad \cdots \cdots (5)$$

212

Also on simplifying (4) we obtain

$$164\,I_1 - 20\,I_B = 16 \quad \cdots \quad (6)$$

Thus we have to simultaneous equations, equations (5) and (6), which we may solve for I_1 and I_B. Their solution is

$$I_B = 0{\cdot}0114 \text{ mA}, \; I_1 = 0{\cdot}099 \text{ mA}.$$

Thus the new quiescent value of collector current is

$$I_C = 100 \times 0{\cdot}0114 = 1{\cdot}14 \text{ mA}$$

and the corresponding value of collector-emitter voltage,

$$V_{CE} = V_{CC} - R_L I_C - R_E(I_B + I_C)$$
$$= 16 - 5 \times 1{\cdot}14 - 1 \times 1{\cdot}15 = 9{\cdot}15 \text{ V}.$$

Note: The change in quiescent point for this circuit for a change in h_{FE} from 50 to 100 is relatively small, thus demonstrating the good stability properties of the emitter-bias method of automatic biasing. These results should also be compared with those obtained in Example 16.2 for the fixed and collector-to-base bias methods.

Problems: D   Small-signal Amplifiers
12.1 The sinusoidal signal generator connected across the input terminals of the small-signal, common emitter amplifier drawn in fig. P12.1 feeds in a base current which varies the collector current between $\pm 0{\cdot}3$ mA about its quiescent value of $I_C = 2$ mA. If the collector supply voltage $V_{CC} = 20$ V and the load resistor $R_L = 4{\cdot}7$ kΩ calculate the following:

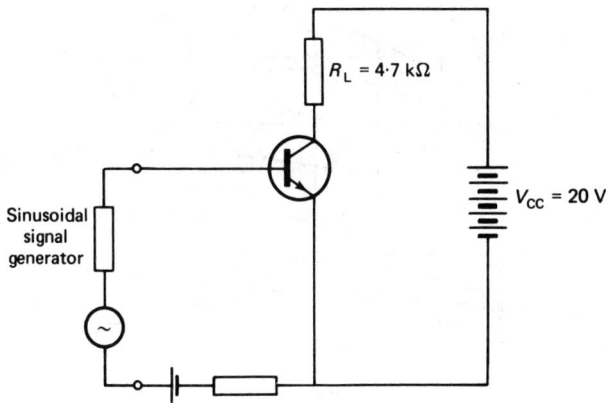

Fig. P12.1. Amplifier circuit for Problem 12.1.

213

(a) The quiescent value of collector-emitter voltage.

(b) The value of collector-emitter voltage when the signal generator produces (i) maximum, (ii) minimum base-emitter forward-bias.

(c) The peak and r.m.s. values of signal voltage developed across R_L.

12.2 The output characteristics of the transistor used in the common emitter amplifier of fig. P12.2(a) are given in (b). The collector supply voltage $V_{CC} = 9$ V. Calculate (a) the value of base current, (b) the value of the load resistor R_L to provide a quiescent operating point of $V_{CE} = 6$ V, $I_C = 15$ mA. Estimate also the power dissipated in the transistor at the quiescent operating point.

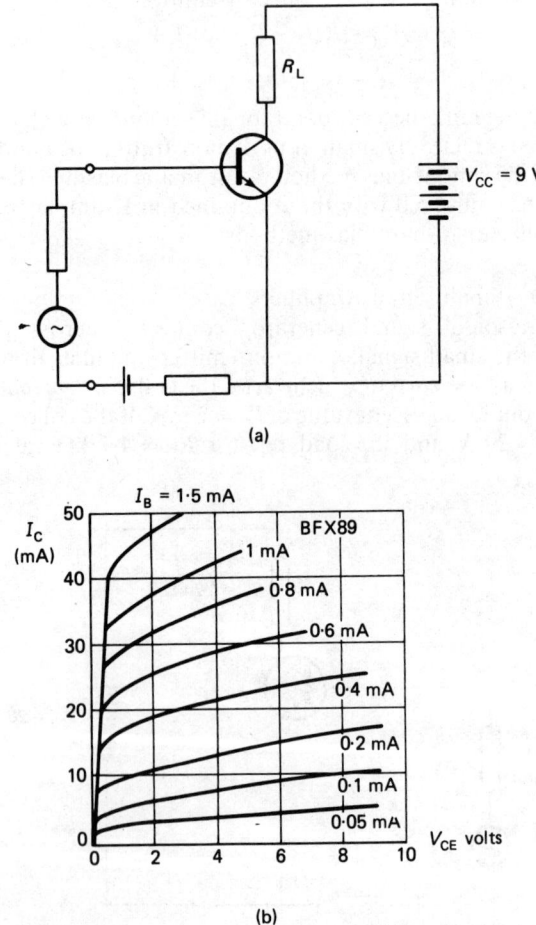

(a)

(b)

Fig. P12.2. Circuit and Characteristics for Problem 12.2.

12.3 Draw the circuit diagram of a common emitter amplifier employing a bias network which supplies a fixed base-bias current. Show that the value of the resistor R_B used in the network is given to a good degree of approximation by

$$R_B = V_{CC}/I_B$$

where V_{CC} = collector supply voltage, I_B = required base-bias current.
State the major drawbacks of this method of bias.

A common emitter amplifier employing a 'fixed' bias network has the following components:

load resistor, $R_L = 2$ kΩ,
collector supply of voltage $V_{CC} = 12$ V,
a transistor with a d.c. current gain, $h_{FE} = 100$ (at $I_C = 2$ mA).

If a quiescent collector current of $I_C = 2$ mA is required calculate a suitable value for the bias resistor R_B. If the transistor were replaced by one with an h_{FE} value of 200, R_B remaining unchanged, calculate the new quiescent values of collector current and collector-emitter voltage.

12.4 Describe the main factors which influence the choice of a quiescent operating point in a small-signal, common-emitter amplifier.

Explain also, with the aid of waveform diagrams, the effect of a small sinusoidal current input on the quiescent condition.

12.5 (a) Calculate the values of base current I_B, collector current I_C, and collector-emitter voltage V_{CE}, in the transistor circuit of fig. P12.5(a). The transistor has a d.c. current gain of $h_{FE} = 100$, and it may be assumed that the base-emitter voltage is 0·6 V.

(b) Calculate the values of R_L and R_B in the amplifier circuit of fig. P12.5(b) to achieve a quiescent operating point of $V_{CE} = 8$ V, $I_C = 1$ mA. The transistor has a d.c. current gain of $h_{FE} = 50$ at $I_C = 1$ mA. State any assumptions made.

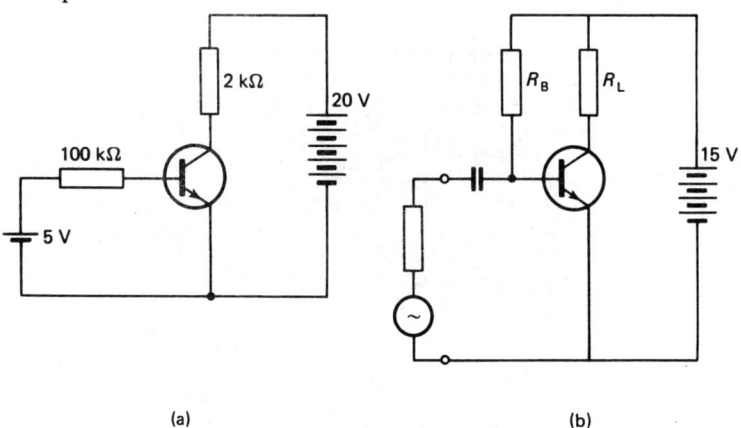

(a)   (b)

Fig. P12.5. Circuits for Problem 12.5.

If the generator connected in the base circuit is sinusoidal and produces a base current variation of peak value 10 μA, estimate, stating any assumptions made, the peak amplitude of signal output voltage across R_L, given that the small-signal current gain of the transistor $h_{fe} = 60$.

Sketch also the waveforms of signal input, base-emitter voltage, and output voltage and comment on their phase relationship.

13.1 Fig. 13.1(a) shows the circuit diagram of a common emitter amplifier with a collector supply voltage $V_{CC} = 8$ V and a load resistor $R_L = 2$ kΩ; (b) shows the output characteristics of the transistor used in the circuit. Write down the load line equation and draw this line on the output characteristics. Hence determine:

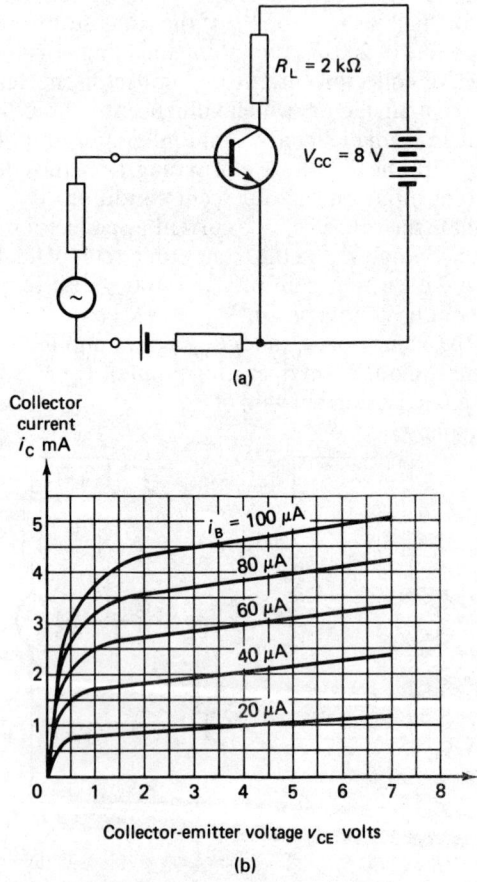

(a)

Collector current i_C mA

$i_B = 100$ μA

80 μA

60 μA

40 μA

20 μA

Collector-emitter voltage v_{CE} volts

(b)

Fig. P13.1. for Problem 13.1.

216

(a) The quiescent operating point if the base current is 40 μA.

(b) The variation in collector current and collector-emitter voltage if the signal source varies the base current by ± 20 μA about the quiescent value of 40 μA.

13.2 The following results were obtained by taking measurements on a small-signal common emitter amplifier:

   a.c. input base current = 10 μA r.m.s.

   a.c. base-emitter voltage = 20 mV r.m.s.

   a.c. component of collector current = 1·5 mA r.m.s.

The load resistance used in the amplifier was $R_L = 2 \cdot 2$ kΩ.

   Use these results to calculate the following:

(a) The small signal input resistance to the amplifier.

(b) The current and voltage gains.

(c) The power gain; express your answer in dB.

13.3 (a) The power gain of an amplifier is quoted as 40 dB. Calculate the ratio of the signal output to input power. If the r.m.s. signal input current to the amplifier is 50 μA and the input resistance 4 kΩ, calculate the signal input and output powers.

(b) The voltage gain of an amplifier is 100. The input resistance is 2 kΩ and the load resistance is 3·3 kΩ. If the r.m.s. signal input current is 10 μA, calculate the following:

(i)   The r.m.s. values of signal input and output voltages.

(ii) The current gain of the amplifier.

(iii) The power gain of the amplifier in dB.

13.4 Fig. P13.4 (overleaf) shows the output characteristics of a transistor employed in a common emitter amplifier. The collector supply voltage and load resistor used in the circuit have the values:

$$V_{CC} = 20 \text{ V}, R_L = 500 \text{ }\Omega.$$

The base bias current is set at $I_B = 200$ μA.

   Draw the load line of the amplifier on the output characteristics and hence determine the values of collector current and collector-emitter voltage at the quiescent operating point.

   If an a.c. signal of peak current amplitude of 100 μA is fed into the base and the small signal input resistance $R_i = 1 \cdot 2$ kΩ, determine:

(a) The peak values of a.c. output current and voltage.

(b) The voltage, current, and power gains of the amplifier.

(c) The a.c. power developed in the load.

13.5 Describe the thermal runaway effect in a transistor.

   Draw diagrams showing two practical types of heat sinks used in transistor circuits and explain why they are normally necessary.

14.1 Draw a circuit diagram of a single-stage thermionic triode amplifier. Explain its small-signal operation.

   The load resistor and H.T. supply in the circuit of fig. P14.1 have the values $R_L = 5 \cdot 6$ kΩ, $V_{HT} = 300$ V. The following measurements were

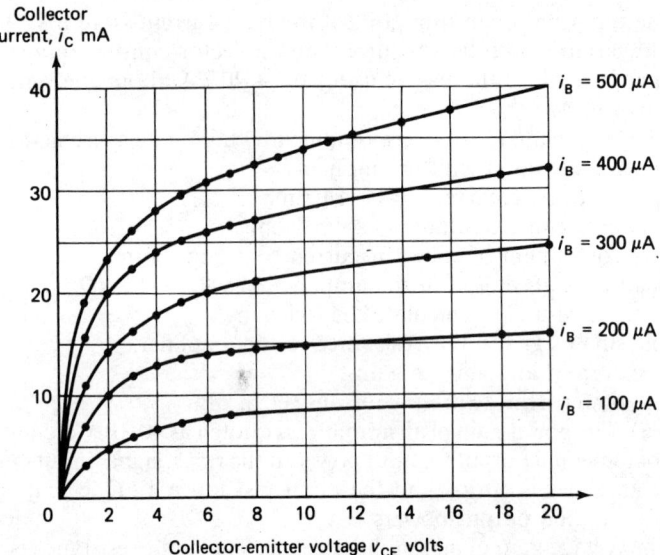

Fig. P13.4. Output Characteristics for Problem P13.4.

recorded for the anode current I_A when the control grid voltage V_G was varied:

$V_G = -1$ V, $I_A = 10$ mA
$V_G = -3$ V, $I_A = 7$ mA
$V_G = -5$ V, $I_A = 4$ mA
$V_G = -10$ V, $I_A = 0$ mA.

Calculate the corresponding values of anode voltage V_A.

If the potentiometer in the grid circuit is set so as to provide a grid bias

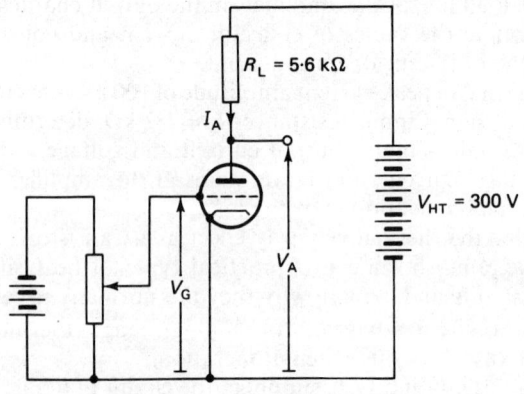

Fig. P14.1. for Problem 14.1.

voltage of $V_G = -3$ V and an a.c. signal generator of peak amplitude 2 V is applied in series in the grid circuit, calculate the peak-to-peak voltage swing across the triode and the voltage gain of the stage. Sketch also the waveforms of grid signal input voltage and output load voltage and comment on their phase relationship.

14.2 Describe the main factors which dictate the choice of a quiescent operating point in a small signal thermionic triode amplifier.

Explain, also, with the aid of waveform diagrams, the effect of a small sinusoidal voltage input on the quiescent condition.

14.3 The output characteristic of the triode used in the amplifier of fig. P14.3(a) are given in (b). The H.T. supply voltage is 240 V. Calculate (a) the value of grid bias, (b) the value of the load resistor R_L to provide a quiescent point of $V_A = 200$ V, $I_A = 3\cdot6$ mA.

The dashed curve in fig. P14.3(b) represents the maximum permitted power dissipation in the anode. The value quoted for this particular triode is $P_A = 1\cdot4$ W. If V_G is set at -1 V, determine the maximum anode current that could be allowed to flow without exceeding the maximum anode dissipation.

15.1 Fig. P15.1(a) shows the circuit diagram of a triode amplifier and (b) the output characteristics of the triode used. Write down the load line equation and draw this line on the output characteristics. Determine the following:

(a) The quiescent operating point of the amplifier if the grid bias voltage is set at -2 V.

(b) The maximum to minimum variation of anode voltage and current when an a.c. voltage generator of 1 V peak amplitude is applied in series in the grid input circuit.

(c) The voltage gain of the stage.

15.2 The output characteristics of a triode used in an amplifier having a resistive load $R_L = 50$ kΩ and an H.T. anode supply of 300 V are given in fig. P15.2. Draw the load line on the characteristics and hence construct the dynamic $i_A - v_G$ characteristic for the amplifier. Use this characteristic to determine the value of grid bias required to select a quiescent point of $V_A = 215$ V, $I_A = 1\cdot7$ mA. Use the dynamic characteristic to make an accurate sketch of the anode current waveform when a sinusoidal signal of peak amplitude 1 V is applied to the grid circuit and when the above quiescent point is used.

15.3 The manufacturers' output characteristics of a triode used in an amplifier are given in fig. P15.3. The amplifier has a resistive load of 1 kΩ and an H.T. supply voltage of 150 V. Determine the quiescent operating point when the grid bias supply is set to -2 V.

If a sinusoidal generator of peak amplitude 1 V is applied in the grid circuit determine (a) the peak and r.m.s. values of the a.c. components of anode voltage and current about the quiescent point, and (b) the voltage gain of the amplifier.

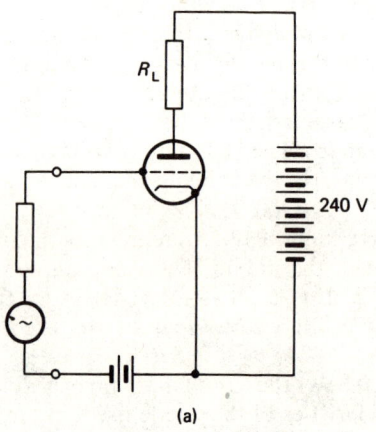

(a)

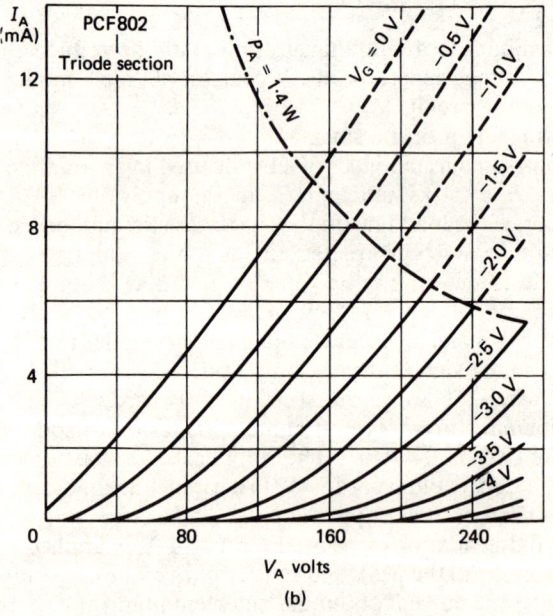

(b)

Fig. P14.3. for Problem 14.3.

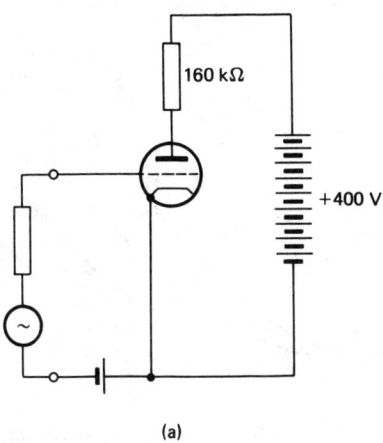

(a)

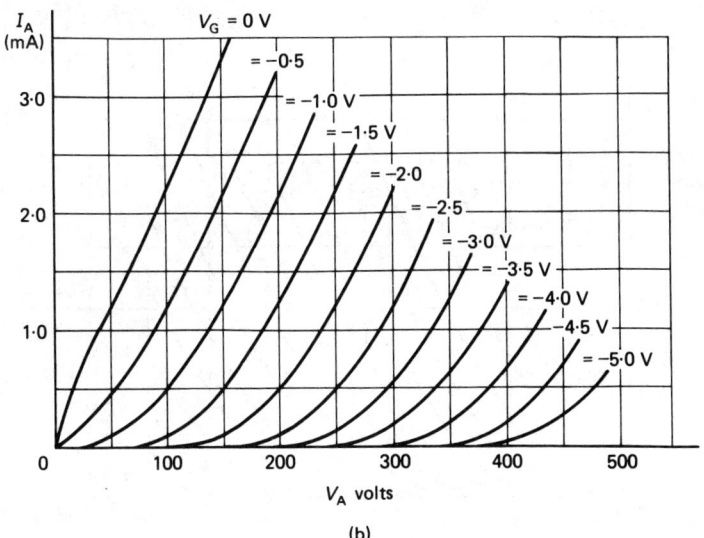

(b)

Fig. P15.1. Circuit and Characteristics for Problem 15.1.

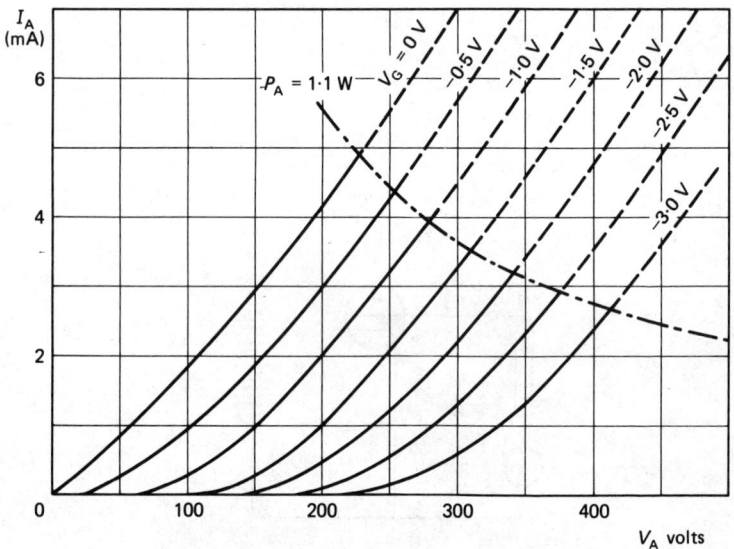

Fig. P15.2. Characteristics for Problem 15.2.

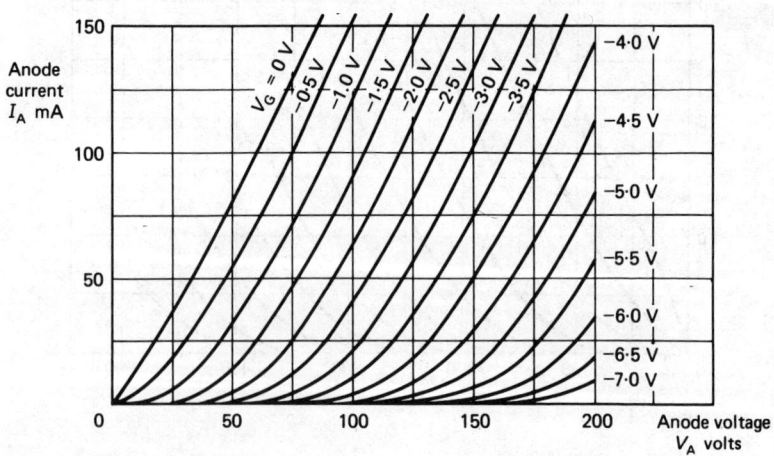

Fig. P15.3. Characteristics for Problem 15.3.

16.1 Explain how automatic bias is achieved in the thermionic triode amplifying circuit in fig. P16.1. What factors determine the value of the capacitor C_C.

Calculate the value of R_C to provide a grid bias voltage of -4 V, given that it is known, from drawing the load line on the anode characteristic, that when $V_G = -4$ V, the anode current is 10 mA.

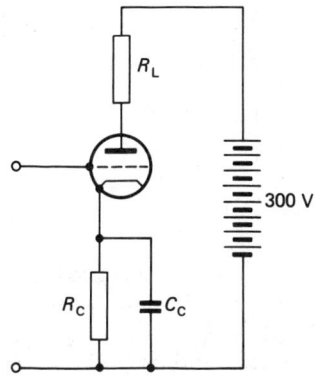

Fig. P16.1. Circuit for Problems 16.1 and 16.2.

16.2 The output characteristics of the triode used in the amplifier of fig. P16.1 are given in fig. P16.2. Calculate the values of R_L and R_C to achieve a quiescent operating point of $V_A = 156$ V, $I_A = 12$ mA.

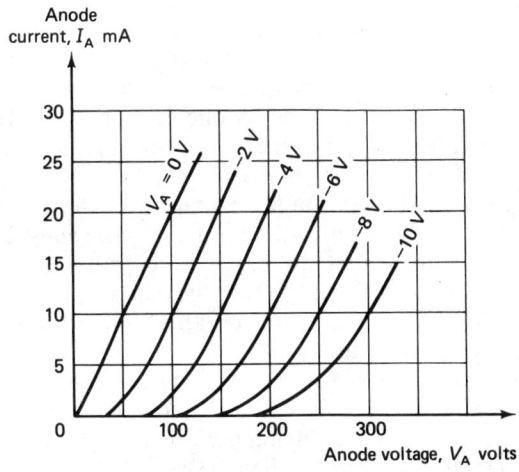

Fig. P16.2. Characteristics for Problem 16.2.

223

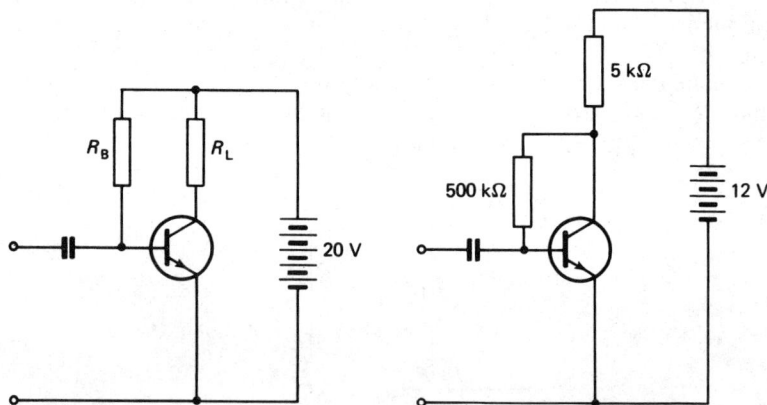

Fig. P16.3. for Problem 16.3. Fig. P16.4. for Problem 16.4.

16.3 Explain how automatic bias is achieved in the amplifier circuit of fig. P16.3. Calculate the values of R_L and R_B, such that the amplifier has a quiescent operating point of $V_{CE} = 12$ V, $I_C = 10$ mA. The transistor used in the amplifier is silicon and has a d.c. current gain of $h_{FE} = 100$.

16.4 Explain the action of the bias network in the circuit of fig. P16.4. Calculate, stating any assumptions made, the values of the base and collector currents and the collector-emitter voltage for this circuit, given that the transistor is silicon and has an $h_{FE} = 100$. Calculate also the change in collector current if the transistor is replaced by one with an h_{FE} value of 50.

16.5 The transistor amplifier circuit shown in fig. P16.5 is to be designed to have a quiescent collector current of 1 mA. Calculate the value of the resistor R_B, given that the h_{FE} of the transistor is 50. Calculate also, with the same value of R_B, the new quiescent collector current when the transistor in the circuit is replaced by one of value $h_{FE} = 100$. Assume that the base-emitter voltage of the transistor remains approximately constant at 0·6 V.

16.6 The circuit diagram of an amplifier employing emitter bias is shown in fig. P16.6. The transistor used in this circuit has a value of $h_{FE} = 200$ and it may be assumed that the base-emitter voltage drop is 0·6 V. Resistors R_1 and R_2 are selected so that the quiescent value of collector current is 1 mA, and the bias supply current flowing through R_1 is 0·2 mA. Calculate the following:

(a) The quiescent values of base and emitter current.
(b) The voltage V_2 across R_2 and the required values of R_1 and R_2.
Describe the effect on the amplifying performance of the circuit when a capacitor of large value (greater than 500 μF) is connected across the emitter resistor R_E.

224

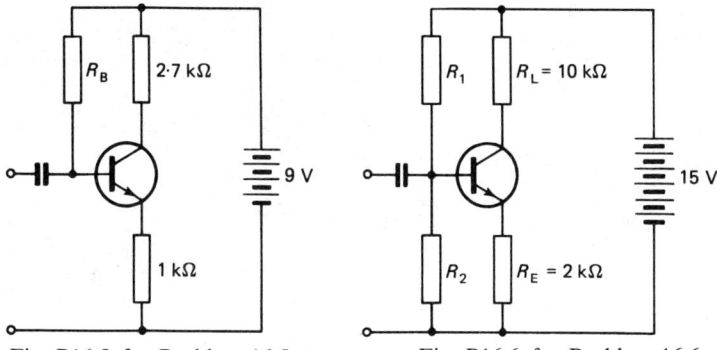

Fig. P16.5. for Problem 16.5.   Fig. P16.6. for Problem 16.6.

Answers

12.1 (a) 10·6 V (b) (i) 9·19 V (ii) 12·01 V (c) 1·41 V, 1·0 V

12.2 (a) 0·2 mA (b) 200 Ω; 90 mW

12.3 $R_B = V_{CC}/I_B$ gives, with $I_B = 0.02$ mA, $R_B = 600$ kΩ; the nearest $\pm 10\%$ preferred value would be 560 kΩ.
$I_C = 4$ mA, $V_{CE} = 4$ V

12.5 (a) $I_B = 44$ μA, $I_C = 4.4$ mA, $V_{CE} = 11.2$ V
(b) $R_L = 7$ kΩ; nearest preferred value is 6·8 kΩ.
   $R_B = 750$ kΩ (neglecting base-emitter voltage drop) or 720 kΩ (if V_{BE} taken as 0·6 V); nearest preferred value is 680 kΩ.
   Peak output signal amplitude = 4·2 V, assuming $R_o \gg R_L$

13.1 $v_{CE} = 8 - 2i_C$ (v_{CE} in V, i_C in mA)
(a) $V_{CE} = 3.9$ V, $I_C = 2.0$ mA
(b) i_C varies from 1·1 mA to 2·8 mA,
   v_{CE} varies from 2·4 V to 5·8 V

13.2 (a) 2 kΩ (b) $A_i = 150$, $A_V = 165$ (c) 44 dB

13.3 (a) 10 000, 10 μW and 100 mW
(b) (i) 20 mV and 2 V (ii) 60·6 (iii) 37·8 dB

13.4 $I_C = 15$ mA, $V_{CE} = 12.5$ V
(a) 7 mA, 3·2 V
(b) $A_V = 27$, $A_i = 70$, $A_P = 1900$ or 33 dB
(c) 12 mW

14.1 244 V, 260·8 V, 277·6 V, 0; 33·6 V, 8·4

14.3 (a) -2 V (b) 11·1 kΩ (12 kΩ for nearest $\pm 10\%$ preferred value);
7·2 mA

15.1 (a) $V_A = 240$ V, $I_A = 1$ mA
(b) 304 V to 165 V, 1·46 mA to 0·6 mA
(c) 70 (note all values are approximate)

15.2 -1.3 V

15.3 $V_A = 94$ V, $I_A = 56$ mA (a) 18·5 V, 13·1 V r.m.s.; 18·5 mA, 13·1 mA
r.m.s. (b) 18·5

16.1 $R_C = 400\ \Omega$

16.2 $R_L = 12\ k\Omega$, $R_C = 330\ \Omega$

16.3 $R_L = 800\ \Omega$, $R_B = 200\ k\Omega$ (neglecting V_{BE}) or 194 kΩ (taking V_{BE} = 0·6 V)

16.4 $I_B \approx 12\ \mu A$, $I_C \approx 1·2\ mA$, $V_{CE} \approx 6$ V (neglecting V_{BE}).
When $h_{FE} = 50$, I_C changes to 0·8 mA

16.5 $R_B = 369\ k\Omega$, $I_C \approx 1·8\ mA$

16.6 (a) $I_B = 5\ \mu A$, $I_E \approx 1\ mA$

(b) $V_2 \approx 2.6$ V, $R_2 = 13·3\ k\Omega$, $R_1 = 62\ k\Omega$
($R_2 = 12\ k\Omega$, $R_1 = 56\ k\Omega$ if $\pm 10\%$ preferred values used).

E Waveform Generators

Section 17: *The expected learning outcome of this section is that the student should know typical oscillator waveforms.*

17.1
Sketches of output waveforms of oscillators in common use: sinusoidal, rectangular, sawtooth

An electronic oscillator is a device capable of generating and sustaining at its output terminals a time varying signal, i.e. a voltage or current waveform.

Fig. 17.1 shows a block diagram and a simple equivalent circuit of an oscillator connected across the input terminals of a circuit. v_s represents the oscillator voltage, R_s the internal resistance of the oscillator, and R_i the input resistance of the circuit we are driving; v_i is the output voltage of the oscillator which, of course, equals the input voltage to the circuit.

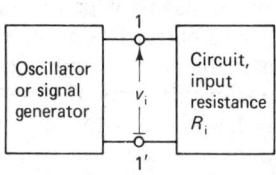
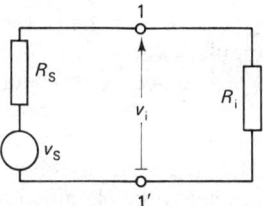

(a) Block diagram showing an oscillator connected across the input terminals of a circuit

(b) Simple equivalent circuit of (a) where oscillator is represented as a voltage generator, v_s and R_s, and input effect of the circuit by R_i; $v_i = v_s R_i/(R_s + R_i)$

Fig. 17.1.

Perhaps the best known oscillator is the sinusoidal oscillator. Typical sketches of the output voltage waveform for such an oscillator are drawn in fig. 17.2. We may express this waveform mathematically as

$$v_i = V_i \sin \omega t$$

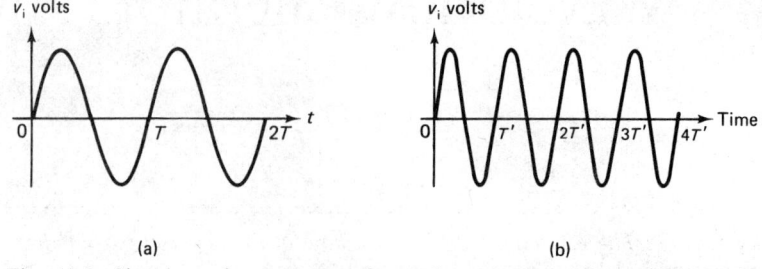

(a)   (b)

Fig. 17.2. Sketches of output waveforms from a sinusoidal oscillator. The waveforms shown have equal amplitudes but are of different frequency. (a) frequency $f = \frac{1}{T}$, (b) $f' = \frac{1}{T'}$; T and T' are the respective periodic times (time of one cycle) of the waveforms.

where V_i = peak amplitude of the waveform (r.m.s. amplitude = $V_i/\sqrt{2}$),
$\omega = 2\pi f$ = the angular or radian frequency of the waveform, units radians per second (rad/s),
f = frequency of waveform, units hertz (Hz),
T = periodic time of waveform, units seconds (s); remember $f = 1/T$.

Sinusoidal oscillators are available commercially with frequencies from less than 1 hertz up to 100 or so gigahertz (1 GHz = 10^9 Hz) corresponding to periodic times from seconds to thousand-millionths of seconds.

Fig. 17.3 shows the output waveforms of oscillators which produce rectangular waves. (a) is a sketch of a symmetrical square wave, symmetrical because the voltage excursions in the positive and negative directions are equal, square because the positive and negative voltage durations of the wave are equal, both being $\frac{1}{2}T$, where T = the repetition period of the wave; (b) shows a sketch of a rectangular wave, whilst (c) would tend to be classified as a pulse train since the duration τ of the

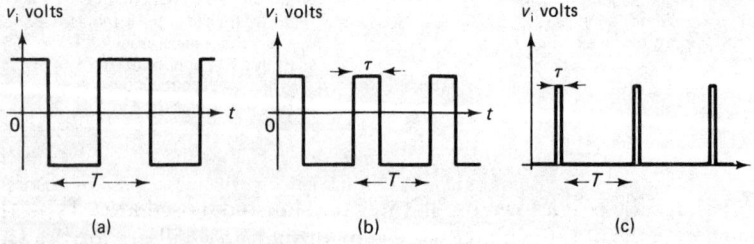

(a)   (b)   (c)

Fig. 17.3. Sketches of output waveforms from an oscillator generating rectangular waves. (a) Symmetrical square wave; (b) rectangular wave (alternating positive and negative); (c) pulse train of positive rectangular pulses.

228

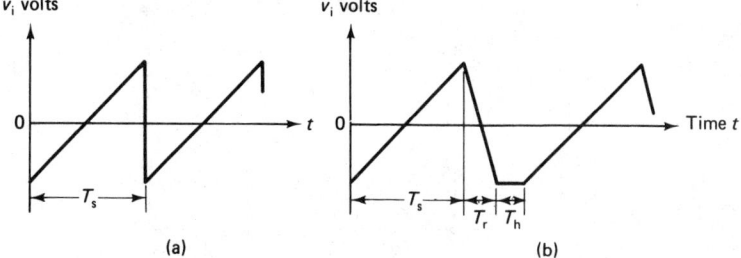

Fig. 17.4. Sketches of output waveforms from an oscillator generating sawtooth waves. (a) ideal sawtooth waveform with zero flyback time; (b) a practical sawtooth waveform. T_s = sweep time, T_r = flyback or retrace time, T_h = hold time.

rectangular pulses is very short compared to the repetition period T. The ratio $\tau : T$ is often known as the mark-space ratio; the repetition frequency of the wave is defined as $f = 1/T$. Oscillators which produce square waves are normally called square-wave generators, whilst oscillators which produce rectangular waves or pulse trains are known as pulse generators. Pulse generators usually have the facility to alter the mark-space ratio and so may be used to generate a variety of rectangular waveforms.

Fig. 17.4 shows two sketches of sawtooth waveforms. We have previously met sawtooth waveforms and used them to drive the X plates of an oscilloscope so that we may display signal waveforms. Oscillators which generate sawtooth waveforms are also known as time-base generators and ramp generators.

17.2
The common uses of sinusoidal, rectangular, and sawtooth waveforms
The sinusoidal oscillator is used as a basic signal generator in both test and communication applications. Fig. 17.5 illustrates a test application; Fig. 17.6 a communications application.

In fig. 17.5, the sinusoidal input voltage to the amplifier is displayed on one pair of Y plates and the output voltage displayed on the second pair of Y plates of a double beam (dual trace) cathode ray oscilloscope.* Most test generators have a variable frequency control and so we may use this to investigate how the voltage gain of our amplifier varies with frequency. Sinusoidal oscillators are available to cover various bands of frequencies; an audio frequency oscillator, for example, would typically provide a sinusoidal signal whose frequency could be varied from a few hertz to about 20 kHz; a low frequency oscillator typically from 10 Hz to

*Most modern test oscilloscopes can display two signals simultaneously. They have facilities for the connection of two separate signal inputs and thus allow us to display on the screen two signal waveforms.

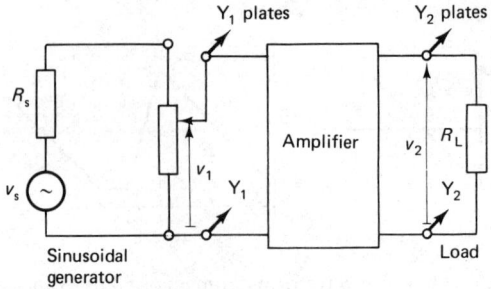

(a) A typical circuit for testing an amplifier

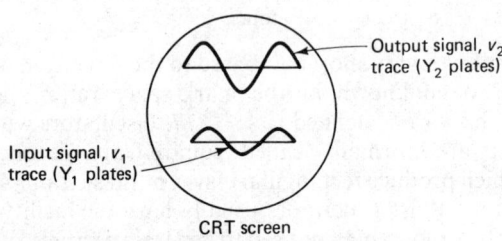

(b) Input and output signal traces on
screen of a dual trace oscilloscope

Fig. 17.5.

the order of 100 kHz; a radio frequency oscillator from 10 kHz to the order of 70 MHz. Sinusoidal oscillators are also available at very much higher frequencies up to the order of 100 GHz.

In the communication system shown in diagrammatic form in fig. 17.6, the sinusoidal oscillator—known as the carrier generator—produces a constant amplitude, constant frequency sine-wave output. Special stabilization circuits are incorporated in the carrier generator to ensure this. The output, known as the carrier wave is then fed, via an amplifier, into a modulator. The signal we wish to transmit is first converted into an electrical waveform, amplified, and then fed into the modulator, where it is used to control the carrier in some definite way. In an amplitude modulated system, the signal varies the amplitude of the carrier so that this is made to follow exactly the amplitude variation of the signal. In frequency modulated systems, the signal controls the frequency of the carrier so that the frequency of the carrier is increased and decreased as the signal increases and decreases. Typical output waveforms for an amplitude modulated wave and a frequency modulated wave are sketched in fig. 17.6(b).

Rectangular, square, and pulse generators are used extensively in all branches of electronics and allied areas: in switching, triggering, control,

230

counting, computing, coding, and communications applications. For example, a rectangular wave could be used to bias a valve or transistor on or off, i.e. by applying the wave to the input grid-cathode circuit of a valve or by applying the wave to the base-emitter circuit of a transistor (the use of a transistor as a switch is discussed in Section 21). A rectangular wave could also be applied to the grid of a CRT to brighten or switch on the electron beam during its scan time and to cut off the beam during flyback. Rectangular and pulse waveforms are also used to communicate information (we discuss in Section 19 some examples of these applications); rectangular and pulse waveforms are used to operate and control logic circuits—logic circuits being the basic building blocks used in modern digital electronics.

Rectangular waveform generators are also used in test applications. The output voltage or current waveforms of a circuit fed with a square wave or pulse train input provide valuable information regarding the

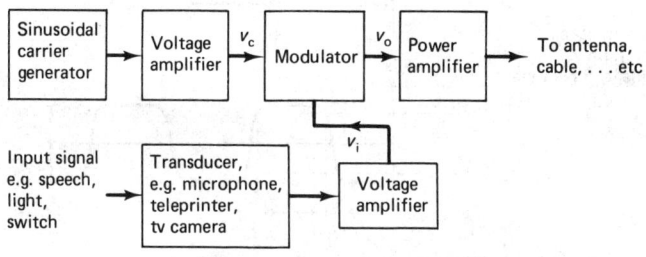

(a) A basic block diagram of a communications transmitter

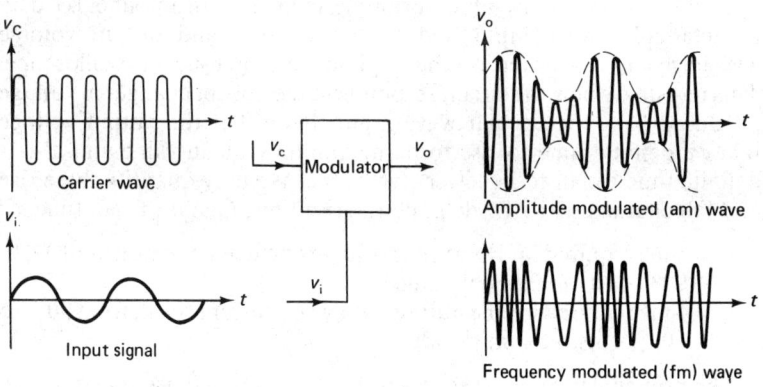

(b) Waveform sketches showing output from transmitter for the cases of amplitude and frequency modulation. For am, the input signal varies the amplitude of the carrier; for fm, the input signal varies the frequency of the carrier

Fig. 17.6.

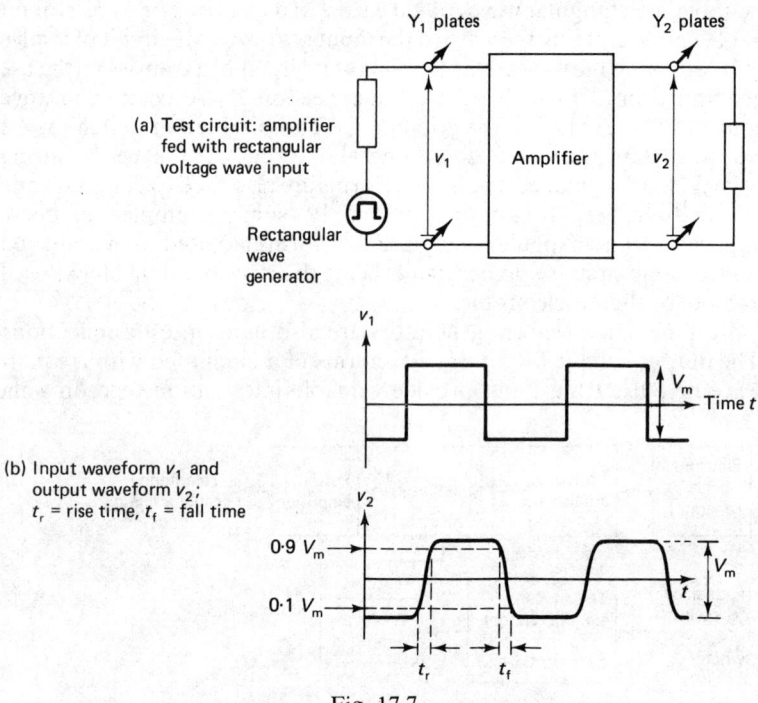

(a) Test circuit: amplifier fed with rectangular voltage wave input

Y_1 plates $\qquad$ Y_2 plates

Amplifier

v_1 $\qquad$ v_2

Rectangular wave generator

(b) Input waveform v_1 and output waveform v_2; t_r = rise time, t_f = fall time

v_1

Time t

V_m

v_2

$0.9\ V_m$

$0.1\ V_m$

V_m

t

t_r $\qquad$ t_f

Fig. 17.7.

characteristics of the circuit; for example, in fig. 17.7 an amplifier is fed by a rectangular wave input and both the input and output voltage waveforms are displayed on the Y plates of a dual-beam oscilloscope. This test shows how the circuit responds to the abrupt changes in voltage produced by a rectangular wave input. Note that the output voltage takes a definite time to rise to its maximum plateau value and also a definite time to fall to its lower level value. We may quantify these rise and fall characteristics by defining the rise time, t_r, and the fall time, t_f:

t_r = time required for the output voltage or current to rise from 10% to 90% of its maximum value.

t_f = time required for the output voltage or current to fall from 90% to 10% of its maximum value.

The principal use of sawtooth waveforms is as time-base generators to drive the X-plates of cathode ray oscilloscopes and for the deflection of the electron beam in television tubes. Sawtooth waveforms may also be used to effect linear changes in many other applications. For example, a sawtooth voltage may be used to change the frequency of a sinusoidal

232

oscillator linearly with time. Such oscillators are known as sweep-frequency oscillators and find important applications in the testing of electronic circuits.

Section 18: *The expected learning outcome of this section is that the student should know the principles of simple sinusoidal oscillators.*

18.1

A sine-wave oscillator is an amplifier with positive feedback sufficient to maintain its own output

Let us first define the meaning of feedback in an amplifier. In general, we may state that when part of the output signal of an amplifier is combined with the input signal, feedback is said to exist. If the resultant effect of the feedback is to increase the input at the amplifier input terminals, then the feedback is known as positive or regenerative feedback. If the resultant input signal is reduced, the feedback is called negative or degenerative feedback.

Fig. 18.1(a) shows an amplifier with no external feedback. The voltage gain of the amplifier

$$A = \frac{\text{signal voltage amplitude across output terminals}}{\text{signal voltage amplitude across input terminals}} = \frac{V_o}{V_i}$$

Note that (a) defined above contains no sign. If the voltage gain is a

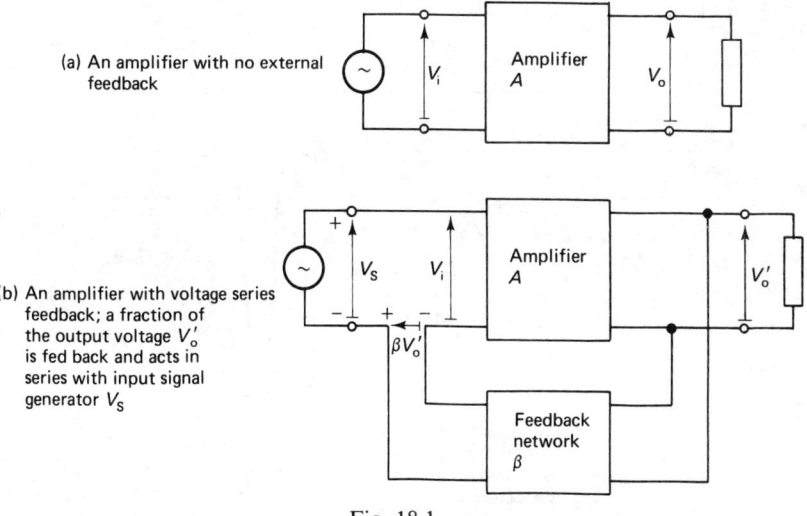

(a) An amplifier with no external feedback

(b) An amplifier with voltage series feedback; a fraction of the output voltage V_o' is fed back and acts in series with input signal generator V_S

Fig. 18.1.

positive number, this means that the output is in phase with the input. If the voltage gain is a negative number, the negative sign is included to indicate that phase inversion occurs in the amplifier (phase inversion was discussed in Sub-sections 12.6 and 14.4).

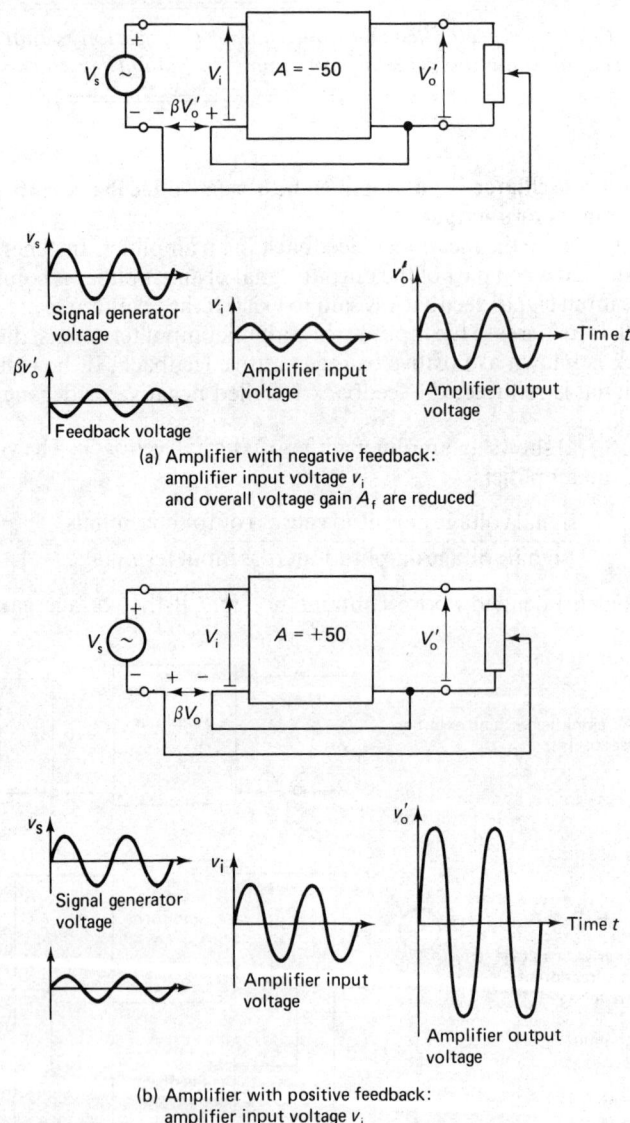

(a) Amplifier with negative feedback: amplifier input voltage v_i and overall voltage gain A_f are reduced

(b) Amplifier with positive feedback: amplifier input voltage v_i and overall voltage gain A_f are increased

Fig. 18.2.

Fig. 18.1(b) shows a schematic diagram illustrating one method of how feedback might be applied. A fraction β of the output voltage V_o', that is voltage of $\beta V_o'$, is applied in series with the signal generator voltage V_S in such a way that the resultant signal appearing at the input terminals of the amplifier is

$$V_i = V_S + \beta V_o'$$

If $\beta V_o'$ is in phase with the signal generator voltage V_S, the feedback voltage increases the effective input signal and we have positive feedback. If $\beta V_o'$ is in anti-phase with V_S, the input signal is decreased, $\beta V_o'$ subtracts from V_S, and we have negative feedback. The situation of negative feedback occurs, for example, if phase inversion occurs in the amplifier (and hence the voltage gain A is a negative quantity) and we feed back from the output through a resistive network so that no further phase shift occurs. Remember that phase inversion occurs in a single-stage common emitter transistor amplifier and in a common cathode triode amplifier. The amplifier itself sees $V_i = V_S + \beta V_o'$ at its input and hence it is this voltage that is amplified. Thus, the output voltage of the amplifier is

$$V_o' = A V_i = A(V_S + \beta V_o')$$

and using this equation to solve for V_o' we have

$$V_o' = A V_S + A\beta V_o'$$

$$V_o' - A\beta V_o' = A V_S$$

$$V_o'(1 - A\beta) = A V_S$$

So
$$V_o' = \frac{A V_S}{1 - A\beta}$$

and the overall voltage gain of the amplifier with feedback is defined as

$$A_f = \frac{\text{output voltage}}{\text{signal generator voltage}} = \frac{V_o'}{V_S} = \frac{A}{1 - A\beta}$$

We will now investigate how feedback changes the overall voltage gain of an amplifier by substituting some values in the above formula for A_f.
(a) Suppose the gain of the amplifier shown in fig. 18.2(a) is $A = -50$ (i.e. phase inversion occurs in the amplifier) and the fraction $\beta = 0.01$ of the output voltage is fed back. We have negative feedback since the input signal amplitude to the amplifier is effectively decreased, i.e.

$$V_i = V_s + \beta V_o' = V_s + \beta A V_i$$

$$= V_s + 0.01 \times -50 \times V_i = V_s - 0.5 \ V_i$$

So
$$1.5 \ V_i = V_s, \ V_i = V_s/1.5 = 0.67 \ V_s.$$

The overall voltage gain,

$$A_f = \frac{V_o'}{V_s} = \frac{A}{1-A\beta} = \frac{-50}{1-(-50 \times 0.01)} = \frac{-50}{1+0.5}$$

$$= \frac{-50}{1.5} = -33.3$$

showing that the magnitude of the overall gain is decreased from 50 to 33.3.

You may wonder at this stage why we may use negative feedback, especially if its effect is to reduce the overall amplifier gain. Let us briefly comment on this. Negative feedback is used since it stabilizes the voltage and current gains of an amplifier. It makes them less dependent on the characteristics of the active components (valves, transistor, or integrated circuits) and on temperature, supply voltage, signal frequency, and other variations. It also reduces distortion and noise effects. In practical amplifiers, negative feedback is invariably employed.

(b) Suppose now the 'normal' gain of the amplifier shown in fig. 18.2(b) is $A = +50$ and a fraction $\beta = 0.01$ of the output voltage is fed back. We have, in this case, positive feedback since the input of the amplifier is increased, i.e.

$$V_i = V_s + \beta V_o' = V_s + \beta A V_i$$
$$= V_s + 0.01 \times 50 \ V_i = V_s + 0.5 \ V_i$$

So $\quad 0.5 \ V_i = V_s, \ V_i = V_s/0.5 = 2 \ V_s.$

The overall voltage gain,

$$A_f = \frac{A}{1-A\beta} = \frac{50}{1-50 \times 0.01} = \frac{50}{0.5} = 100,$$

is also increased—in fact, it is doubled. Thus, positive feedback increases the overall voltage gain of an amplifier. From an amplification point of view, positive feedback is seldom used. Although it increases the overall voltage gain, it decreases the amplifier stability. Indeed, the amplifier may break into oscillation when sufficient positive feedback occurs. For amplifier applications, this is extremely undesirable.

If we increase the fraction of output voltage fed back to $\beta = 0.02$, we obtain a very interesting situation:

$$A\beta = +50 \times 0.02 = 1.$$

So the voltage gain with feedback, $A_f = \dfrac{A}{1-A\beta} = \dfrac{50}{1-1} = \dfrac{50}{0} \to \infty$

i.e. the voltage gain approaches infinity. We know that, in practice, we cannot achieve infinite gain in an amplifier, so what does this result infer

236

in physical terms? It means that a vanishingly small input voltage would give rise to a finite (that is, a definite amount of) output voltage even when no signal generator is present in the input circuit. Thus, if we were to remove the signal generator from the input circuit and apply sufficient positive feedback so that $A\beta$ was equal to, or even greater than, 1 our amplifier would become an oscillator, generating a continuous waveform at its output terminals.

This principle is applied in all positive feedback oscillators. We can thus state that a sine-wave oscillator is an amplifier with positive feedback sufficient to maintain its own output, the output being a continuous sine-wave. Fig. 18.3(a) shows a schematic diagram of a positive feedback oscillator.

Finally, let us give a brief explanation as to how the circuit shown in fig. 18.3(a) acts as an oscillator. When we switch on the amplifier, steady currents flow and voltages are established, and our amplifier is capable of producing gain. Superimposed on these steady values are small components of alternating voltages and current which cover all frequencies. These components are known as noise and are always present in electronic circuits. It is these noise components which provide

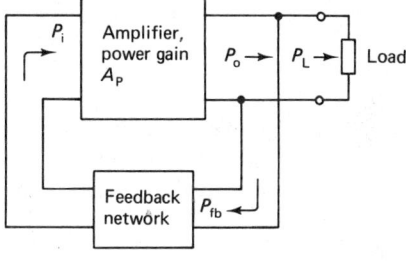

(a) Block diagram of a basic oscillator: the oscillator is an amplifier with positive feedback sufficient to maintain its own output

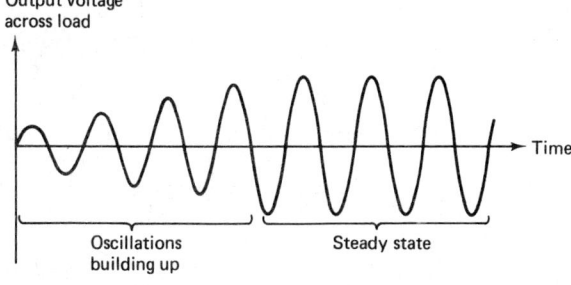

(b) Sinusoidal output voltage waveform from the oscillator

Fig. 18.3.

237

an initial input which may start the build up of oscillation. The frequency at which the sinusoidal oscillator operates is the frequency at which the total phase shift of a signal, which proceeds from input to output through the amplifier and then through the feedback network back to the input, is 2π radians or 360°. This condition gives positive feedback since the signal fed back is in phase with the original input signal. Oscillation will be sustained at this frequency provided that $A\beta = 1$. Thus, if the condition that $A\beta = 1$ can be met for a given frequency, oscillation will be obtained at this frequency.

In practice, the oscillation amplitude builds up, as shown in fig. 18.3(b), until a constant amplitude sinusoidal waveform is obtained. When this condition, known as the steady state, is reached, the power fed back at the oscillation frequency is equal to the input power (sufficient to maintain a constant output) plus the loss incurred in the feedback network. Thus, if A_p = the power gain of the amplifier, P_fb = the feedback power, P_i = input power, P_loss = the total loss in the feedback network, P_o = the output power from the amplifier, and P_L = the useful oscillator power supplied to the external load, we have

$$P_\mathrm{fb} = P_\mathrm{i} + P_\mathrm{loss}$$

$$P_\mathrm{o} = A_\mathrm{p} P_\mathrm{i}$$

$$P_\mathrm{L} = P_\mathrm{o} - P_\mathrm{fb}$$

18.2
A sine-wave oscillator requires both a frequency determining circuit and a method of self stabilization

A sine-wave oscillator requires a frequency determining circuit, so that we may design our oscillator to give power at a given frequency, and also a method of self-stabilization so that the oscillator produces a continuous sine wave of constant amplitude.

Fig. 18.4(a) shows an example of an oscillator in which the frequency determining circuit is a parallel L-C circuit. This circuit acts as the load of the amplifier. Energy is extracted from this circuit and is fed back to the amplifier input by means of magnetic (inductive) coupling. Coil L_1 is inductively coupled to coil L of the L-C circuit. The relative direction of the windings of L_1 and L are such that they introduce a phase shift of 180°. Thus, if we employ an amplifier which produces phase inversion, i.e. a phase shift of 180°, the total phase shift from input to output through the amplifier and back again via the feedback circuit is 360°. This means our feedback signal effects positive feedback and hence, if the amplifier produces sufficient gain, the condition $A\beta \geq 1$ is satisfied and oscillation will build up and be sustained. The parallel L-C circuit possesses frequency selective properties and as a result the output from our

238

(a) An oscillator employing a
parallel $L - C$ circuit to
determine its frequency.

Oscillation frequency $f = \dfrac{1}{2\pi\sqrt{LC}}$ Hz

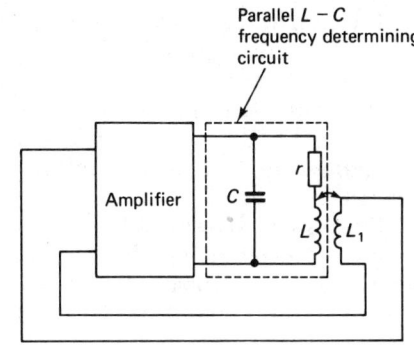

Parallel $L - C$
frequency determining
circuit

(b) An oscillator employing three
$R - C$ networks which determine
its oscillation frequency

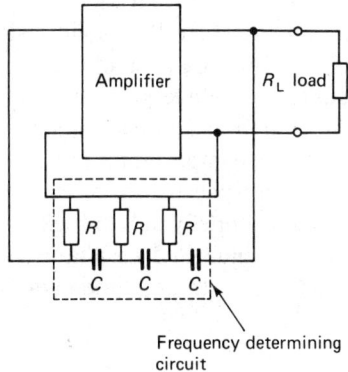

Frequency determining
circuit

Fig. 18.4.

amplifier reaches a maximum, or very close to a maximum, at the
frequency $f_o = 1/[2\pi\sqrt{(LC)}]$ hertz.*

*Provided that a coil of high Q-factor is used, that is

$$Q = \frac{\omega L}{r} \gg 1 \text{ (typically 50 to several 100)}$$

where L = coil inductance, r = loss resistance of coil including winding resistance and any
other associated losses, $\omega = 2\pi f$, f = frequency;
the parallel L-C circuit acts as a very high impedance load to signal frequencies at or close

to $f = \dfrac{1}{2\pi\sqrt{(LC)}}$.

The voltage gain of most amplifiers increases as the load value increases and thus at f_o the
gain of the amplifier tends to a maximum. As the frequency is increased or decreased about
f_o the load impedance presented by the L-C circuit, and therefore the gain, falls off very
rapidly. In fact the magnitude of the L-C circuit impedance falls to one-half its value at f_o at
the frequencies $f = f_o(1 \pm 1/2Q)$. Thus if the Q-factor is large the circuit is very selective.

239

It is also at this frequency, and only at this frequency, that the total input to output phase shift in our amplifier is $180°$. At all other frequencies the phase shift will be different. Hence, the parallel L-C circuit not only produces maximum gain at the frequency f_0 $= 1/[2\pi\sqrt{(LC)}]$, it also supplies the correct phase condition. Circuit diagrams of two oscillators employing a parallel L-C circuit are drawn in fig. 18.5.

Fig. 18.4(b) shows another example of an oscillator in which the frequency determining circuit consists of three R-C networks. This circuit produces a phase shift of $180°$ at a frequency determined by the products of $R \times C$. In fact the actual value is given by

$$f = \frac{1}{2\pi RC\sqrt{6}} \text{ hertz, for a triode amplifier;}$$

or $f = \dfrac{1}{2\pi RC(6 + 4R_L/R)}$ hertz, for a common emitter transistor amplifier, where R_L = load resistor in collector lead.

Thus, if we use an amplifier which produces phase inversion and has sufficient voltage gain to satisfy the criterion $A\beta = 1$, we will obtain oscillation at a given frequency determined by the value of RC.

It is also important that an oscillator is stabilized so that it produces a continuous sine wave of constant amplitude. Automatic bias circuits, discussed briefly in Sub-section 18.5, may be used to effect this. In high quality sinusoidal oscillators, such as those used for communication purposes, both amplitude and frequency must by controlled to a high degree.

18.3
The approximate frequency of oscillation of most L-C sine-wave oscillators is $f_0 = \dfrac{1}{2\pi\sqrt{(LC)}}$ hertz

Most sine-wave oscillators which employ an L-C circuit have a frequency of oscillation given, to a good degree of approximation, by

$$f_0 = \frac{1}{2\pi\sqrt{(LC)}} \text{ hertz}$$

where C = capacitance of circuit in farads,
L = inductance of coil in henrys.

Note that the coil should have a high Q-factor, i.e. $Q = \dfrac{\omega L}{r} \gg 1$.

Normally, a parallel L-C circuit is used although sometimes a series L-C circuit is employed. The actual position of the L-C circuit may be in the output circuit of the amplifier, as shown in figs. 18.4 and 5, or it may be located in the input circuit of the amplifier with feedback power coupled from the output into this circuit. Sometimes two similar L-C circuits are used in an oscillator, one in the input, the other in the output circuit of the amplifier, with coupling so that positive feedback occurs from output to input.

Example 18.1

A frequency determining circuit in an L-C sine-wave oscillator consists of a high Q inductor of inductance $L = 200 \mu H$ in parallel with a capacitor of capacitance $C = 450$ pF. Determine the frequency of oscillation. Take $\pi = 3\cdot142$.

A variable capacitor C_1 is connected in parallel with the L-C circuit. Determine the frequency range of the oscillator if this capacitor can be varied from 20 pF to 150 pF.

Solution

Since
$$L = 200 \ \mu H = 200 \times 10^{-6} \ H$$
$$C = 450 \ pF = 450 \times 10^{-12} \ F$$
then
$$LC = 200 \times 10^{-6} \times 450 \times 10^{-12} = 9 \times 10^{-14}$$
and
$$\sqrt{(LC)} = 3 \times 10^{-7},$$
so the frequency of oscillation is

$$f_o = \frac{1}{2\pi\sqrt{(LC)}} = \frac{1}{2\pi \times 3 \times 10^{-7}} = \frac{10^7}{6 \times 3\cdot142} \ Hz$$

$$= 530\cdot4 \ kHz.$$

When the variable capacitor C_1 is connected in parallel with the L-C circuit the total capacitance in parallel with the inductor is increased to $C + C_1$. Thus the corresponding oscillation frequency changes to

$$f = \frac{1}{2\pi\sqrt{[L(C + C_1)]}} \ hertz.$$

The lowest value of $C_1 = 20$ pF and so the oscillation frequency when the variable capacitor is at its minimum setting is

$$f = \frac{1}{2\pi\sqrt{[200 \times 10^{-6}(450 + 20) \times 10^{-12}]}}$$

$$= \frac{10^7}{2\pi\sqrt{9\cdot4}} = 519\cdot0 \ kHz.$$

The highest value of $C_1 = 150\,\text{pF}$ and thus the corresponding oscillation frequency is

$$f = \frac{1}{2\pi\sqrt{[200 \times 10^{-6}(450 + 150) \times 10^{-12}]}}$$

$$= \frac{10^7}{2\pi\sqrt{12}} = 459.4 \text{ kHz}.$$

18.4
The circuit diagrams of a tuned-anode thermionic triode oscillator and a tuned-collector transistor oscillator

Circuit diagrams for a tuned-anode thermionic triode oscillator and a tuned-collector transistor oscillator are drawn in figs. 18.5(a) and (b) respectively.

In (a) the L-C circuit is located in the anode lead and acts as the load of the triode. Coil L_1 is inductively coupled to coil L and feeds back energy from the anode output circuit into the grid input circuit. At the oscillation frequency, $f_o = 1/[2\pi\sqrt{(LC)}]$, positive feedback is effected so that the voltage fed back to the grid-cathode terminals is 'in phase'. Thus, when the circuit is switched on, oscillations will build up at this frequency provided, of course, that the gain of the amplifier is sufficient. The R_G-C_G parallel combination acts as an automatic bias network. It also effects a degree of amplitude stabilization of the oscillator voltage developed across the L-C circuit. We will explain the action of this network in the next sub-section. In the diagram shown, the oscillator output is taken across terminals 2-2'. Alternatively, we may couple out power from the oscillator by employing another coil which is also inductively coupled to the parallel circuit coil L. Our external load, in this case, would be connected across the terminals of this second coil. This method of coupling is known as transformer coupling.

In (b) the L-C circuit is located in the collector lead. Oscillations build up and are sustained at the frequency $f_o = 1/[2\pi\sqrt{(LC)}]$ by coupling energy from the L-C circuit back into the base circuit. The oscillator output is taken across the terminals 2-2'. Alternatively, we may use transformer coupling and take power from the oscillator by employing a second coil inductively coupled to L. The other components in the circuit—R_1, R_2-C_2, and R_E-C_E—provide automatic bias for the transistor. We will consider their action in the next sub-section.

18.5
Methods of applying bias in tuned-circuit oscillators

The automatic bias in the tuned-anode triode oscillator is provided by the R_G-C_G network in the grid circuit. Let us explain the action of this circuit by reference to fig. 18.6. When the oscillator is first switched on, the grid bias voltage is zero and the operating point is at Q_o, this point

242

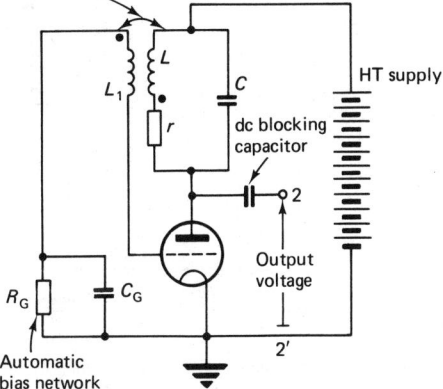

(a) Circuit diagram of a tuned-anode triode oscillator

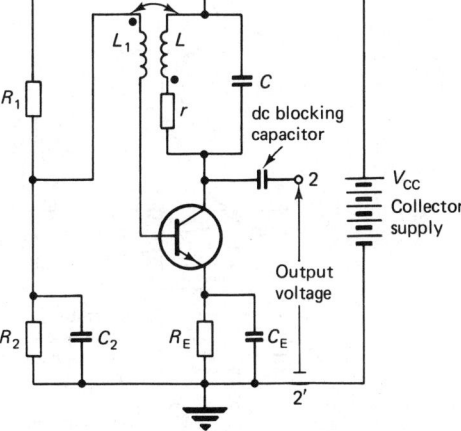

(b) Circuit diagram of a tuned-collector transistor oscillator

Fig. 18.5.

being marked in on the dynamic i_A-v_G characteristic of the triode shown in fig. 18.6(b). Switching on also initiates oscillations in the L-C circuit and energy will be coupled into the grid circuit via L_1. We have represented this effect in fig. 18.6(a) by marking in the voltage v_{fb} which acts in series with the C_G-R_G network. v_{fb} is the effective positive feedback voltage at the oscillation frequency, $f_o = 1/[2\pi\sqrt{(LC)}]$. As the initial oscillation starts, the grid will go positive on the first positive half-cycle of v_{fb}. The grid will therefore collect electrons and a current will flow in the grid circuit charging up the capacitor C_G, with the terminal closest to the

243

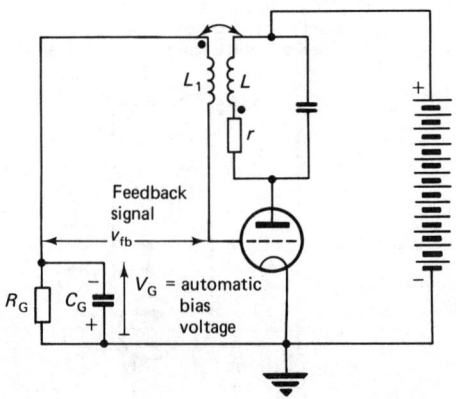

(a) Tuned-anode oscillator showing $R_G - C_G$ automatic bias network. Note that the signal fed back to the grid circuit v_{fb} acts in series with the bias voltage V_G

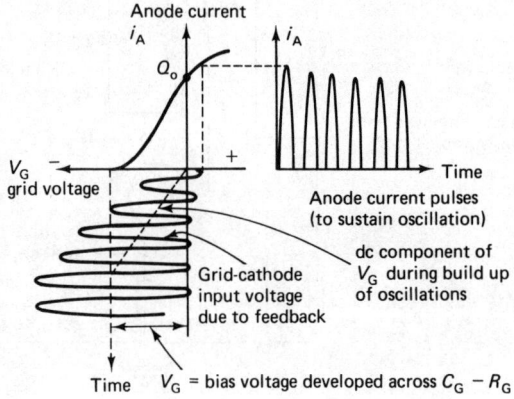

(b) Waveform sketches showing the build up of oscillation in the grid circuit and the pulses of anode current produced to sustain oscillations. Also shown is the $i_A - v_G$ dynamic characteristic of the triode, which is used to determine the pulses produced by the grid-cathode input voltage

Fig. 18.6.

grid negative with respect to the terminal connected to the cathode. Provided that the condition $C_G R_G > T$, where $T = 1/f_o$ = period of the oscillation voltage is satisfied, the capacitor will hold this charge over the period T, and hence the voltage developed across the C_G-R_G network acts as a steady d.c. bias voltage. This bias voltage displaces the operating point to the left as the oscillator amplitude grows, as shown in fig. 18.6(b). The amplitude of the oscillations in both grid and anode circuits soon stabilizes and when equilibrium is established, a relatively small grid current flows only during positive peaks of v_{fb} to maintain C_G charged.

The magnitude of the bias voltage may be controlled by varying the values of R_G and C_G. This process is normally carried out experimentally. In the case shown in fig. 18.6(b), the grid bias is chosen so that the triode is operated beyond cut-off for most of each cycle. Anode current flows for only a short duration on the positive peaks of v_{fb}. The anode current is thus in the form of pulses, as shown in (b). These pulses serve to maintain voltage oscillations of constant amplitude across the L-C circuit. Any variation that tends to reduce the amplitude of the oscillations will cause reduced feedback and thus reduce the bias. Reduction in bias causes the amplification of the triode to increase and hence the oscillations will build up again automatically. Thus, the C_G-R_G network acts so as to stabilize the oscillator output. The mode of operation described above, where anode current only flows for a short proportion of each cycle, is known as class C operation and is used in oscillators to increase their efficiency. Class C operation will be discussed in the Electronics III Unit.

In the tuned-collector oscillator shown in fig. 18.7(a), bias is provided using an emitter bias network. Remember, we discussed this method of providing bias of an amplifier stage in Section 16.1. The voltage developed across R_2 of the R_1-R_2 network by the collector voltage supply acts to forward-bias the base-emitter junction of the transistor, while the voltage developed across the emitter resistor R_E due to emitter current flowing in R_E acts in the opposite sense. Capacitors C_E and C_2 act as by-pass capacitors and provide an effective short-circuit for a.c. currents at the oscillation frequency in the emitter lead and base input circuit respectively. C_E also plays an important role when the oscillator works under class C conditions. In this respect, C_E supplies a similar function to C_C of the automatic bias network used in the tuned-anode triode circuit.

In fig. 18.7(b), we have illustrated the build up of oscillations when the oscillator operates under 'small-signal' linear conditions. In this case, the amount of feedback and/or the gain of the transistor are sufficient to build up and sustain oscillation, but they are below the values which would drive the transistor into non-linear operation (i.e. when the transistor is driven so that the base current is cut off on one part of the cycle and driven into saturation on another part of the cycle). Thus, the feedback signal produces sinusoidal variations in both base and collector currents about the operating points Q set by the emitter bias network. Point Q is shown on the dynamic i_B-i_C characteristic in the diagram of fig. 18.7(b). This form of operation is known as class A. Class A is the form of operation used in small-signal amplifiers.

However, most oscillators work under class C conditions. Class C

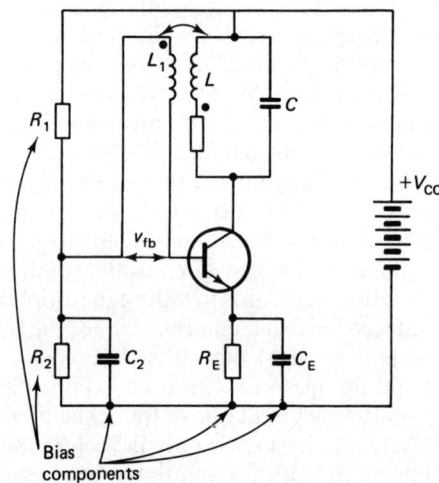

(a) Tuned-collector oscillator showing bias network components

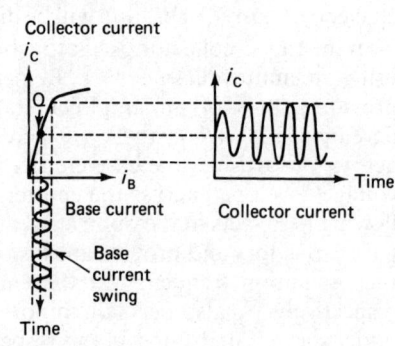

(b) Build up of oscillation when circuit operates under class A conditions

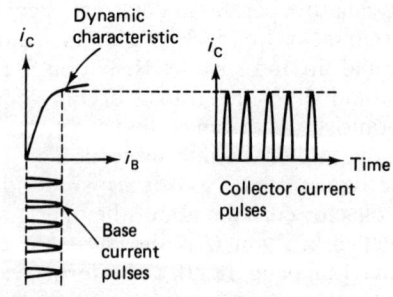

(c) Sketches of waveforms of base and collector current when the oscillator operates under class C conditions

Fig. 18.7.

operation gives very much increased efficiency and much greater power output from our oscillator. The base and collector current waveforms for class C operation are sketched in fig. 18.7(c). In the steady state, the (n-p-n) transistor conducts strongly only for a short duration on positive peaks of the feedback signal v_{fb} and thus only pulses of base, and therefore collector and emitter current flow once each cycle. The collector current pulses supply energy to the L-C circuit to sustain oscillations* of constant voltage amplitude. The emitter current pulses keep C_E charged and thus maintain the d.c. voltage across the R_E-C_E network. This voltage acts essentially to reverse-bias the base-emitter junction and thus the transistor only conducts when the feedback signal is at such a magnitude as to forward-bias this junction. This situation occurs on positive peaks of the feedback signal for an n-p-n transistor.

Problems: E Waveform Generators

17.1 Draw to scale the following oscillator waveforms, labelling both voltage and time axes:

(a) A sinusoidal voltage of 5 V peak amplitude and 10 kHz frequency.

(b) A symmetrical square wave varying between ± 5 V and of 1 μs period.

(c) A sawtooth voltage rising from 0 to 5 V with a sweep time of 9 ms and falling linearly to 0 V in a fly-back time of 1 ms.

17.2 Describe two common uses of each of the following oscillators: sinusoidal oscillators, pulse generators, sawtooth generators.

18.1 Explain, with the aid of suitable diagrams, the following statements:

(a) A sine-wave oscillator is an amplifier with positive feedback sufficient to maintain its own output.

(b) A sine-wave oscillator requires a frequency determining element and a method of self-stabilization.

18.2 Draw a circuit diagram of a tuned anode thermionic triode oscillator and describe the following:

(a) How the oscillator frequency is determined.

(b) How positive feedback is effected.

(c) How automatic bias and amplitude stabilization may be achieved.

18.3 Fig P18.3 shows a circuit diagram of a tuned collector transistor oscillator. Explain the action of the circuit and estimate its frequency of oscillation.

*At first sight it might be difficult to understand why such unidirectional pulses of collector current (and anode current in the tuned-anode oscillator) lead to a sinusoidal output voltage from the tuned circuit. However, if we regard an L-C circuit as one which, given an incentive, will oscillate freely at its natural frequency of $1/[2\pi\sqrt{(LC)}]$ we can then consider the current pulses as providing the 'pushes' once per cycle to keep the sinusoidal oscillations going. We can compare the mechanism to giving pushes to a swing once per its natural period.

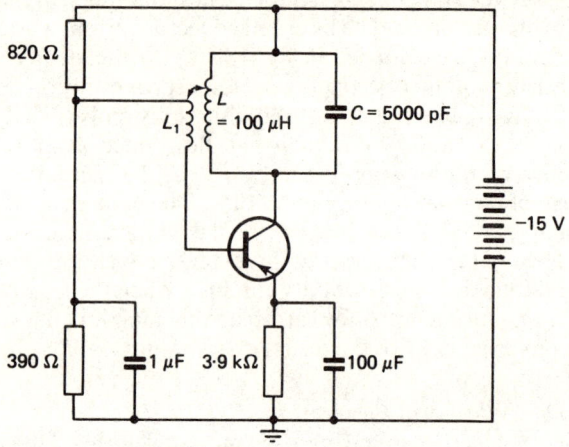

Fig. P18.3. Oscillator circuit for Problem 18.1

Answers
18.3 225 kHz

F Logic Elements and Circuits

Section 19: *The expected learning outcome of this section is that the student should know that information can be communicated by two-state signals.*

19.1
Simple examples of two-state devices

Three 'everyday' examples of two-state devices—although you probably would not have thought of them as two-state devices—are shown in fig. 19.1. Each 'device' has two states: (a) the gate has a 'closed' state and an 'open' state; (b) a simple two-state traffic light indicates a 'go' state when the green light is on, and indicates a 'stop' state when the red light is on; (c) the gate valve has two states, one 'closed', one 'open'.

Although these devices seem far removed from the field of electronics, they do act in an analogous way to a simple switch in an electrical circuit. For example, the switch S in the circuit of fig. 19.2 is a two-state device: the switch has an 'on' and 'off' state. In the 'on' state, S is closed and a conducting connection is made between its contacts A and B. Current flows in the circuit when S is 'on'. In the 'off' state, S is opened and the connection between A and B is broken. No current then flows in the circuit. Thus, when the switch is in the 'on' state, the battery voltage is impressed across the resistor, current flows in the circuit and the voltage developed across the resistor is $V_o = V_S$. When the switch is off, no current flows and $V_o = 0$. Fig. 19.2(b) shows a graph of the output voltage (voltage across R) against time. It is assumed that the switch is 'on' for 10 seconds, 'off' for 10 seconds, and so on. The circuit provides an putput voltage at either of two levels or states:

$$\text{switch 'on', output voltage} = V_o = V_S$$
$$\text{switch 'off', output voltage } V_o = 0$$

Fig. 19.3 shows diagrams of some practical switches used to effect the 'on' and 'off' states in electrical circuits. (b) is a simple manually-operated switch, the metal bar C is used to make ('on' state) or break ('off' state) contacts A and B. (c) shows a simplified diagram of a magnetic relay switch. When energizing current I' flows in the solenoid L, it attracts the

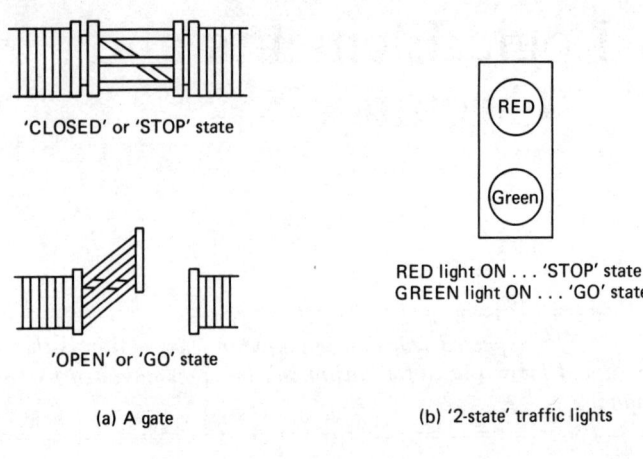

'CLOSED' or 'STOP' state

'OPEN' or 'GO' state

(a) A gate

RED light ON . . . 'STOP' state
GREEN light ON . . . 'GO' state

(b) '2-state' traffic lights

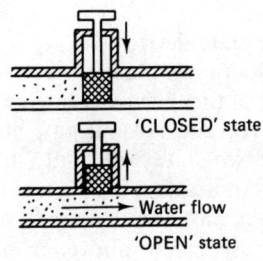

'CLOSED' state

Water flow
'OPEN' state

(c) A simple 'gate' water valve

Fig. 19.1 Three 'everyday' examples of 2-state devices.

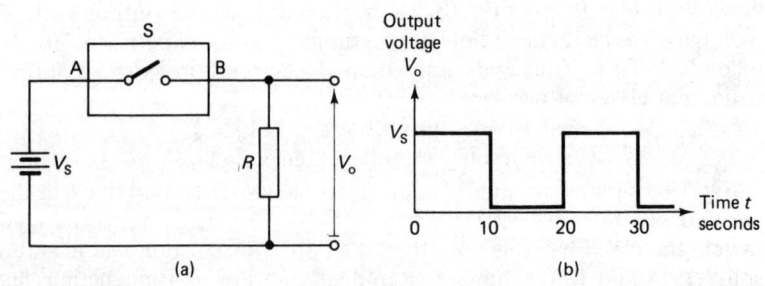

Fig. 19.2.

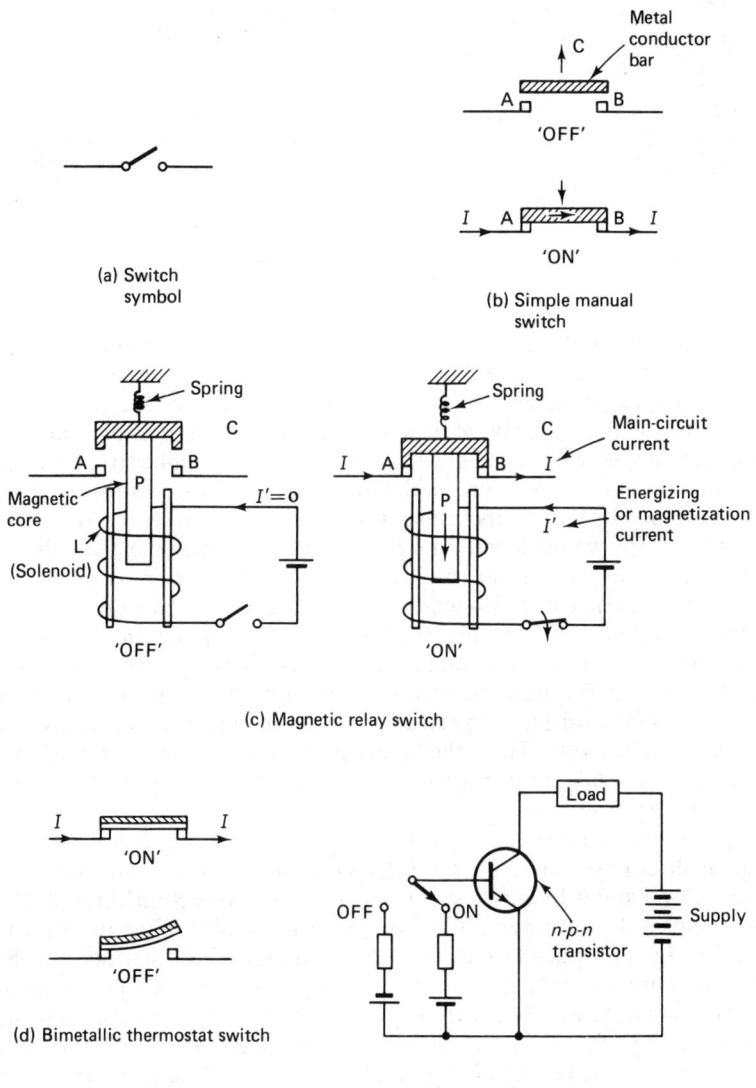

(a) Switch symbol

(b) Simple manual switch

(c) Magnetic relay switch

(d) Bimetallic thermostat switch

(e) Simple transistor switch

Fig. 19.3. Diagrams of some basic switches for use in electrical circuits: the simple switch is a 2-state device for 'ON'–'OFF' control of current and power in electrical circuits.

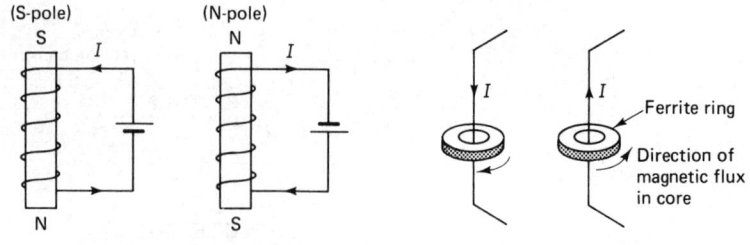

(a) Soft iron core magnetized in 2 states

(b) Ferrite ring (used in some computer memory stores) magnetized in 2 states

Fig. 19.4.

magnetic core P downwards and the conducting bar C joins contacts A and B, thus producing an 'on' condition in the main circuit. When no energizing current flows, the core P is no longer attracted and is pulled upwards by the spring, breaking the circuit, producing the 'off' state. (d) shows a simple switch used in a thermostat. The bimetallic strip expands with increasing temperature, and above a certain temperature it breaks the circuit. If the temperature is subsequently lowered, the strip 'unbends' and below a certain temperature it will make the circuit. (e) shows an example of a simple electronic switch; the action of this switch will be explained in Sub-section 21.3. However, we can offer a brief explanation now. In the 'off' state, the base-emitter junction of the n-p-n transistor is reverse-biased and cuts of current flowing in the transistor, and therefore through the load. In the 'on' state, the base-emitter junction is forward-biased by an amount sufficient to cause the transistor to conduct strongly. Thus, the circuit is switched on and current flows through the load and transistor switch, allowing power to be fed to the load from the supply.

Examples of two-state devices also occur in magnetic circuits. The magnetic core shown in fig. 19.4(a) may be magnetized in one direction (set in one state) by passing current in a given direction through the solenoid, and in the reverse direction (set in a second state) by passing the current in the opposite direction. In computers, information may be stored in tiny rings of ferrite by magnetizing these rings and placing them in one of two states. The ring is magnetized by passing a small current, as shown in fig. 19.4(b), in one direction, or in the second state by passing the current in the reverse direction. The rings remain magnets even when the current is reduced to zero. The two directions of magnetization represent coded information or binary numbers. One state of magnetization is used to represent binary 1, the other state binary 0. Binary numbers are explained briefly in the following sub-section.

In this sub-section, we have introduced the concept of a two-state device. We gave some common examples of two-state devices, showing that each could exist in one or other of two states. We showed that a switch is a two-state device and that switching in an electrical circuit can be controlled manually, magnetically, thermostatically, or electronically. Finally, we showed that a ferrite ring can exist in two states of magnetization.

19.2
Simple examples of information being communicated by two-state devices

Two simple examples demonstrating how information may be communicated by a two-state device (the switch S) are shown in fig. 19.5(a) and (b). In (a) the switch is closed for a period of duration T seconds, opened for T, and then closed for $3T$. The resulting output voltage across the resistor is in fact the electrical signal representation of the letter A in the international morse code system. In (b) the switch is

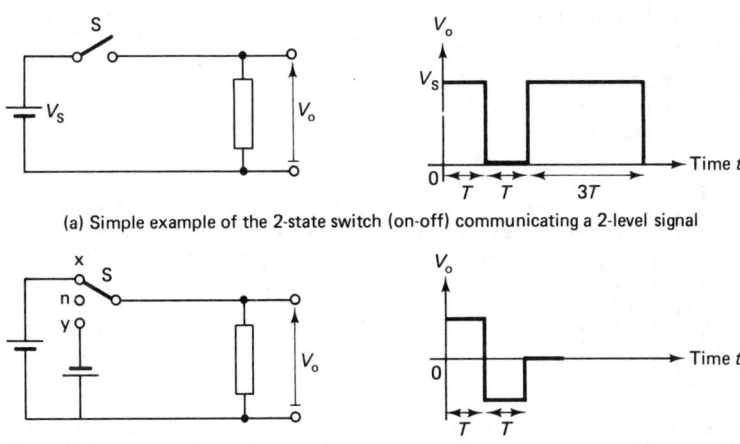

(a) Simple example of the 2-state switch (on-off) communicating a 2-level signal

(b) Simple example of a 2-state device communicating a 2-level (positive-negative) signal

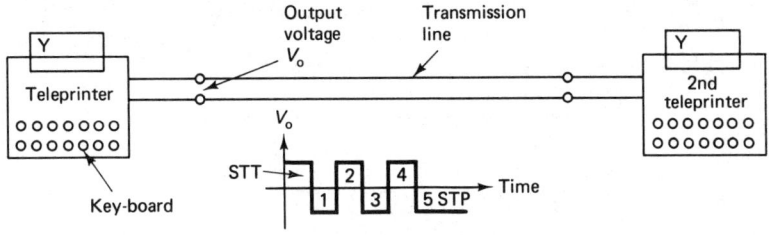

(c) Teleprinter communicating 2-level coded information

Fig. 19.5.

placed in position x for T seconds, and then transferred to position y for time T, and finally to position n. The resulting output voltage is the letter A in the international cable code. Thus the operation of a switch, even in very simple circuits, can be used to communicate information. The concept of transmitting information in a simple 2-level coded form was, of course, and still is the basis of telegraphy. For example, if we were to depress the Y letter key on a teleprinter, the waveform shown in (c) would be generated: the first pulse (labelled STT) enacts a start command, the next five pulses (labelled 1, 2, 3, 4, 5) are the code for Y, the final pulse (labelled STP) is a stop command. If this pulse train were transmitted down a line (a pair of conducting wires) to a second teleprinter it would be 'recognized' by the internal circuits of the latter as information for a Y to be printed and Y would be printed.

The majority of logic systems and most digital communications systems convey information by voltage or current waveforms consisting of trains of pulses, the pulse height taking one of two levels. The pulse duration (we used the symbol T above) is normally fixed. For manually operated systems values of T used vary from ~ 10 to 100 ms or even seconds, whilst for very high speed data transmission systems T may be fractions of microseconds.

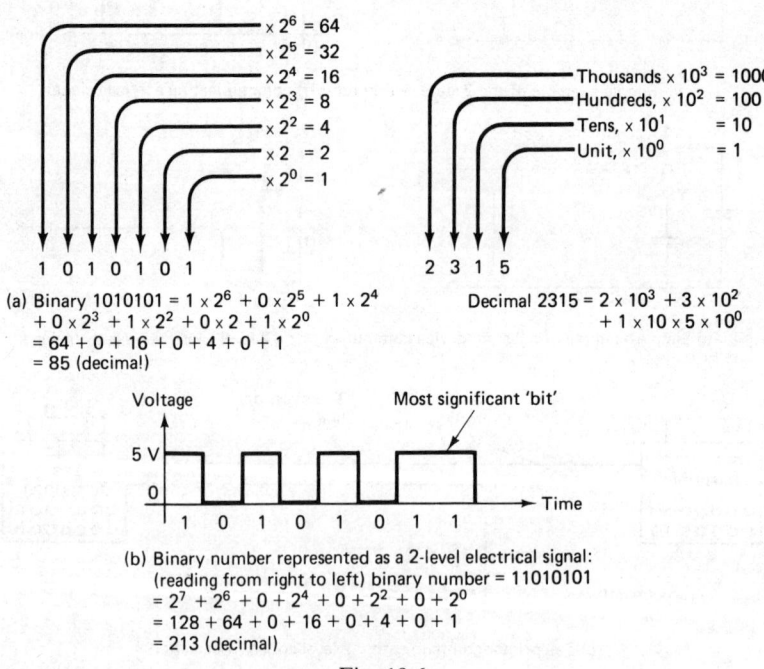

(a) Binary 1010101 = $1 \times 2^6 + 0 \times 2^5 + 1 \times 2^4$
    $+ 0 \times 2^3 + 1 \times 2^2 + 0 \times 2 + 1 \times 2^0$
    $= 64 + 0 + 16 + 0 + 4 + 0 + 1$
    $= 85$ (decimal)

Decimal 2315 = $2 \times 10^3 + 3 \times 10^2$
    $+ 1 \times 10 \times 5 \times 10^0$

(b) Binary number represented as a 2-level electrical signal:
    (reading from right to left) binary number = 11010101
    $= 2^7 + 2^6 + 0 + 2^4 + 0 + 2^2 + 0 + 2^0$
    $= 128 + 64 + 0 + 16 + 0 + 4 + 0 + 1$
    $= 213$ (decimal)

Fig. 19.6.

254

Table 19.1 *Numbers in decimal (base* 10) *and binary (base* 2) *notation*

Decimal number	Binary number	Decimal number	Binary number
0	00000	9	01001
1	00001	10	01010
2	00010	11	01011
3	00011	12	01100
4	00100	13	01101
5	00101	14	01110
6	00110	15	01111
7	00111	16	10000
8	01000		

In digital computers—perhaps the best known of all logic-digital systems—information is communicated, stored, and processed in 2-level or binary form. Let us consider briefly how numerical information is represented in binary form. Binary numbers are written as a sequence of 1's and 0's and Table 19.1 gives the decimal numbers 1 to 16 in binary form. Fig. 19.6(a) illustrates the meaning of the relative position of a 1 or 0 in a binary number showing also the significance of the figure positions in a decimal number with which we are very much more familiar. Fig. 19.6(b) shows an electrical representation of the binary number 11010101, the binary 1 being represented by a pulse of level $+5$ V and a binary 0 by 0 V.

Most logic systems operate using two-state devices communicating information by 2-level signals. In logic terminology we describe one of these levels as a 1 state and the other level as a 0 state.

In positive logic systems the more positive of the 2 levels is defined as the 1 state and the lower level as the 0 state. Normally a reasonably generous tolerance may be allowed in the actual voltage magnitude of a 1 and a 0 state, since the logic elements in the system essentially only have to distinguish between 2 levels. Thus if we are working in a positive logic system where a 1 state is nominally 5 V and the 0 state 0 V, a typical specification might be as follows:

signals levels 4 to 6 V recognized as 1 state signals
signals levels 0 to 0·4 V recognized as 0 state signals.

Obviously the system must be designed to cater for such variations and indeed a logic system employing, for example, diodes and transistors (DTL logic) does.

In a negative logic system the more negative (or less positive) voltage level is defined as the 1 state. For example in a negative logic system using

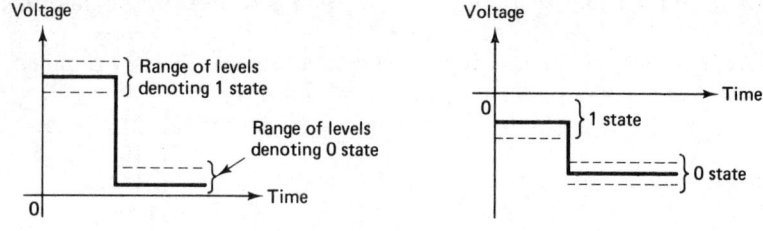

(a) Positive logic: the more positive voltage level is the 1 state

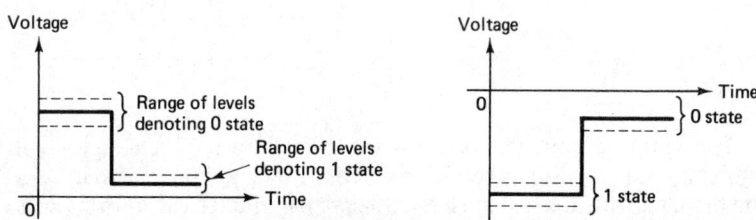

(b) Negative logic: the more negative (or least positive) voltage level is the 1 state

Fig. 19.7.

integrated circuits (P-MOS logic), the logic 1 state is represented by -11 V (minimum) and the 0 state 0 to -3 V (maximum). Thus 1 state signals must be at least -11 V or more negative, and 0 signals within the range 0 to -3 V. Diagrams illustrating the definitions of positive and negative logic 1 and 0 states are drawn in fig. 19.7.

Our specific learning objective of this sub-section was to give simple examples of information being communicated by 2-state devices. We showed that a simple switching circuit can generate 2-state electrical signals and that by using a known code information can be communicated. We cited telegraphy as an example. Other examples in which information is communicated essentially in 2-state pulse form are: telemetry systems (physical data such as temperature, speed, pressure, flow rate, from a removed or remote position, e.g. from a furnace, rocket, or chemical reaction, may be converted into an electrical signal and this signal may then be coded and transmitted in 2-state form); pulse code modulation telephony (in PCM telephony speech and even television signals are coded and transmitted in 2-state form); digital computers (e.g. numbers are communicated and processed in binary form). The majority of logic systems and the logic elements making up the systems which we shall be considering next are essentially 2-state systems and devices. The elements act on or communicate 2-level signals (logic 1's or 0's).

256

Section 20: *The expected learning outcome of this section is that the student should understand the function of 'AND', 'OR', and 'NOT' gates: three basic logic elements.*

20.1
The logical function of the AND gate

Let us first introduce the meaning of the logical AND function by investigating the action of the simple switching circuit of fig. 20.1. If the

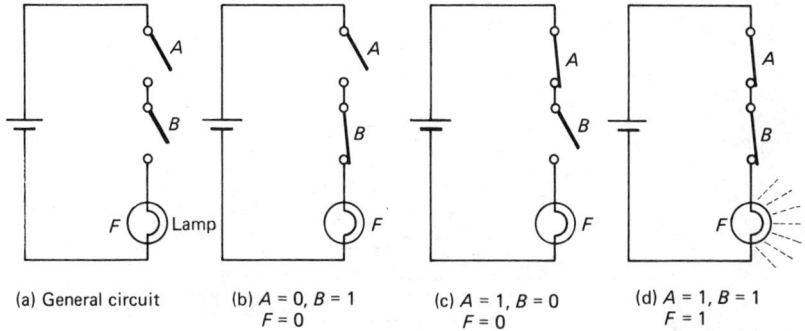

(a) General circuit   (b) $A = 0, B = 1$ (c) $A = 1, B = 0$ (d) $A = 1, B = 1$
$F = 0$ $F = 0$ $F = 1$

Fig. 20.1. A simple circuit illustrating the meaning of the 'AND' function.

lamp is to be switched on, both switches A AND B must be closed. If either switch is opened, the lamp will be switched off. Thus, adopting logic terminology, we will denote the closed and opened states of switches A and B and the on and off states of the lamp as 1 and 0, i.e.

$A = 1$ defines switch A closed, $A = 0$ defines switch A open;
$B = 1$ defines switch B closed, $B = 0$ defines switch B open;
$F = 1$ defines lamp on, $F = 0$ defines lamp off.
We can therefore state that, in the circuit of Fig. 20.1,
when $A = 0, B = 0$, then $F = 0$;
when $A = 0, B = 1$, then $F = 0$(see fig.20.1(b));
when $A = 1, B = 0$, then $F = 0$ (see fig.20.1(c));
when $A = 1$ AND $B = 1$, then $F = 1$ (see fig.20.1(d)).

The circuit is in fact a form of 2-input AND gate and the above statements define the AND function for 2 inputs, i.e. A and B are regarded as the input singals and F as the output signal.

More generally, we may state that an AND gate has two or more inputs and a single output, and performs the following logic function:
the output of an AND gate is a 1 state if, and only if, all inputs are in the 1 state.

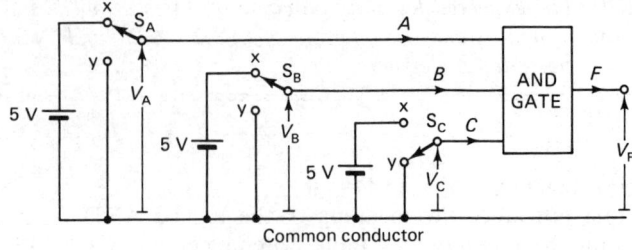

(a) 3-input AND gate
(diagram shows V_A = 5 V, V_B = 5 V,
V_C = 0 so $A = B = 1$, $C = 0$
and for this case output
$V_F = 0$ corresponding to $F = 0$)

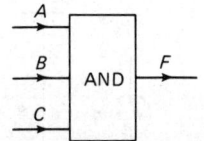

(b) Simplified diagram normally drawn:
common conductor is not drawn nor
are input signal or output details
shown

Fig. 20.2.

Fig. 20.2 shows a diagram of a 3-input AND gate. The AND gate is represented by the rectangular box. We will regard this box as containing the electronic circuits (plus power supplies to operate them) necessary to effect the logical AND function. This is a simple example of a systems approach where we are concerned with the function of an electronic element rather than its detailed circuitry. The 3-input signals are denoted by A, B, and C and the gate output by F. In the diagram, the input signals are simulated by the switching in or out of a 5 V battery between an input line and a common conductor. In practice, these signals are generated by actual devices or by other gates in or feeding the logic system.

We will assume that we are working in a positive logic system so that the 1 state corresponds to an input or output voltage level of 5 V, and the 0 state to a voltage level 0 V. Thus, with switch S_A in position x, the input voltage to line A, $V_A = 5$ V, and the input signal assumes a 1 state. We will denote this by $A = 1$. When S_A is moved to position y, the input voltage $V_A = 0$ V corresponds to a 0 state, i.e. $A = 0$. Similarly, with switches S_B and S_C in position x, $V_B = V_C = 5$ V and so B and C assume the 1 states, i.e. $B = C = 1$; when S_B and S_C are switched to position y, V_B

258

$= V_C = 0$ V corresponding to $B = C = 0$. The 3-input AND gate performs the following logical function:

(1) When input signals $A = 0$, $B = 0$, $C = 0$, the output signal $F = 0$,
i.e. when all three switches are in y position $V_A = V_B = V_C = 0$ V corresponding to $A = B = C = 0$, the AND gate produces zero output voltage, $V_F = 0$ corresponding to $F = 0$.
(2) When $A = 0$, $B = 0$, $C = 1$, then $F = 0$,
i.e. when $V_A = 0$, $V_B = 0$, but $V_C = 5$ V and the AND gate output voltage $V_F = 0$ corresponding to $F = 0$.
(3) When $A = 0$, $B = 1$, $C = 1$, then $F = 0$,
i.e. when $V_A = 0$, $V_B = V_C = 5$V, the AND gate output voltage $V_F = 0$, so $F = 0$.
(4) when $A = B = C = 1$, then $F = 1$,
i.e. when all three switches are in the x position, $V_A = V_B = V_C = 5$ V, the AND gate output voltage $V_F = 5$ V, so $F = 1$.

The AND gate output only assumes the 1 state when all inputs are 1; otherwise the output assumes the 0 state.

20.2
The construction of a truth table for a 3-input AND gate
As well as defining a logical function or operation in words we can also construct a table which defines the function explicitly. Such a table contains all possible combinations of input signal states and lists against each the corresponding output signal state. This table is called a truth table.

Our object in this sub-section is to construct a truth table for a 3-input

Table 20.1   The truth table for a 3-input AND gate

Inputs			Output
A	B	C	F
0	0	0	0
0	0	1	0
0	1	0	0
0	1	1	0
1	0	0	0
1	0	1	0
1	1	0	0
1	1	1	1

AND gate. Actually we have gone a long way to accomplishing this in the last sub-section, when considering the AND gate of fig. 20.2. In all there are 8 possible combinations of input signal states: we start with $A = B = C = 0$ and then systematically list the other 7 combinations on the left-hand side of the truth table, Table 20.1. By definition the output of an AND gate is only 1 when all inputs are 1. Hence $F = 0$ for all input combinations other than $A = B = C = 1$ and thus the F output column contains 0s except for the last row when $F = 1$, as $A = B = C = 1$.

20.3
The Boolean symbol for 'AND'

Logic function operations, such as the 'AND' and the 'OR' and 'NOT' which we will be meeting later in this section, are called 'Boolean' after a nineteenth century mathematician Boole. Boole developed an algebra to handle logic operations and although we will not be concerned with this algebra, except to express logic gate relations, you will certainly meet it in your subsequent studies of logic circuits.

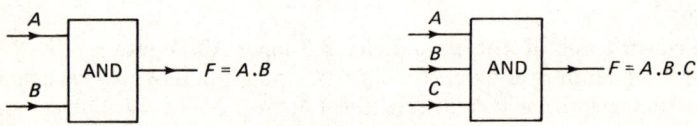

2 and 3-input AND gates showing the Boolean 'dot' symbol for 'AND'

Fig. 20.3.

The Boolean symbol for 'AND' is denoted by a dot, e.g. $A.B$ reads A AND B. Thus the logic equation expressing the output F of the 2- and 3-input AND gates shown in fig. 20.3 may be written in Boolean algebra notation as

$$F = A.B$$
$$F = A.B.C$$

the latter equation being read as F equals A AND B AND C.

The dot is frequently missed out and it is understood that its absence means the AND operation. Thus the output from a 4-input AND gate, for example, with inputs A, B, C, D may be written as

$$F = ABCD \text{ as well as } F = A.B.C.D$$

and is read as F equals A AND B AND C AND D. By definition of the AND gate,

$$F = 1 \text{ when } A = B = C = D = 1$$
$$= 0 \text{ for all other input signal combinations.}$$

Note also that other symbols are used to denote AND. Occasionally a cross ' × ' or an inverted v are used, e.g. $A \times B$ and $A \wedge B$ mean A AND B: whilst in set theory the cap symbol $\cap$ is used, so $A \cap B$ means A AND B.

20.4
The BS circuit symbol for an AND gate
The present BS symbol (British Standard BS 3939, Section 21, issued March 1969) for the AND gate is drawn in fig. 20.4(a).

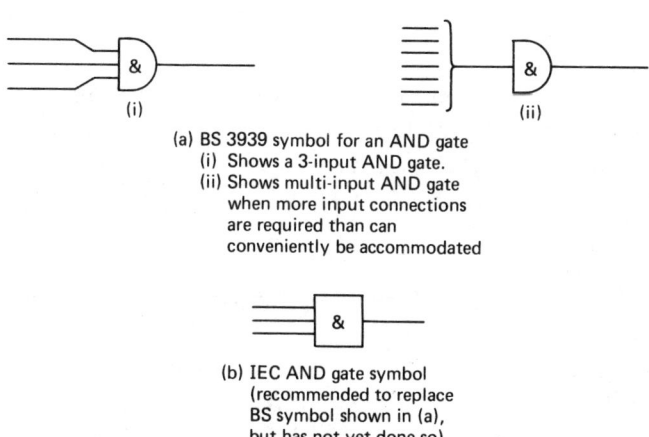

(a) BS 3939 symbol for an AND gate
(i) Shows a 3-input AND gate.
(ii) Shows multi-input AND gate when more input connections are required than can conveniently be accommodated

(b) IEC AND gate symbol (recommended to replace BS symbol shown in (a), but has not yet done so)

Fig. 20.4. AND gate symbols.

20.5
The recognition of the superseded BS circuit symbol for an AND gate
The AND gate symbol drawn in fig. 20.4(a) has, at the time of writing (Spring 1977), not been superseded and is in current use.

However, to facilitate easier drawing and drawing by automated equipment there has been a considerable movement to replace the semi-circular or D-shaped logic gate symbols by rectangles. In 1972 the International Electrotechnical Commission (I.E.C.) published No. 117, Part 15, Binary Logic Elements, in which it recommended that the D symbol be replaced by a rectangle. Thus the AND gate symbol is likely to become that drawn in fig. 20.4(b). The British Standards Institution have indicated that they intend to modify BS 3939, Section 21, to conform with the I.E.C. symbol.

Example 20.1
The input-output voltage levels for two gates which are used in a system employing positive logic are:

Inputs		Output		Inputs		Output
A	B	F		A	B	F
0 V	0 V	0·2 V		0 V	0 V	0·1 V
0 V	5 V	4·5 V		0 V	5 V	0·2 V
5 V	0 V	4·6 V		5 V	0 V	0·2 V
5 V	5 V	4.7 V		5 V	5 V	5 V

Gate P   Gate Q

In both cases the maximum-minimum voltage levels corresponding to the 1 and 0 states are given by:

1 state · · · 4 V (min) to 6 V
0 state · · · 0 V to 0·4 V (max)

Write down the truth tables for gates P and Q and identify which gate is an AND gate.

Determine also the output from gate P and gate Q if the following signal inputs are applied to each gate:

(a) $A = 101001$
    $B = 100101$

(b) $A = 001000$
    $B = 100111$

Note. These signals are shown graphically in fig. 20.5(a) and (b), respectively.

Solution

The truth tables for gates P and Q are given below. All voltage levels from 0 to 0·4 V become '0' logic states, whilst all levels between 4 and 6 V become '1' logic states.

Input		Output		Input		Output
A	B	F		A	B	F
0	0	0		0	0	0
0	1	1		0	1	0
1	0	1		1	0	0
1	1	1		1	1	1

Truth table for gate P   *Truth table for gate Q*

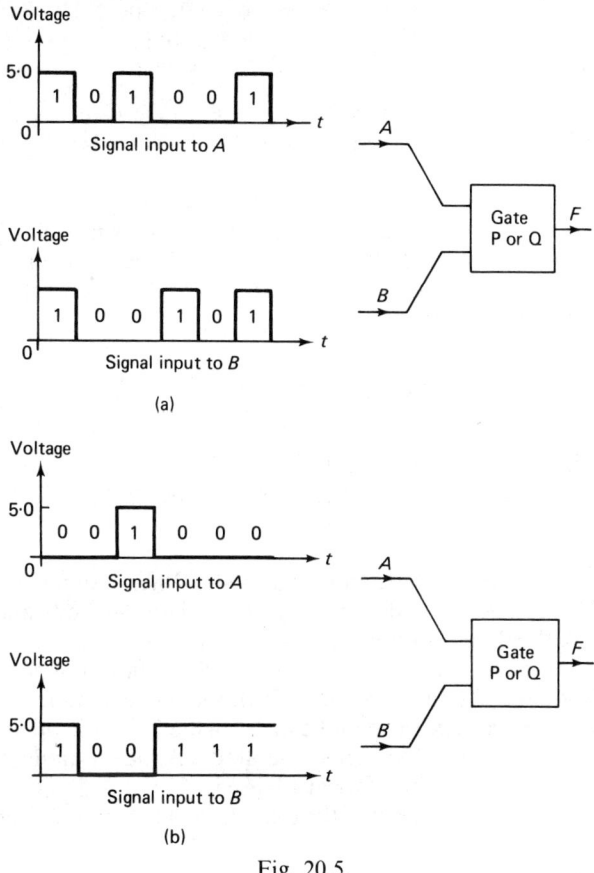

Fig. 20.5.

Gate Q is an AND gate. Its truth table shows its output $F = 1$ if, and only if the two inputs A and B are both 1.
Outputs from gate P:

(a) $A = 101001$ (b) $A = 001000$
      $B = 100101$ $B = 100111$
      $F = 101101$ $F = 101111$

F is obtained for each input signal pair from the truth table of gate P. For example, in (a) consider first input signals $A = 1, B = 1$: look along truth table row $A = 1, B = 1$ to find $F = 1$; second pair of input signals $A = 0$, $B = 0$; look along truth table row $A = 0, B = 0$ to find $F = 0$.
Output from gate Q (an AND gate):

(a) $A = 101001$　　　　　　　　(b) $A = 001000$
　　$B = 100101$　　　　　　　　　　$B = 100111$
　　$F = 100001$　　　　　　　　　　$F = 000000$

F is obtained either from the truth table of gate Q or using the definition of an AND gate, i.e. $F = 1$ only if $A = B = 1$, otherwise $F = 0$.

Example 20.2
Determine the output voltage V_F in the circuit shown in fig. 20.6 ((a) shows the full diagram including means for supplying signal inputs; (b) is a simplified diagram, the common conductor line and input signal details being missed out), when

(a) $V_A = 6$ V, $V_B = 6$ V
(b) $V_A = 0$ V, $V_B = 6$ V
(c) $V_A = 0$ V, $V_B = 0$ V.

Assume the two diodes are ideal.
Show also that the circuit acts as an AND gate.

Solution
First let us recap on the action of a diode. When a diode is reverse-biased, i.e. the diode cathode is more positive than the diode anode (see fig. 20.7(a)), the diode conducts a minutely small current. This reverse-bias current is normally neglected, and in an ideal diode it is zero. When a diode is forward-biased (see fig. 20.7(b)) the diode conducts strongly and only a small voltage drop of the order of 0·6 V for silicon and 0·2 V for germanium is produced across the diode. An ideal diode has zero voltage drop across it when forward-biased.

Diode action is at the heart of the operation of the circuit given in our problem.

(a) When $V_A = 6$ V and $V_B = 6$ V both diodes DA and DB are reverse-biased and therefore cannot conduct current. Thus no current can flow through resistor R and there is zero voltage drop across R. Hence the voltage at F is the same as the battery supply voltage, so

$$V_F = 5 \text{ V}$$

(b) If $V_A = 0$ V, then diode DA becomes forward-biased and current flows from the battery supply through R and DA. Since diode DA is ideal there is zero voltage drop across it and all the 5 V battery voltage is dropped across R. Thus F is at zero volts and

$$V_F = 0$$

The state of diode DB does not affect $V_F = 0$ in this case. When $V_B = 6$ V diode DB is reverse-biased by $V_B - V_F = 6 - 0 = 6$ V and therefore conducts no current.

264

(c) If $V_A = 0$ and $V_B = 0$, both diodes are forward-biased and conduct current. Current flows from the battery via R and then through DA and DB, the diodes sharing the current, see fig. 20.7(c). Since no voltage is developed across an ideal diode in the forward-biased mode,

$$V_F = V_A = V_B = 0$$

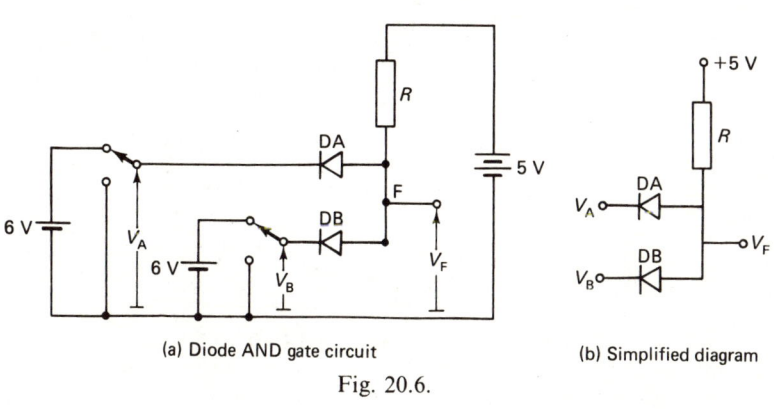

(a) Diode AND gate circuit

(b) Simplified diagram

Fig. 20.6.

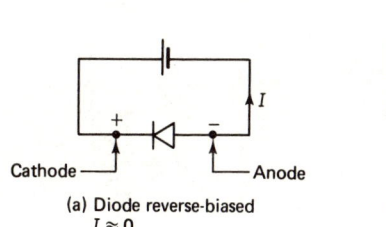

Cathode — + — — Anode

(a) Diode reverse-biased
$I \approx 0$

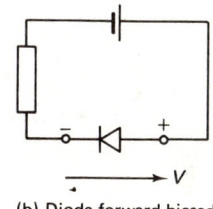

(b) Diode forward-biased
$V \approx 0$

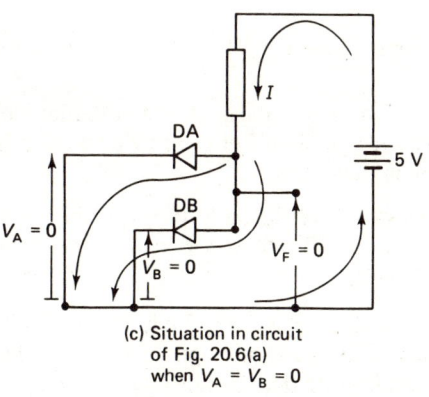

(c) Situation in circuit
of Fig. 20.6(a)
when $V_A = V_B = 0$

Fig. 20.7.

265

20.6
The logical function of the OR gate

In Sub-section 20.1 we introduced the meaning of the AND function by noting the action of a simple circuit containing two switches in series. Following along similar lines let us introduce the meaning of the logical OR function by investigating the circuit shown in fig. 20.8, where in this

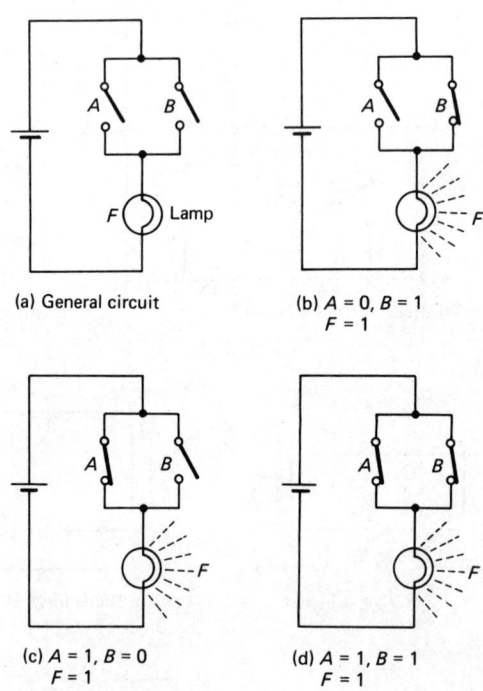

(a) General circuit

(b) $A = 0, B = 1$
$F = 1$

(c) $A = 1, B = 0$
$F = 1$

(d) $A = 1, B = 1$
$F = 1$

Fig. 20.8. A simple circuit illustrating the meaning of the 'OR' function.

case we have two switches in parallel. We note that if either switch A OR B is closed the lamp will light. Let us draw up a truth table for the circuit using the following notation:

$A = 1$ defines switch A closed, $A = 0$ defines switch A open;
$B = 1$ defines switch B closed, $B = 0$ defines switch B open;
$F = 1$ defines lamp on, $F = 0$ defines lamp off.

The truth table, Table 20.2, defines the function of a 2-input OR gate. A and B are regarded as the input signals and F as the output signal.

In general, we can state that an OR gate has two or more inputs and a single output and performs the following logic function:

266

Table 20.2 *The truth table for a 2-input OR gate*
(derived using the switching circuit of fig. 20.8)

Input		Output
A	B	F
0	0	0
0	1	1
1	0	1
1	1	1

the output from an OR gate is a 1 state if one or more inputs are in the 1 state.

Thus output $F = 1$ if one or more inputs are 1,
$\quad\quad\quad = 0$ if all inputs are 0.

20.7
The construction of a truth table for a 3-input OR gate

The truth table for a 3-input OR gate, shown diagrammatically in fig. 20.9, is given in Table 20.3. There are 8 different combinations for the input signal states A, B, and C. These are listed systematically beginning with $A = B = C = 0$ on the left-hand side of the table. By definition of the OR gate function the output state F is a 1 state if one or more inputs are in the 1 state. Thus $F = 1$ for all input signal combinations other than $A = B = C = 0$ when $F = 0$. Thus the output F column contains 1s except for the first row when all input signals are in the 0 state.

Table 20.3　*The truth table for a 3-input OR gate*

Inputs			Output
A	B	C	F
0	0	0	0
0	0	1	1
0	1	0	1
0	1	1	1
1	0	0	1
1	0	1	1
1	1	0	1
1	1	1	1

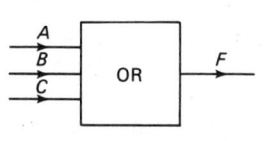

Fig. 20.9. Diagram of a 3-input 'OR' gate.

20.8
The Boolean symbol for 'OR'

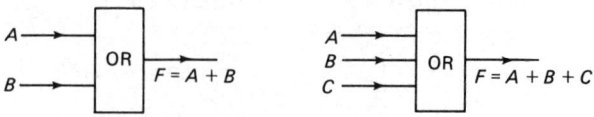

2 and 3-input OR gates showing the Boolean '+' symbol for 'OR'

Fig. 20.10.

The Boolean symbol for the 'OR' function is denoted by $+$, e.g. $A+B$ reads A OR B. Thus the logic equation expressing the output F of the 2 and 3-input OR gates shown in fig. 20.10 may be written in Boolean algebra notation as

$$F = A+B$$
$$F = A+B+C,$$

the latter equation being read as A OR B OR C.

Very occasionally the symbol v is used, so AvB reads A OR B; whilst in set theory the cup symbol $\cup$ is often employed, so $A\cup B$ reads A OR B.

20.9
The BS circuit symbol for an OR gate
The BS 3939 symbol for an OR gate is drawn in fig. 20.11(a).

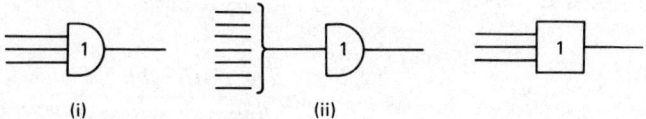

(i)   (ii)

(a) BS3939 symbol for an OR gate
  (i) shows a 3-input OR gate
  (ii) shows a multi-input OR gate where more
    input connections are required than can
    conveniently be accommodated.

(b) I.E.C. OR gate symbol (recommended
  to replace BS symbol shown in
  (a), but has not yet done so)

Fig. 20.11. 'OR' gate symbols.

20.10
The recognition of the superseded BS circuit symbol for an OR gate
The OR gate symbol drawn in fig. 20.11(a) has at the time of writing not been superseded and is in current use. However, as previously stated in Section 20.5, the I.E.C. recommended in 1972 that the semi-circle gate symbol be replaced by a rectangle. Thus the OR gate symbol is likely to become that drawn in fig. 20.11(b).

268

Example 20.3

Draw the BS symbol for a 5-input OR gate and a 5-input AND gate. If the input signals to both gates are A, B, C, D, E write down the Boolean expressions for the respective gate output function. State also the output state of the OR and the AND gate when

(i) $A = B = C = D = E = 0$,
(ii) $A = 1, B = 0, C = 1, D = 0, E = 1$,
(iii) $A = B = C = D = E = 1$.

Solution

(a) 5-input OR gate   (b) 5-input AND gate

Fig. 20.12.

The BS symbol for a 5-input OR gate is drawn in fig. 20.12(a) and for a 5-input AND gate in fig. 20.12(b).

The output function F of the OR gate in Boolean form is written as

$$F = A + B + C + D + E$$

The output function F of the AND gate is written as

$$F = A.B.C.D.E$$

To find the output state of the OR gate we apply the definition of the OR gate: $F = 1$ if one or more inputs are 1. Thus for

case (i) $F = 0$ (since all inputs are 0)
   (ii) $F = 1$ (at least one input is 1, in fact $A = C = E = 1$)
   (iii) $F = 1$ (all inputs are 1).

To find the output state of the AND gate we apply the AND gate definition: $F = 1$ if and only if all inputs are 1. Thus for

case (i) $F = 0$ (since all inputs are 0)
   (ii) $F = 0$ (since not all inputs are 1)
   (iii) $F = 1$ (all inputs are 1).

Example 20.4

Fig. 20.13 shows a circuit containing 3 switches. If a switch open is denoted by a 0 state and a switch closed by a 1 state, and the lamp off by a 0 state and the lamp on by a 1 state, construct a truth table for the circuit. That is regard switches A, B, C as input signals and the state of the lamp as the output signal F.

Write down also the Boolean expression for F.

269

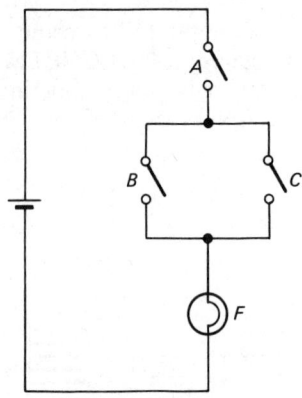

Fig. 20.13.

Solution

Let us systematically list all possible combinations of the switches and see whether or not the lamp lights.

(1) When all switches are open the lamp is off, so in logic notation when $A = 0$, $B = 0$, $C = 0$, then $F = 0$.

(2) When switches A and B are open and C is closed, the lamp is still off, so when $A = 0$, $B = 0$, $C = 1$, then $F = 0$.

(3) When switch A is open, B is closed, and C is open, the lamp is still off, so when $A = 0$, $B = 1$, $C = 0$, then $F = 0$.

(4) When switch A is open and B and C are closed, the lamp is still off, so when $A = 0$, $B = 1$, $C = 1$, then $F = 0$.

(5) When switch A is closed and B and C are open, the lamp is still off, so when $A = 1$, $B = 0$, $C = 0$, then $F = 0$.

(6) When A is closed, B is open, and C is closed we have a complete circuit so the lamp lights, hence when $A = 1$, $B = 0$, $C = 1$, then $F = 1$.

(7) Likewise with A closed, B closed, C open, the lamp lights, so when $A = 1$, $B = 1$, $C = 0$, then $F = 1$.

(8) When all switches are closed the lamp lights, so when $A = 1$, $B = 1$, $C = 1$, then $F = 1$.

We now have all the data to construct the truth table of the circuit. The truth table is given below:

The 'function' of the circuit can be stated as follows:

when switches A AND B OR A AND C OR A AND B AND C are closed (i.e. 1 states) the lamps light (i.e. $F = 1$). Thus we may write down the Boolean expression for F as

$$F = A.B + A.C + A.B.C$$

You may have realized that the last term is in effect redundant. The circuit will provide a 1 output when A AND B OR A AND C are closed. It

270

Inputs			Output
A	B	C	F
0	0	0	0 · · · (1)
0	0	1	0 · · · (2)
0	1	0	0 · · · (3)
0	1	1	0 · · · (4)
1	0	0	0 · · · (5)
1	0	1	1 · · · (6)
1	1	0	1 · · · (7)
1	1	1	1 · · · (8)

does not require both B AND C to be closed. Thus we could simplify the Boolean expression for F and write

$$F = A.B + A.C$$

Note. An equal sign in Boolean algebra does not mean 'equal' in the conventional algebraic sense (if it did the above expressions are obviously not equal). In Boolean algebra two expressions are equal if they have the same truth table. If you are familiar with Boolean algebra you may like to check and show that the above two expressions are equal in the logic sense. You can also check that both have identical truth tables and are therefore equal. In any event please do not worry. Our specific learning objective is that you should recognize the Boolean symbols for AND and OR and know the functions of AND and OR gates.

Example 20.5
Fig. 20.14 shows a diagram of a logic system consisting of two OR gates feeding an AND gate. Determine, for the inputs shown on the diagram, the outputs F_1 and F_2 from the OR gates and the system output F_3.

Solution

OR gate 1:
$$A = 1010$$
$$B = 1011$$
$$F_1 = 1011$$

F_1 is obtained using the OR gate definition: if one or more inputs is a 1 then the output is a 1.

OR gate 2:
$$C = 0010$$
$$D = 1110$$
$$F_2 = 1110$$

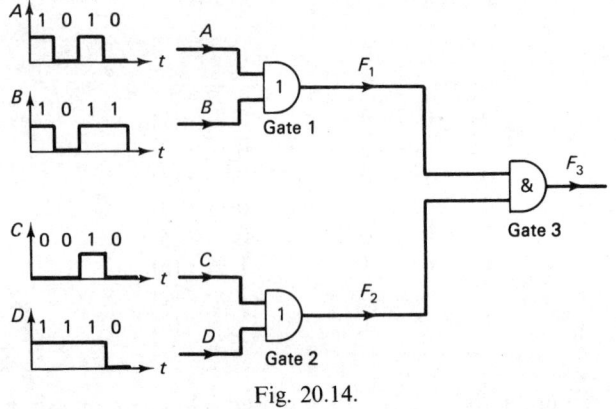

Fig. 20.14.

AND gate 3:

$$F_1 = 1011$$
$$F_2 = 1110$$ inputs
$$F_3 = 1010 \cdots \text{system output}$$

F_3 is obtained using the AND gate definition: output is a 1 if and only if all inputs are 1.

20.11
The logical function of the NOT gate

The NOT gate has a single input and a single output and performs the logical function of negation:

the output of a NOT gate is in the 1 state when the input signal is in the 0 state and vice versa.

Thus if the input to a NOT gate is

$A = 0$, the NOT gate output $F = 1$,
if $A = 1$, the NOT gate output $F = 0$.

A NOT gate is also known as an inverter (since it inverts a 0 input to a 1 input and vice versa) and sometimes as a negator (since it performs the logic operation of negation). The action of a simple transistor NOT gate circuit is explained in Section 21.4. A simple diagrammatic representation of its NOT function is illustrated in fig. 20.15.

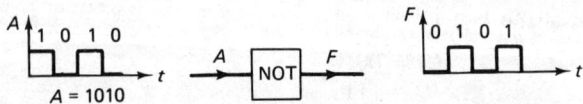

Fig. 20.15. Diagram illustrating function of a 'NOT' gate.

20.12

The construction of a truth table for a NOT gate

The truth table for a NOT gate is constructed in Table 20.4. When $A = 0, F = 1$; when $A = 1, F = 0$.

Table 20.4
Truth table for
a NOT gate

Input A	Output F
0	1
1	0

20.13

The Boolean symbol for 'NOT'

The Boolean symbol for 'NOT' is denoted by a bar. Thus $\bar{A}$ reads NOT A. The logic equation for a NOT gate is

$$F = \bar{A} \quad (F \text{ 'equals' NOT } A)$$

Sometimes a prime is used instead of a bar to denote the NOT operation, so in this case A' would denote NOT A.

20.14

The BS circuit symbol for a NOT gate

The BS 3939 symbol for a NOT gate is drawn in fig. 20.16(a). The standard convention used to indicate logic negation (that is, to indicate that a 0 is changed to a 1 and vice versa) is a small circle drawn at the point where a signal line joins the logic gate symbol. In fig. 20.16(a) this circle is drawn at the output of the OR gate symbol to denote the NOT gate.

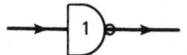

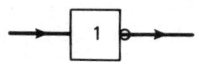

(a) BS 3939 symbol for a NOT gate

(b) I.E.C. (International Electrotechnical Commission) NOT gate symbol—recommended to replace the BS symbol in (a) but has not yet done so

Fig. 20.16.

20.15

The recognition of the superseded BS circuit symbol for a NOT gate

The NOT gate symbol drawn in fig. 20.16(a) has at the time of writing

(Spring 1977) not been superseded and is in current use. However, the I.E.C. recommended in 1972 that the semi-circle gate symbol be replaced by a rectangle. Thus the NOT gate symbol is likely to become that drawn in fig. 20.16(b).

Example 20.6
(a) An input signal $A = 01010011$ is applied to a NOT gate. Determine the output signal.
(b) Input signals $A = 101011$, $B = 100001$ are applied at the input lines of the logic system shown in fig. 20.17. Determine the corresponding output signals F_1 and F_2.

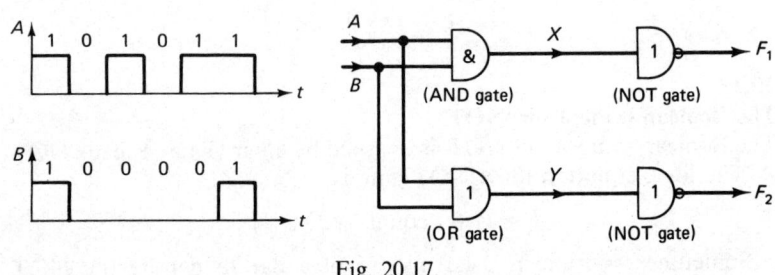

Fig. 20.17.

Solution
(a) The output of the NOT gate is $F = \bar{A}$, so on interchanging 0 for 1 and 1 for 0 we have,

$$F = \bar{A} = 10101100$$

(b) The output X from the AND gate is a 1 when A AND B are both 1, otherwise it is 0, so

with $A = 101011$
$$\updownarrow \quad \updownarrow, \quad X = 100001$$
$B = 100001$

The output from the NOT gate,

$$F_1 = \bar{X} = 011110$$

The output Y from the OR gate is a 1 if either A or B is a 1 or when A and B are 1, so

with $A = 101011$, $Y = 101011$
$\quad\quad B = 100001$

and the final output from the NOT gate,

$$F_2 = \bar{Y} = 010100$$

274

Example 20.7
Construct truth tables for the logic systems shown in fig. 20.18(a) and (b). Hence show that both systems perform the same logic function.

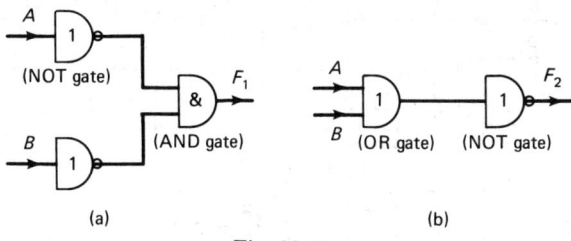

(a)   (b)

Fig. 20.18.

Solution
The truth tables are given below. Note that to help construct the tables the intermediate gate outputs are also given. To obtain the input figures remember that A may be either 1 or 0, and so may B. Thus the possible inputs to the NOT gates can only be $A = 0, B = 0; A = 0, B = 1; A = 1, B = 0$; and $A = 1, B = 1$.

Inputs		Outputs from NOT gate		Output from AND gate
A	B	$\bar{A}$	$\bar{B}$	F_1
0	0	1	1	1
0	1	1	0	0
1	0	0	1	0
1	1	0	0	0

Truth table for circuit (a)

Inputs		Output from OR gate	Output from NOT gate
A	B	$A+B$	F_2
0	0	0	1
0	1	1	0
1	0	1	0
1	1	1	0

Truth table for circuit (b)

It will be seen that the outputs from both circuits are the same in every input case. Thus we may conclude that the two circuits perform the same logic function.

Example 20.8
This example is included to demonstrate a practical application of the use of logic gates in a simple interlock system to control the 'on-off' operation of a machine.

The operation of the system, shown diagrammatically in fig. 20.19, can be explained as follows:

275

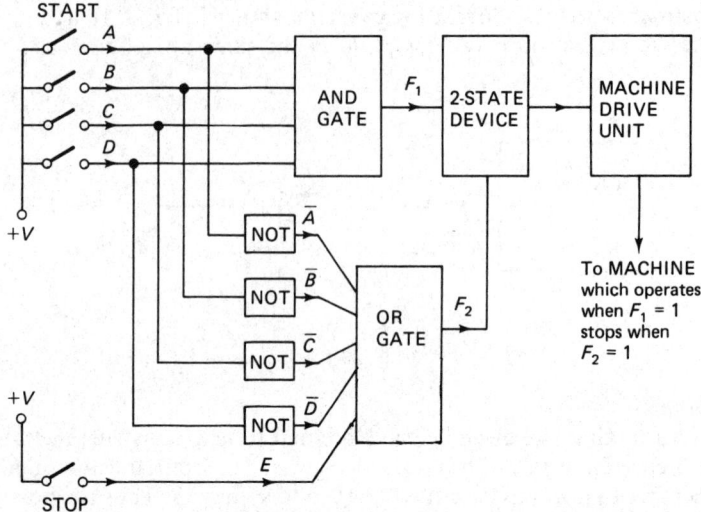

Fig. 20.19. A simple interlock control system using logic gates.

(a) The machine will only start if
    (i) $A = 1$, i.e. the start switch is closed,
AND (ii) $B = 1$, i.e. switch B is closed.
  In a practical system $B = 1$ could represent a safety condition being satisfied; e.g. a safety door being closed, if the door were open then B would be a 0.
AND (iii) $C = 1$, i.e. switch C closed.
$C = 1$ could represent a physical requirement being met, e.g. switch C could be a thermostatic switch which is closed above a certain temperature and opens if the temperature falls below the required minimum operating temperature.
AND (iv) $D = 1$, i.e. switch D is closed.
$D = 1$ could represent a feed condition being satisfied; $D = 0$ would then represent insufficient or no feed materials.
(b) Thus if $A = 1$ AND $B = 1$ AND $C = 1$ AND $D = 1$, the AND gate output $F_1 = 1$. The $F_1 = 1$ output is used to 'set' the state of the 2-state device so that the device provides a signal to the machine drive unit which in turn switches on the machine.
(c) If one or more of the input conditions are changed so that A OR B OR C OR D are switched to a 0 state, then the machine will be switched off. For example if $C = 0$ (indicating, for example, that the temperature has fallen outside the operating range), a 0 state will be fed to the NOT gate via the paralleled input C line. The OR gate will then produce an output $F_2 = 1$. This output is used to 'reset' the 2-state device into a

276

second state. This state acts to disconnect the drive unit and the machine stops.

(d) Similarly if the stop switch is closed, an $E = 1$ state is fed into the OR gate. Thus $F_2 = 1$, the 2-state device is switched into its second state, and the drive unit is disconnected.

Problem

(a) Draw a logic circuit diagram of the system of fig. 20.19 using BS symbols for the gates.

(b) Write down the outputs, F_1 and F_2 using Boolean notation.

(c) Determine F_1 and F_2 when

  (i) $A = 1, B = 1, C = 1, D = 0, E = 0$.
  (ii) $A = 1, B = 1, C = 1, D = 1, E = 0$.

Solution

(a) The logic circuit diagram of fig. 20.19 is drawn in fig. 20.20.

(b) $F_1 = A.B.C.D$
   $F_2 = \bar{A} + \bar{B} + \bar{C} + \bar{D} + E$

(c) (i) When $A = B = C = 1, D = E = 0$,
the output from the AND gate, $F_1 = 0$ (*since not all inputs are 1, D = 0*),
the output from the OR gate, $F_2 = 1$ (since $\bar{D} = 1$ and the OR gate gives a 1 output if one or more of its inputs are 1).

(ii) When $A = B = C = D = 1, E = 0$

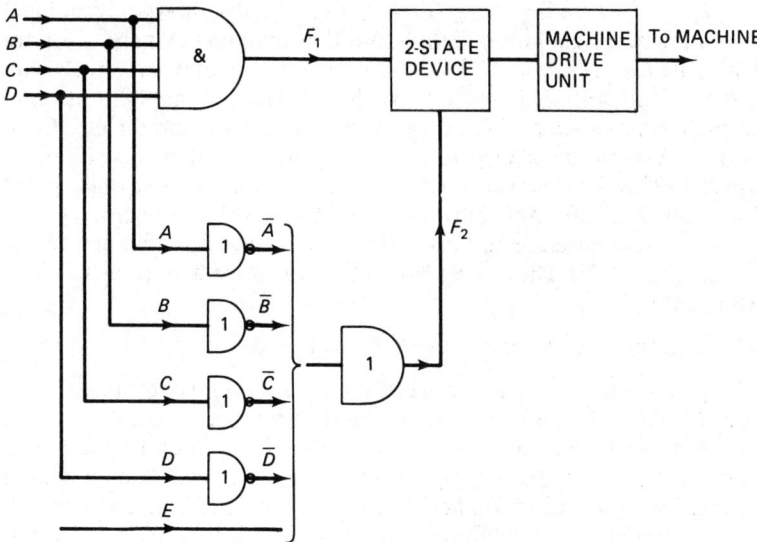

Fig. 20.20. Interlock control system of Fig. 20.19 with gates drawn using BS 3939 symbols.

277

$F_1 = 1$ (all 4 inputs to the AND gate are in the 1 state).
$F_2 = 0$ (all 5 inputs to the OR gate are 0, i.e. $\bar{A} = \bar{B} = \bar{C} = \bar{D} = 0, E = 0$).

Section 21: *The expected learning outcome of this section is that the student should understand the action of simple electronic gates.*

21.1
The action of a 3-input diode AND gate

Fig. 21.1 shows a 3-input diode AND gate for a positive logic system working from a 5-V supply.

Considering next the signal input and output levels, suppose we select:

voltage levels 5 to 6 V to correspond to a 1 state,
voltage levels 0 to 1 V to correspond to a 0 state.

These ranges allow for non-ideal diode action and the fact that in a practical system there is always some variation in the voltage levels of 1 and 0 signals. In fig. 21.1(a), 1 and 0 input states are effected by switches and the 6-V batteries. Thus, when a switch is in a position x, 6 V and therefore a 1 state is fed into an input line; when a switch is in a position y, the input line is connected to a common line at 0 V and this situation corresponds to feeding in a 0 state. Fig. 21.1(b) shows a simplified diagram where signal input details and the common 0 V conductor are deliberately not shown. This is normal practice, and when circuits are drawn in this way it is understood that all voltages at the input and output terminals are measured with respect to a 0-V common line. Thus, with a 6-V input connected to A ($A = 1$), we know that this voltage is represented at A with respect of a 0-V conductor; if $A = 0$ we understand that input A is automatically connected to the 0-V conductor.

Let us now describe the operation of the gate by reference to fig. 21.2(a), (b), and (c). These three cases illustrate the basic action of a diode AND gate.

(a) All inputs are at 0 volts, so $A = B = C = 0$.

Thus, all three diodes are forward-biased and conduct current. In the forward-biased mode there is zero voltage drop across an ideal diode. Hence the voltage at output F must be 0 and the output state $F = 0$. Current flows from the supply through resistor R and is then shared amongst the three diodes. The 5-V supply is dropped entirely across R. For practical diodes, a small voltage of about 0·6 V for silicon and about 0·2 V for germanium will occur across each diode, so $V_F = 0·6$ V (Si) or V $= 0·2$ V (Ge). In either case this voltage will be interpreted as a 0 state

278

since all logic gate systems work with some tolerance. The tolerance specified in our case is that levels between 0 and 1 V are interpreted as 0 state signals.

(b) $A = 1, B = 1, C = 0$, i.e. $V_A = 6$ V, $V_B = 6$ V, $V_C = 0$ V.

Thus, diodes DA and DB will be reverse-biased, but diode DC will be forward-biased. The last diode carries all the current supplied from the 5-V battery via R. The voltage at F is therefore still 0 V (or approximately so), so $F = 0$.

(c) $A = 1, B = 1, C = 1$, i.e. $V_A = V_B = V_C = 6$ V.

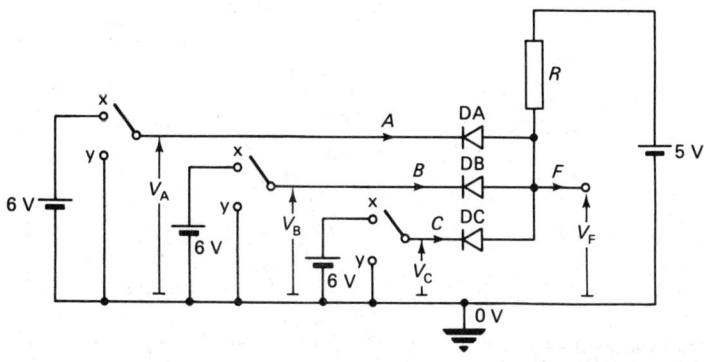

(a) 3-input diode AND gate working with positive logic. Diagram shows input states simulated by 6 V batteries and switches: when $V_A = V_B = V_C = 6$ V, $V_F = 5$ V so when A and B and $C = 1$, output $F = 1$

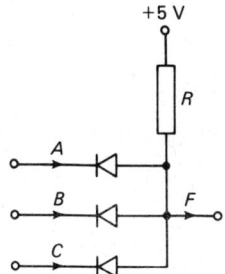

(b) Simplified diagram. Electronic gate circuit diagrams normally miss out common conductor and input signal details

Fig. 21.1. 3-input positive logic diode 'AND' gate.

In this case all diodes are reverse-biased (their cathodes are at a more positive voltage than their anodes) and hence no current can flow in the circuit. Thus, the voltage drop across R is zero and the output at F rises to 5 V, i.e. $V_F = 5$ V and $F = 1$. $V_F = 5$ V is a 1 state since a voltage level

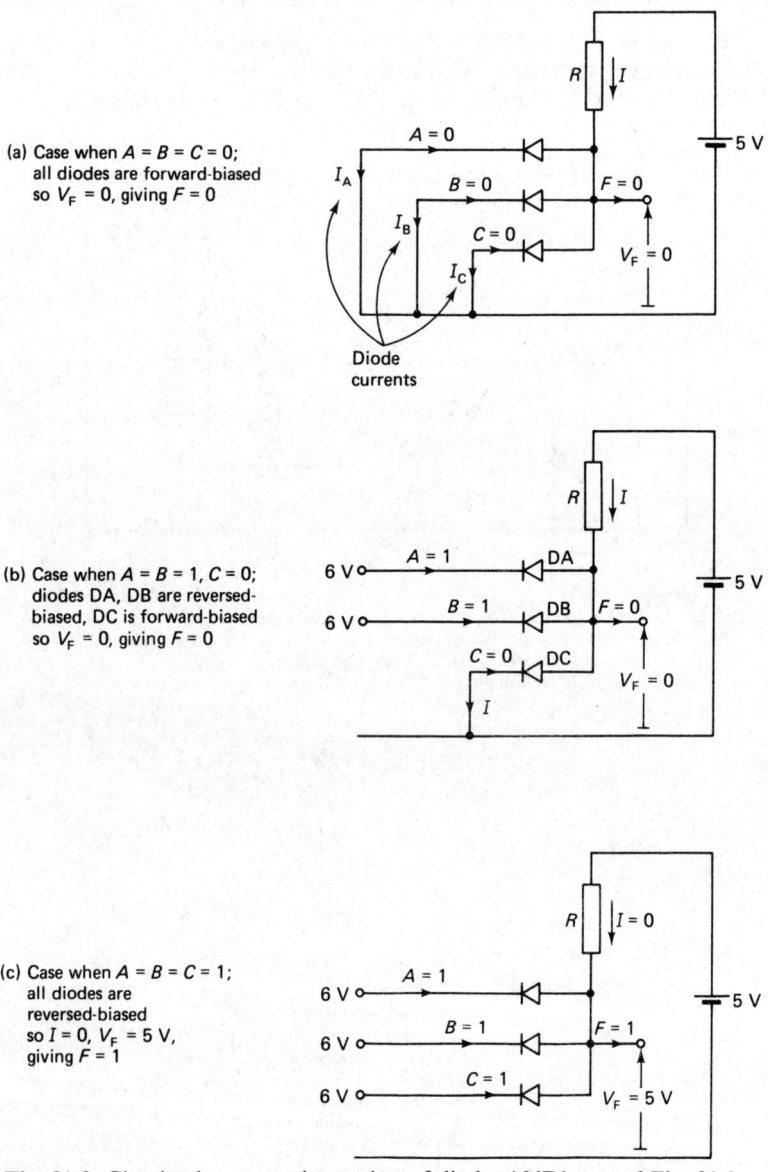

(a) Case when $A = B = C = 0$; all diodes are forward-biased so $V_F = 0$, giving $F = 0$

(b) Case when $A = B = 1$, $C = 0$; diodes DA, DB are reversed-biased, DC is forward-biased so $V_F = 0$, giving $F = 0$

(c) Case when $A = B = C = 1$; all diodes are reversed-biased so $I = 0$, $V_F = 5$ V, giving $F = 1$

Fig. 21.2. Circuits demonstrating action of diode 'AND' gate of Fig. 21.1.

between 5 and 6 V is defined as a 1 state in our original specification.

Although we have considered only three combinations of input signals out of a possible eight, it is apparent that if, and only if, all inputs are 1, the output is 1. You might like to check the circuit for yourselves with other values of A, B, and C and construct a truth table for the circuit. This truth table will show that the circuit of fig. 21.1 acts as an AND gate.

21.2
The action of a 3-input diode OR gate

Fig. 21.3(a) shows a 3-input positive logic OR gate. When one or more diodes are forward-biased by a 1 state (positive voltage) input the diode(s) conduct(s) transmitting a 1 state to the output, which is taken across resistor R. For example, fig. 21.3(b) shows the case when $V_A = 6$ V, $V_B = V_C = 0$ V, i.e. $A = 1$, $B = C = 0$. Diode DA is forward-biased and conduction of current occurs via DA through R. The voltage developed across R,

$$V_F = 6 - \text{voltage drop across diode DA}$$

and since the latter term is small (about 0·6 V for Si and 0·2 V for Ge diodes) V_F is about 5·4 V (Si) or 5·8 V (Ge) and corresponds to a 1 state. Only when $A = B = C = 0$, i.e. all inputs are at zero volts is the output zero. Thus the circuit acts as a positive logic OR gate.

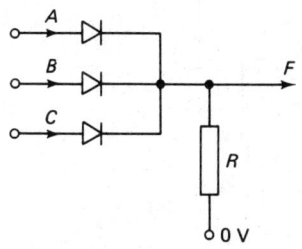

(a) 3-input positive logic
diode OR gate

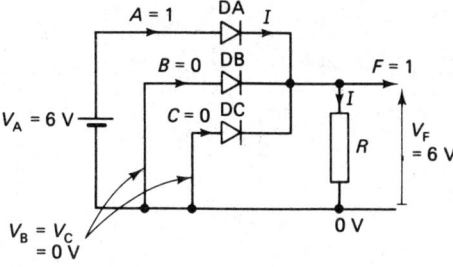

(b) Case when $A = 1$, $B = C = 0$,
diodes DB, BC are reversed-
biased, but DA is forward
biased, so $V_F = 6$ V and $F = 1$

Fig. 21.3.

281

Example 21.1

Sketch the waveforms for the output voltage obtained from the diode circuits of fig. 21.4(a) and (b) when the input signal waveforms shown in (c) are applied at their input lines.

Solution

(a) Circuit (a) is a positive logic AND gate. Only when all 3 diodes are reversed-biased, which occurs when all 3 input signals are 11 V, will the output be equal to the 10-V supply. In all other cases one or more of the diodes will be forward-biased by 0 V inputs and thus the output will be approximately 0 V (allowing for a small voltage drop across a diode). Hence to make a sketch of output voltage we find the times when all 3 inputs are 11 V. This only occurs in the 'fifth pulse' interval and during this interval the output is 10 V. At all other times the output is 0 V. The output voltage waveform is shown in fig. 21.5(a).

(b) Circuit (b) is a positive logic OR gate. If one or more inputs are 11 V the respective diodes are forward-biased and the output becomes 11 V, or approximately so allowing for a small voltage drop across the diode. Only when all 3 input signals are 0 V is the output 0 V. The output voltage waveform is shown in fig. 21.5(b).

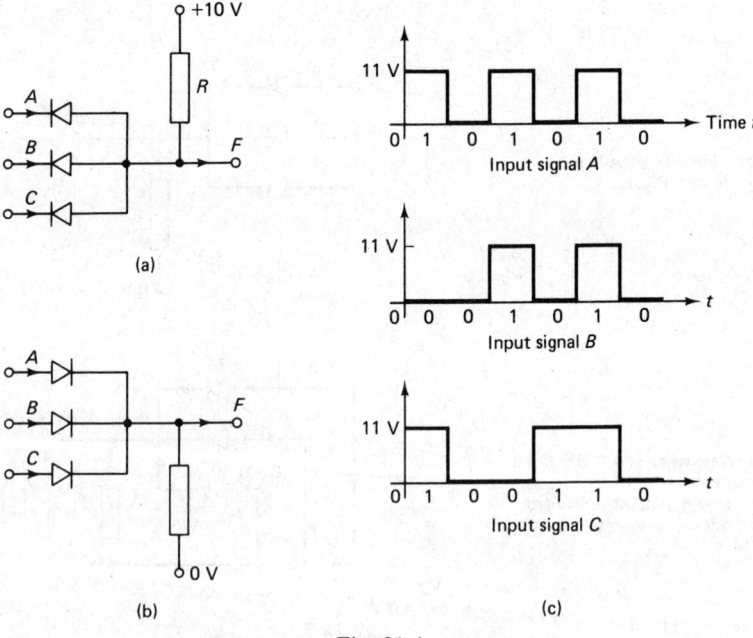

Fig. 21.4.

282

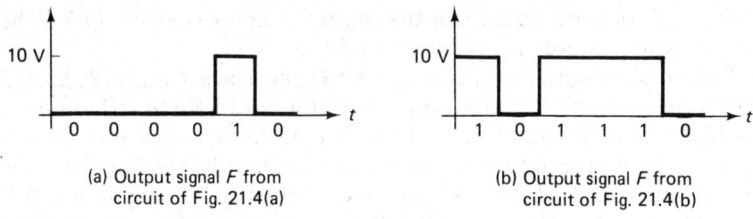

(a) Output signal F from circuit of Fig. 21.4(a)

(b) Output signal F from circuit of Fig. 21.4(b)

Fig. 21.5.

Example 21.2

Sketch the waveforms for the output voltage obtained from the diode circuits of fig. 21.6(a) and (b) when the input signal waveforms shown in (c) are applied at their input lines.

Assuming that a voltage level of between -16 V and -15 V is a 1 state and a voltage level of approximately 0 V is a 0 state, describe the logic function of the circuits.

Solution

(a) The diodes in circuit (a) are reversed-biased by a -16 V signal and forward-biased by a 0-V signal. Thus only when all 3 inputs are -16 V is

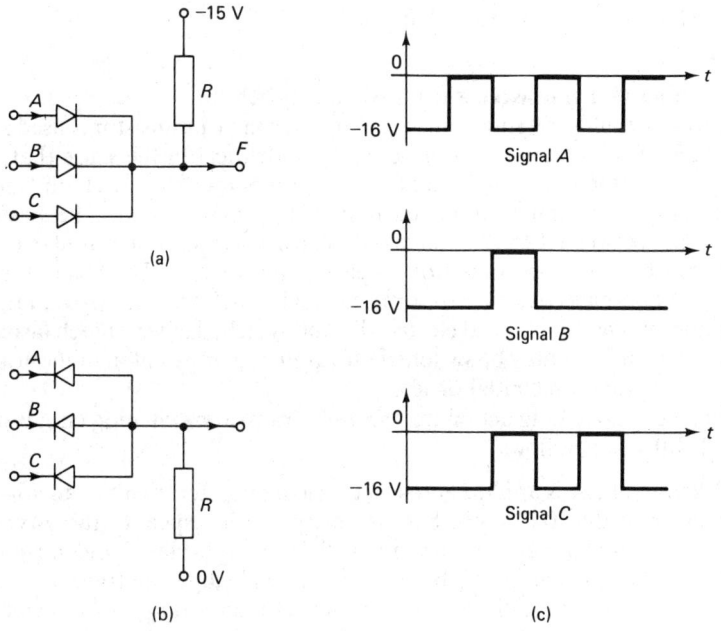

Fig. 21.6.

the output equal to the -15 V supply. In all other cases one or more diodes are forward-biased and the output is approximately O-V. The output voltage waveform is shown in fig. 21.7(a).

The circuit acts as a negative logic AND gate since if, and only if, all 3 inputs are -16 V (i.e. 1 states) is the output -15 V (i.e. a 1 state).
(b) If one or more inputs are -16 V, one or more diodes are forward-biased and thus the output is -16 V 'less' a small voltage drop across the diode, e.g. about $-15\cdot4$ V if silicon diodes are used. If all inputs are O V the output is O V. However, the latter condition does not arise for the signals shown. The output is always about -16 V as shown in fig. 21.7(b).

The circuit acts as a negative logic OR gate, since if one or more inputs equal -16 V (i.e. 1 states), the output is -16 V (i.e. a 1 state).

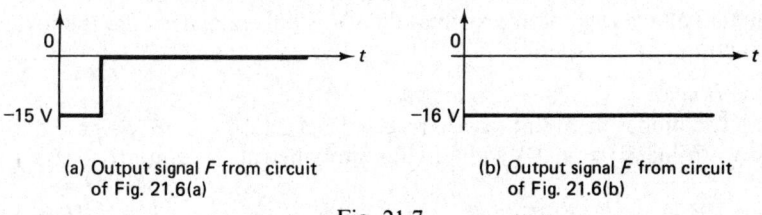

(a) Output signal *F* from circuit of Fig. 21.6(a)

(b) Output signal *F* from circuit of Fig. 21.6(b)

Fig. 21.7.

21.3
The action of a transistor when used as a switch

Figures 21.8 and 21.9 show simple circuits in which a transistor is used as a switch. The operation of both circuits is identical in the sense that a small current fed into the base of a transistor is used to 'switch on' and maintain a very much larger current in the main circuit. The basic difference between the two circuits is that an *n-p-n* transistor is used in fig. 21.8(b) whilst a *p-n-p* transistor is used in fig. 21.9(b). The transistors work in an analogous way to a mechanical switch with the advantage that they may be controlled electrically and switched at very much faster rates. A transistor may be switched off-to-on and on-to-off in millionths of seconds (microseconds) or less.

The basic switching action for the *n-p-n* transistor switching circuit of fig. 21.8(b) is as follows:

1. When a voltage is applied across the base-emitter terminals of an *n-p-n* transistor so that the *p*-type base is positive with respect to the *n*-type emitter the *p-n* base-emitter junction will be forward-biased and current will flow into the base. This base current I_B will cause the transistor to conduct between its collector and emitter terminals. If I_B is of sufficient magnitude (we will consider this value later) it will cause the transistor to conduct strongly, with the current flowing through the transistor limited

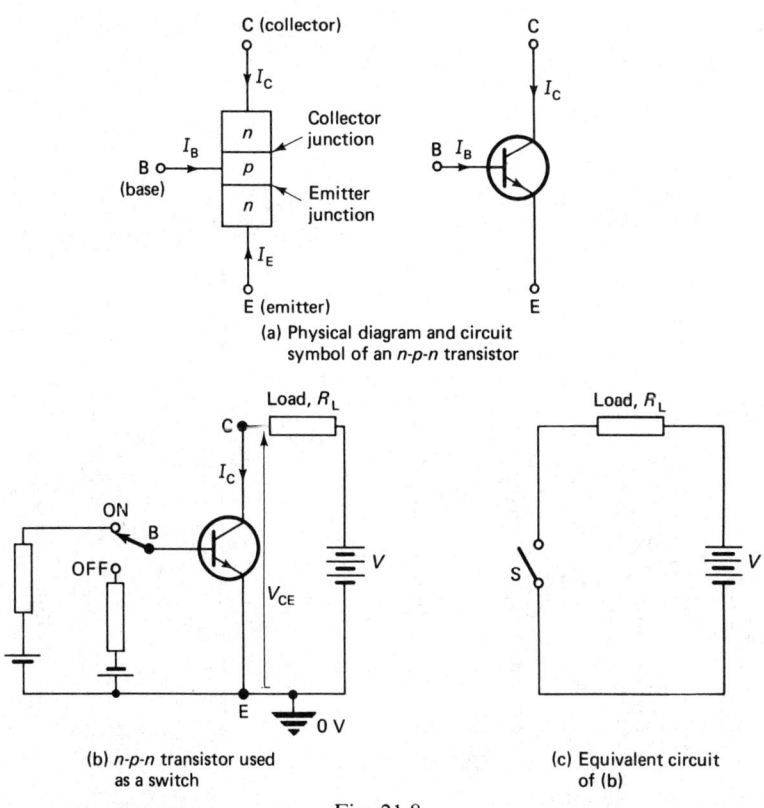

(a) Physical diagram and circuit symbol of an *n-p-n* transistor

(b) *n-p-n* transistor used as a switch

(c) Equivalent circuit of (b)

Fig. 21.8.

primarily by the load resistor R_L and supply voltage V, and virtually independent of I_B. Under these conditions we say that the transistor is in its 'on' or saturated state. In the on state the collector-emitter voltage is very small, typically 0·35 V or less for silicon transistors and even less for germanium, and so nearly all the supply voltage is applied across the load giving the main circuit current (the transistor collector current),

$$I_C = \frac{V - V_{CE}}{R_L} \approx \frac{V}{R_L} \text{ as } V \gg V_{CE}$$

The on state of the transistor is equivalent to the switch S in Fig. 21.8(c) being closed.

2. When a negative voltage is applied across the base-emitter terminals the base-emitter (B-E) *p-n* junction is reverse-biased and no current is fed into the base. Since base current is necessary to produce significant current flow through the transistor, the application of reverse-bias

285

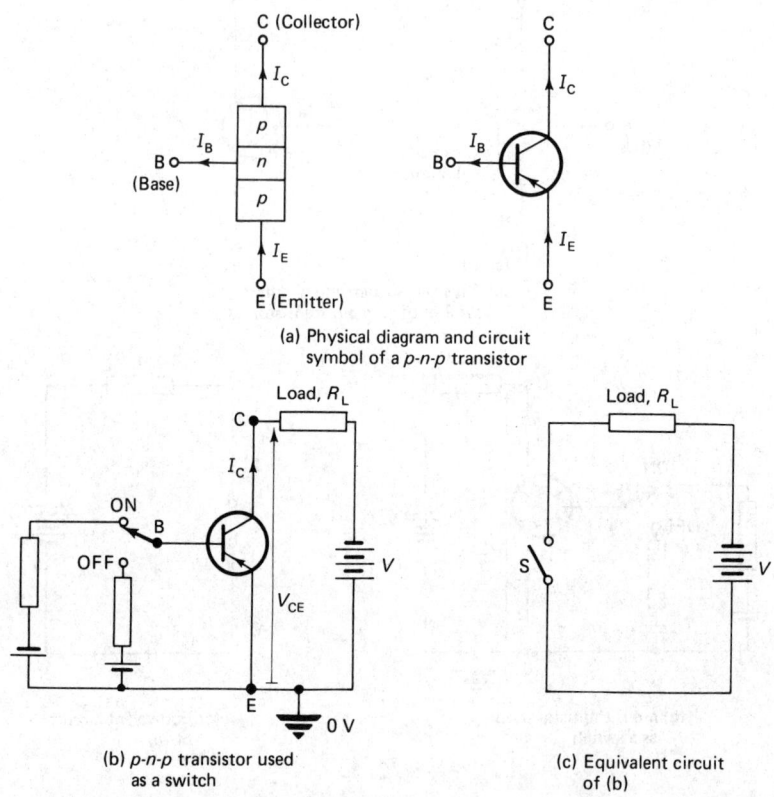

(a) Physical diagram and circuit symbol of a *p-n-p* transistor

(b) *p-n-p* transistor used as a switch

(c) Equivalent circuit of (b)

Fig. 21.9.

results in practically zero collector current. In fact there is a very small collector current, known as the reverse collector saturation current I_{CBO}*, due to the flow of minority carriers across the collector-base junction which acts as a reverse-biased diode. At 25°C, I_{CBO} is of the order of nano-amperes (10^{-9} A) for silicon and micro-amperes (10^{-6} A) for germanium, and although I_{CBO} approximately doubles for every 10°C temperature rise there is little error in taking $I_C = I_{CBO} \approx 0$ when the B-E junction is reverse-biased. Under these conditions we say that the transistor is in its 'off' or 'cut-off' state. In the off state I_C and therefore the

*I_{CBO} is the collector current which flows when $I_E = 0$, i.e. when the emitter lead is either open-circuited or the base-emitter junction is reversed-biased. If the base lead is open-circuited so that $I_B = 0$ the collector current which flows is denoted by I_{CEO}. I_{CEO} is related to I_{CBO} by the equation $I_{CEO} = I_{CBO}/(1-\alpha)$ where normally $\alpha \approx 0$ for silicon but may be of the order of 0·9 or so for germanium at low current values close to cut-off. Thus in the latter case $I_{CEO} \gg I_{CBO}$, e.g. if $\alpha = 0·9$ for a germanium transistor $I_{CEO} = 10 I_{CBO}$.

main circuit current is virtually zero. The off state of the transistor is equivalent to the switch S in fig. 21.8(c) being open.

In practice we may switch the transistor off by either reverse-biasing the B-E junction or by returning the base (usually via a resistor) to the O-V conductor or by open-circuiting the base. In switching circuit applications the first two methods are preferred. Reverse-biasing the B-E junction provides the most rapid turn-off characteristic, whilst returning the base to O-V is simpler and cheaper since it requires no base bias supply. Open-circuiting the base gives $I_B = 0$ and virtually cuts off the transistor. It is however the least satisfactory of the three methods and gives the slowest turn-off times.

The action of the p-n-p transistor switching circuit of Fig. 21.9(b) is virtually identical to the description given above except that the polarity of the voltage supply is reversed in the main circuit, i.e. the p-type collector is connected to the negative supply terminal and:

1. A negative voltage is applied to forward-bias the n-p base-emitter junction. Provided that the base current which then flows is above a certain value the transistor is switched on;

2. A positive voltage is applied to reverse-bias the base-emitter junction and turn off the transistor. The transistor may also be virtually switched off by returning the base to the O-V common conductor or by open-circuiting the base.

Fig. 21.10 shows a typical output characteristic of a transistor. We

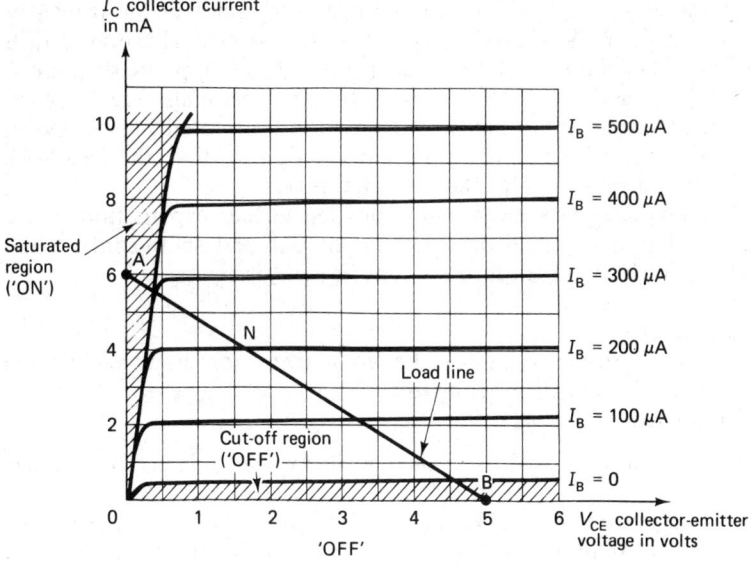

Fig. 21.10.

may use this characteristic to explain the changes that take place when a transistor is switched on and off. Let us consider these changes for the *n-p-n* transistor circuit of fig. 21.8(b). Exactly the same arguments apply for a *p-n-p* transistor. The load line equation for the main circuit of fig. 21.8(b) is

$$V = V_{CE} + R_L I_C$$

This line may be drawn on the output characteristic by first finding two points through which the line must pass. The easiest points to calculate are:

$V_{CE} = 0$, so $V = 0 + R_L I_C$, hence $I_C = \dfrac{V}{R_L}$.

$I_C = 0$, so $V = V_{CE} + 0$, i.e. $V = V_{CE}$.

These two points ($V_{CE} = 0, I_C = V/R_L$) $\cdots$ point A, and ($V_{CE} = V, I_C = 0$) $\cdots$ point B, are plotted on fig. 21.10 and the load line constructed by joining A and B by a straight line.

Now if the B-E junction is reversed-biased, the transistor is cut off so the operating point (i.e. the values of V_{CE}, I_C, and I_B of the transistor) lies at or very close to point B on the load line, i.e. we are at $V_{CE} \approx V, I_C \approx 0$. The shaded region below $I_B = 0$ on the output characteristic is known as the cut-off region and a transistor in its off state should always be in this region.

When we drive current into the base by forward-biasing the B-E junction we move the operating point up the load line towards A. For example, if we set $I_B = 200$ μA the operating point of the transistor moves to point N where $V_{CE} = 1.65$ V, $I_C = 4$ mA. However if I_B is increased beyond $I_B = 300$ μA the operating point approaches point A and $V_{CE} \approx 0$ and $I_C \approx V/R_L$ (in this case 6 mA). The shaded region on the left-hand side of the graph is known as the saturated region and in switching a transistor on we normally design to drive the base with sufficient current to enter the saturated region.

The two examples given below provide further explanation of the action of simple transistor switching circuits and show some of the factors which are taken into account in their design.

Example 21.3
The manufacturer's data for the *n-p-n* transistor employed as the switching element in the circuit of fig. 21.11 give the following information:

When the transistor is switched on, the typical and maximum values of:
base-emitter voltage, $V_{BE(sat)} = 0.95$ V (typical), 1.3 V (maximum);
collector-emitter voltage, $V_{CE(sat)} = 0.15$ V (typical), 0.35 V (maximum);

The static (d.c.) forward current transfer ratio (or gain) (see Sections 6.5–6.9) $h_{FE} = 30$ (minimum), 90 (typical).

Note. In the large-scale production of solid state devices it is impossible to test all devices individually. Manufacturers therefore compile their data sheets from test measurements on a relatively few samples taken from a series of production batches. They specify typical values and usually include minimum and/or maximum values. The values of V_{BE} and V_{CE} quoted above are based on given values of I_B and I_C when the transistor is in its 'on' or saturated state. They are therefore referred to as saturated values and hence the (sat) subscript is included.

The values of h_{FE} are also quoted as minimum and typical values. The minimum value $h_{FE} = 30$ means that it is very unlikely that more than 1 in 1000+ transistors would have a static current gain below 30, whereas a typical transistor would have an $h_{FE} = 90$. In circuit design we use the 'worst-case' values (i.e. minimum h_{FE}, maximum V_{BE} and V_{CE}) to ensure that if we ever have to replace the transistor, our circuit will have an almost 100% probability of functioning correctly.

Problem
Using the given data, calculate for the circuit of fig. 21.11:
   (a) the value of main circuit current when the transistor is switched on,
   (b) the value of base current required to switch the circuit on,
   (c) the value of resistor R_B if the switching base current is effected by a 5-V signal (e.g. by closing the switch shown in fig. 21.11).

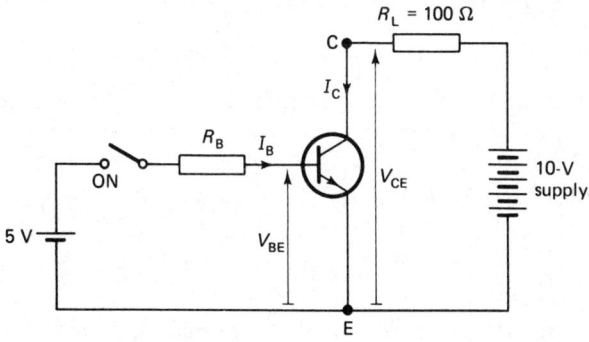

Fig. 21.11. Switching circuit for example 21.3.

Solution
(a) When the transistor is switched on the voltage across its collector-emitter terminals is $V_{CE(sat)}$ and thus the voltage across the $R_L = 100\ \Omega$ load resistor is the supply voltage minus $V_{CE(sat)}$, so the main circuit current,

$$I_C = \frac{10 - V_{CE(sat)}}{100}$$

$$= \frac{10 - 0 \cdot 15}{100} = 98 \cdot 5 \text{ mA (typically, as } V_{CE(sat)} = 0 \cdot 15 \text{ V)}$$

$$= \frac{10 - 0 \cdot 35}{100} = 96 \cdot 5 \text{ mA (using 'worst-case' } V_{CE(sat)} = 0 \cdot 35 \text{ V)}.$$

(b) Let us select our 'on' current as $I_C = 100$ mA (we know that typically the on current is $98 \cdot 5$ mA, so we are only rounding up). We then use $h_{FE(min)}$ to calculate the switching base current. Strictly speaking h_{FE} refers to the ratio of I_C to I_B in the active transistor region (the region we use in employing the transistor as an amplifying rather than a switching element), but it is normal engineering practice to utilize the minimum value of h_{FE} to work out the base current required to produce the 'on' state value of collector current. Thus

$$h_{FE} = \frac{I_C}{I_B}, \text{ so } I_B = \frac{I_C}{h_{FE(min)}} = \frac{100}{30} = 3 \cdot 3 \text{ mA}$$

(c) We can calculate the value of R_B to produce $I_B = 3 \cdot 3$ mA when the 5-V signal is applied as follows:

$$\text{voltage applied to base circuit} = R_B I_B + V_{BE(sat)}$$

i.e. applied voltage = voltage drop across R_B + voltage drop from B to E.

So
$$5 = R_B \times 3 \cdot 3 \times 10^{-3} + 1 \cdot 3$$

where we have used the 'worst-case' value of $V_{BE(sat)}$ to ensure that $I_B = 3 \cdot 3$ mA, even when maximum voltage is dropped across the B-E junction.
On solving for R_B we have

$$R_B = \frac{5 - 1 \cdot 3}{3 \cdot 3 \times 10^{-3}} = 1121 \ \Omega$$

Again, in practice we would take R_B as the nearest preferred resistance value to 1121 Ω. The preferred values* nearest to 1121 Ω are 1 kΩ and 1·2 kΩ. Normally we go 'down' to ensure that I_B is maintained above its switching value, so $R_B = 1$ kΩ would be the design choice.

---

* ±10% (tolerance) preferred values of resistance are: 10, 12, 15, 18, 22, 27, 33, 39, 47, 56, 68, 82, 100×10^n ohms where $n = 1, 2, 3, 4, 5 \ldots$ etc. and such resistors are the ones which would normally be available in an electronics laboratory or workshop.

The points brought out in this example are very useful and may be applied to design simple switching circuits:

1. Transistor on, $I_C = \dfrac{V_{supply} - V_{CE(sat)}}{R_{load}}$

2. Design value for switching current,

$$I_B = \frac{I_C}{h_{FE(min)}}, \text{ i.e. } \frac{\text{required load current}}{\text{minimum value of } h_{FE}}$$

3. Value of R_B included in base circuit to produce required I_B,

$$R_B = \frac{V_{signal} - V_{BE(sat)}}{I_B}.$$

The power dissipated by the transistor in its on state must be less than its rated value. In the above example the power dissipated in the transistor is mainly due to the $I_C \times V_{CE(sat)}$ term, so with $I_C = 100$ mA, $V_{CE(sat)} = 0.35$ V (worst-case value), the power dissipated is $100 \times 0.35 = 35$ mW, well within the rating of an average switching transistor.

Our second switching example demonstrates the use of graphical methods to solve a switching problem. However, although graphical techniques help to explain the operation of a circuit they are only applicable for quantitative deductions provided we have available or have measured the actual characteristics of the transistor we are using. Transistor characteristics vary quite widely so it would be incorrect to base any design calculations on the use of a 'typical' curve.

Example 21.4
The output I_C versus V_{CE} and input I_B versus V_{BE} static characteristics of an n-p-n transistor are given in fig. 21.12(a) and (b) respectively. This transistor is used in the switching circuit of (c).
(a) Write down the load line equation for the main circuit and plot the load line on the (expanded scale) output characteristic given in (a)(i). Hence calculate the main circuit current when $I_B = 10$ mA, 15 mA and 20 mA.
(b) Using the input characteristic, calculate the corresponding values of V_{BE} to produce the above values of I_B.
(c) If a switching current of $I_B = 20$ mA is required, calculate the value of R_B to effect this when the input signal to the base circuit is 6 V.
(d) If R_B is 1 kΩ, calculate V_{BE}, the base and the main circuit currents when switch S is (i) closed, (ii) open.

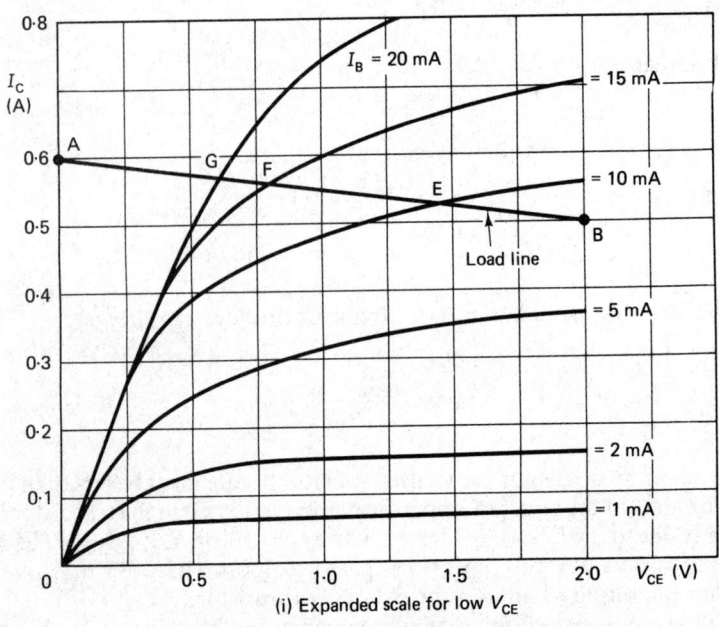

(i) Expanded scale for low V_{CE}

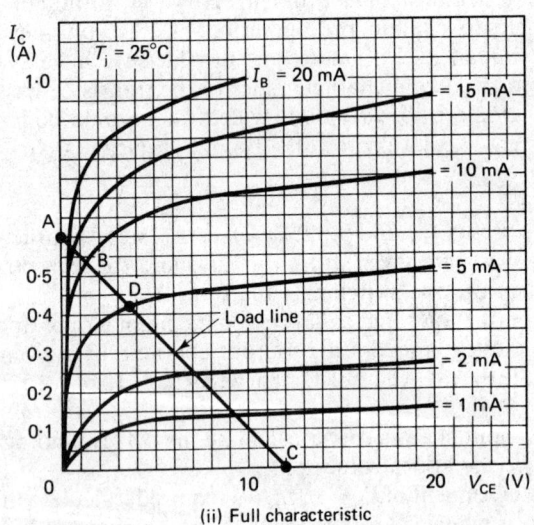

(ii) Full characteristic

(a) Measured static output characteristic for the *n-p-n* silicon transistor used in (c)

Fig. 21.12.

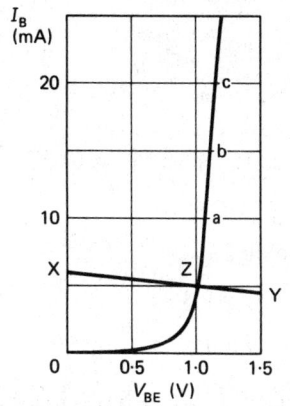

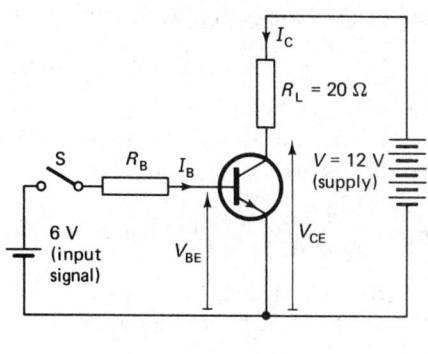

(b) Measured static input characteristic for the *n-p-n* transistor used in (c)

(c) Switching circuit diagram

Fig. 21.12 (*continued*)

Solution

(a) The load line equation is

$$V_{CE} = V_{supply} - R_L I_C$$

i.e. $V_{CE} = 12 - 20I_C$ as $V_{supply} = 12$ V, $R_L = 20$ Ω.

To draw this line on the output characteristic (a)(i) we need first to find two points:

when $V_{CE} = 0$ V, we have $0 = 12 - 20I_C$, so $I_C = \dfrac{12}{20} = 0.6$ A or 600 mA;

when $V_{CE} = 2$ V, we have $2 = 12 - 20I_C$, so $I_C = \dfrac{10}{20} = 0.5$ A or 500 mA.

Thus plotting $V_{CE} = 0, I_C = 600$ mA (point A) and $V_{CE} = 2$ V, $I_C = 500$ mA (point B) on fig. 21.12(a)(i) and drawing a straight line through these points, we have the load line AB superimposed on the static characteristics. The point of intersection E of the $I_B = 10$ mA characteristic and the load line gives the value of I_C and therefore main circuit current when $I_B = 10$ mA. From the graph we have, $I_C = 524$ mA.

Point F corresponds to $I_B = 15$ mA and gives $I_C = 560$ mA.
Point G corresponds to $I_B = 20$ mA and yields $I_C = 568$ mA.

(b) Reading off from the input characteristic of fig. 21.12(b) directly we have:

when $I_B = 10$ mA, $V_{BE} = 1.08$ V (point a);
when $I_B = 15$ mA, $V_{BE} = 1.1$ V (point b);
when $I_B = 20$ mA, $V_{BE} = 1.15$ V (point c).

293

(c) In the base circuit we have:

$$\text{signal voltage applied} = R_B I_B + V_{BE}$$

i.e. $6 = R_B \times 20 \times 10^{-3} + 1\cdot15$
on substituting for signal voltage $= 6$ V, $I_B = 20$ mA, and $V_{BE} = 1\cdot15$ V.
So on solving for R_B we obtain

$$R_B = \frac{4\cdot85}{20 \times 10^{-3}} = 243 \ \Omega \text{ (nearest preferred value 220 } \Omega\text{)}.$$

(d) (i) When switch S is closed
When S is closed the voltage applied to the base circuit of the transistor is 6 V.

To find V_{BE} and I_B when $R_B = 1000 \ \Omega$ we must draw in the base circuit load line on the input characteristic of fig. 21.12(b) and find the point of intersection. This point gives us V_{BE} and I_B.
The base circuit load line equation is

$$6 = R_B I_B + V_{BE} = 1000 \ I_B + V_{BE}.$$

To plot this equation we need 2 points:
when $V_{BE} = 0$ V, $6 = 1000 \ I_B$, so $I_B = 6$ mA;
when $V_{BE} = 1\cdot5$ V, $6 = 1000 \ I_B + 1\cdot5$, so $I_B = 4\cdot5$ mA.
Thus plotting the points $V_{BE} = 0$, $I_B = 6$ mA (point X) and $V_{BE} = 1\cdot5$ V, $I_B = 4\cdot5$ mA (point Y) we can draw the load line. The load line cuts the input characteristic at point Z and thus reading off the values of V_{BE} and I_B at this point we have:
when $R_B = 1$ kΩ $V_{BE} = 1\cdot0$ V, $I_B = 5\cdot0$ mA.
To find the value of I_C corresponding to $I_B = 5$ mA we must draw in the main circuit load line,

$$V_{CE} = 12 - 20 I_C$$

on the full output characteristic of fig. 21.12(a)(ii) and find the point of intersection of the load line with the $I_B = 5$ mA curve. This occurs at D giving $I_C = 420$ mA.

(ii) When switch S is open
When S is open $I_B = 0$. Thus no current is injected into the base and the transistor is virtually cut off. The main circuit current I_C then consists of a tiny leakage current, and for practical purposes may be taken as zero.

21.4
The action of a transistor NOT gate
Fig. 21.13 shows two simple NOT gate circuits. Circuit (a) utilizes an n-p-n transistor and implements the NOT function normally for positive logic systems, whilst circuit (b) employs a p-n-p transistor and

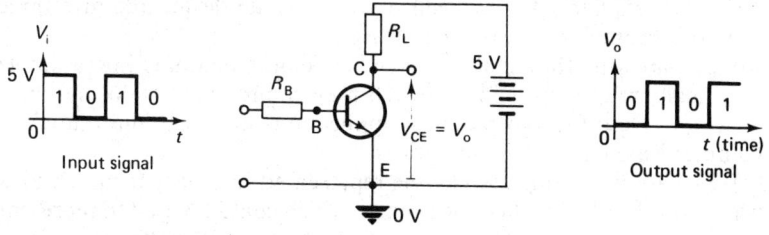

(a) An *n-p-n* transistor NOT gate circuit, $V_o = \overline{V_i}$

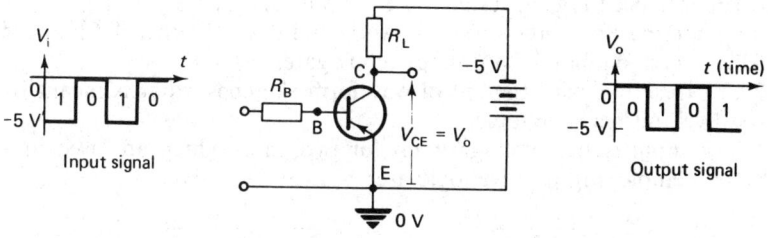

(b) A *p-n-p* transistor NOT gate circuit $V_o = \overline{V_i}$

Fig. 21.13.

implements the NOT function normally for negative logic systems. The action of both circuits is basically a switching operation.

Let us assume in (a) that a 5-V level represents a logic 1 state and O-V a 0 state. Thus when the input signal to the base circuit $V_i = 5$ V the B-E junction of the *n-p-n* transistor is forward-biased. The base resistor R_B is selected so that the correct current is fed into the base to drive the transistor into saturation. A simple design criterion for R_B is given in Example 21.3. The output then falls to $V_o = V_{CE(sat)} \approx$ O V. Thus a 1 state input is inverted to a 0 state output. On the other hand when $V_i = $ O V there is no voltage present to forward-bias the *p-n*, B-E junction and hence $I_B \approx$ O and the transistor is cut-off. In the cut-off condition the collector current is virtually zero, so no voltage drop occurs across the load resistor R_L. The output voltage therefore rises to the supply voltage, i.e. $V_o = 5$ V. Thus a 0 state input produces a 1 state output.

In circuit (b) let us assume -5 V represents a 1 state (in negative logic the more negative voltage level is the 1 state) and O V the 0 state. When $V_i = -5$ V the *n-p*, B-E junction is forward-biased and, assuming R_B is correctly selected, the transistor is switched on, so $V_o = V_{CE(sat)} \approx$ O V. Hence a 1 input produces a 0 output. When $V_i = 0$ the transistor is cut-off. $V_o = $ supply voltage $= -5$ V and thus a 0 input is inverted to a 1 output.

Problems: F Logic Elements and Circuits

19.1 (a) Explain what is meant by a two-state device and give three practical examples of two-state devices.

(b) Explain, with the aid of circuit and waveform diagrams, how a simple electrical circuit containing a switch may communicate information.

19.2 (a) Describe two practical applications in which information is communicated by two-state devices.

(b) Explain how numbers may be represented in binary form. Draw a waveform sketch of an electrical signal which could be used to represent the decimal number of 242 in binary form. State which voltage levels you have used to define your 1 and 0 states.

20.1 (a) State the logical function of an AND gate.

(b) Construct the truth table for a 3-input AND gate.

(c) State the Boolean symbol for AND and draw the British Standard (BS) circuit symbol for a 3-input AND gate.

20.2 Describe, with the aid of waveform sketches, what is meant by positive and negative logic.

The input-output voltage levels for two gates which are used in a system employing positive logic are:

Inputs		Output	Inputs		Output
A	B	F	A	B	F
0 V	0 V	0·1 V	0 V	0 V	0·1 V
0	10 V	9·5 V	0	10 V	0·15 V
10 V	0 V	9·6 V	10 V	0 V	0·2 V
10 V	10 V	9·5 V	10 V	10 V	9·8 V

GATE X	**GATE Y**

In both cases, the maximum-minimum voltage levels corresponding to the 1 and 0 states are given by:

1 state ... 9 V (min) to 10·5 V
0 state ... 0 V   to 1·0 V (max)

Construct the truth tables for each gate and identify which one is an AND gate.

Determine, also, the output from each gate if the following input signals are applied:

(a) $A = 1010101$ (b) $A = 001101$
    $B = 0011001$ $B = 110110$

20.3 Determine the output voltage V_F in the circuit of fig. P20.3 when (a) $V_A = 10$ V, $V_B = 10$ V; (b) $V_A = 10$ V, $V_B = 0$ V; (c) $V_A = 0$ V, $V_B = 0$ V.

It may be assumed that the diodes are ideal.
Write down the truth table for the circuit and hence show that it acts as an AND gate.

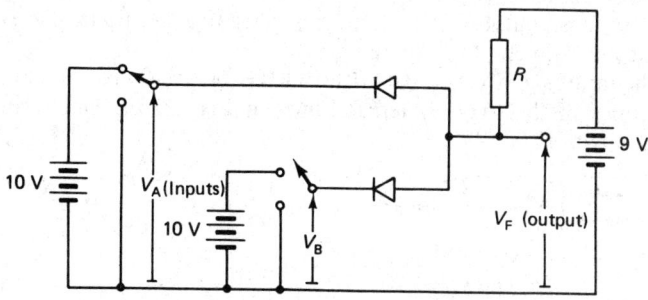

Fig. P20.3. for Problem 20.3.

20.4 (a) State the logical function of an OR gate.
(b) Construct a truth table for a 3-input OR gate.
(c) State the Boolean symbol for OR and draw the BS symbol for a 4-input OR gate.

20.5 Draw the BS symbols for a 5-input AND and a 5-input OR gate. If the input signals to both gates are A, B, C, D, E, write down the Boolean expression for the respective gate output functions.

State, also, the output of the AND and the OR gates when:

(a) $A = B = C = D = E = 1$.
(b) $A = 0, B = C = D = 1, E = 0$.
(c) $A = B = 1, C = 0, D = E = 1$.
(d) $A = B = C = D = E = 0$.

20.6 Describe the action of the circuit shown in fig. P20.6 in logic terms by drawing up a truth table. Write down also the Boolean expression which defines the circuit action.

[*Hint*: take switches A, B, C, D as the inputs and lamp F as the output; define a switch closed as a 1-state and a switch open as an 0-state; when the lamp is on define $F = 1$; when the lamp is off define $F = 0$].

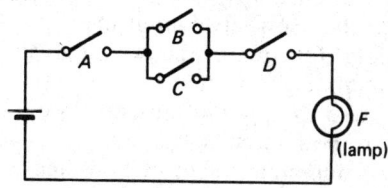

Fig. P20.6. for Problem 20.6.

20.7 (a) State the logical function of a NOT gate and construct its truth table.

(b) State the Boolean symbol for NOT and draw the BS circuit symbol for a NOT gate.

(c) An input signal $A = 101110101$ is applied to a NOT gate. Determine the output from the NOT gate.

(d) The input signals $A = 1011101$ and $B = 0010110$ are applied to the input lines of the logic systems shown in Fig. P20.7. Determine the output signal F.

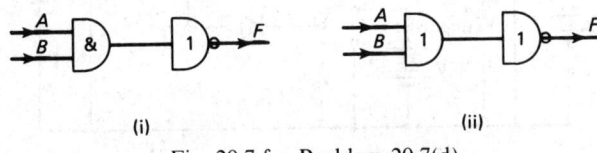

(i)　　　　　　　　　　　　(ii)

Fig. 20.7 for Problem 20.7(d).

20.8 Fig. P20.8 shows a logic system with all gates represented by BS symbols. State the logic function of gates 1, 2, and 3, and determine
(a) the output F_1 from gate 1,
(b) the output F_3 from gate 3,
(c) the system output F,
when the input signals $A = 101011$, $B = 110101$ are applied at the input lines.

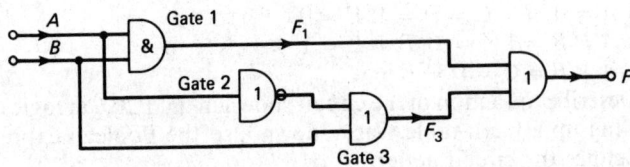

Fig. P20.8. Logic system for Problem 20.8.

21.1 Draw the circuit diagram of (a) a 3-input diode AND gate, and (b) a 3-input diode OR gate, and explain their action.

21.2 Sketch the waveforms of the output voltage obtained from the diode circuits of fig. P21.2(a) and (b) when the input signals A, B, C shown in (c) are applied at their respective input lines.

Describe also the logic function of the two circuits.

21.3 Draw the circuit diagrams of a simple transistor switch when (a) an n-p-n transistor, and (b) a p-n-p transistor, is used.

Explain the action of both circuits.

21.4 Calculate the minimum value of V_S which will switch on the transistor in the circuit of fig. P21.4, given that the base-emitter voltage

298

and the collector emitter voltage of the transistor in its saturation region are $V_{BE(sat)} = 0.8$ V, $V_{CE(sat)} = 0.2$ V, and the value of the d.c. current gain, $h_{FE} = 20$.

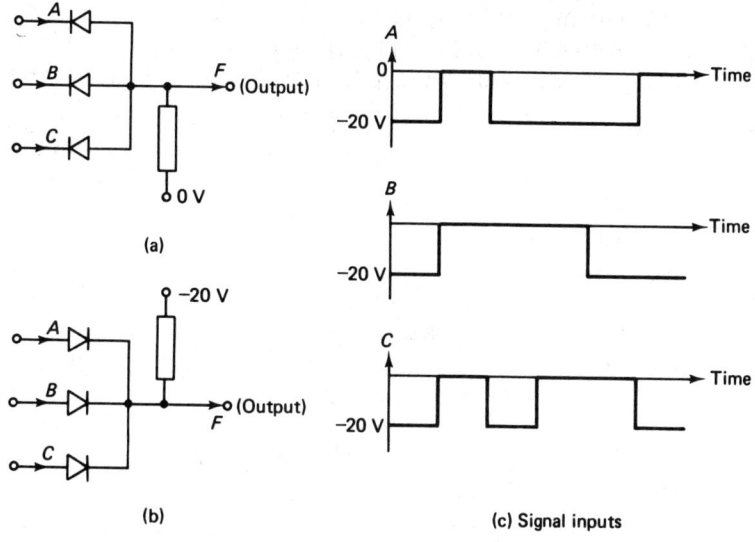

(a)

(b)

(c) Signal inputs

Fig. P21.2. Circuits and input signals for Problem 20.2.

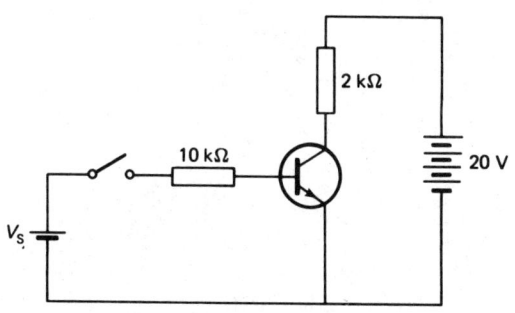

Fig. P21.4. Circuit for Problem 21.4.

21.5   Draw the circuit diagrams of a simple transistor NOT gate when (a) an *n-p-n* transistor, and (b) a *p-n-p* transistor, is used.
Explain, with the aid of waveform diagrams, the action of circuit (a).

Answers
20.2 (a) 1011101   (b) 000100
20.3 (a) 9V   (b)0 V   (c) 0 V

20.5 (a) 1 (AND), 1 (OR) (b) 0 (AND), 1 (OR) (c) 0 (AND), 1 (OR) (d) 0 (AND), 0 (OR)

20.6 $A.B.D + A.C.D + A.B.C.D$

20.7 (c) $\bar{A} = 010001010$
(d) (i) 1101011 (ii) 0100000

20.8 (a) $F_1 = 100001$ (b) $F_3 = 110101$ (c) $F = 110101$

21.2 Circuit (a) acts as an OR gate for negative logic, (b) as an AND gate.

21.4 5·75 V